GEOLOGICAL SOCIETY SPECIAL PUBLICATION NO. 251

Alluvial Fans:
Geomorphology, Sedimentology, Dynamics

EDITED BY

A.M. HARVEY

University of Liverpool, UK

A.E. MATHER

University of Plymouth, UK

and

M. STOKES

University of Plymouth, UK

2005
Published by
The Geological Society
London

THE GEOLOGICAL SOCIETY

The Geological Society of London (GSL) was founded in 1807. It is the oldest national geological society in the world and the largest in Europe. It was incorporated under Royal Charter in 1825 and is Registered Charity 210161.

The Society is the UK national learned and professional society for geology with a worldwide Fellowship (FGS) of 9000. The Society has the power to confer Chartered status on suitably qualified Fellows, and about 2000 of the Fellowship carry the title (CGeol). Chartered Geologists may also obtain the equivalent European title, European Geologist (EurGeol). One fifth of the Society's fellowship resides outside the UK. To find out more about the Society, log on to www.geolsoc.org.uk.

The Geological Society Publishing House (Bath, UK) produces the Society's international journals and books, and acts as European distributor for selected publications of the American Association of Petroleum Geologists (AAPG), the American Geological Institute (AGI), the Indonesian Petroleum Association (IPA), the Geological Society of America (GSA), the Society for Sedimentary Geology (SEPM) and the Geologists' Association (GA). Joint marketing agreements ensure that GSL Fellows may purchase these societies' publications at a discount. The Society's online bookshop (accessible from www.geolsoc.org.uk) offers secure book purchasing with your credit or debit card.

To find out about joining the Society and benefiting from substantial discounts on publications of GSL and other societies worldwide, consult www.geolsoc.org.uk, or contact the Fellowship Department at: The Geological Society, Burlington House, Piccadilly, London W1J 0BG: Tel. +44 (0)20 7434 9944; Fax +44 (0)20 7439 8975; E-mail: enquiries@geolsoc.org.uk.

For information about the Society's meetings, consult *Events* on www.geolsoc.org.uk. To find out more about the Society's Corporate Affiliates Scheme, write to enquiries@geolsoc.org.uk.

Published by The Geological Society from:
The Geological Society Publishing House
Unit 7, Brassmill Enterprise Centre
Brassmill Lane
Bath BA1 3JN, UK

Orders: Tel. +44 (0)1225 445046
Fax +44 (0)1225 442836

Online bookshop: www.geolsoc.org.uk/bookshop

The publishers make no representation, express or implied, with regard to the accuracy of the information contained in this book and cannot accept any legal responsibility for any errors or omissions that may be made.

British Library Cataloguing in Publication Data
A catalogue record for this book is available from the British Library.

ISBN 1-86239-189-0

Typeset by Servis Filmsetting Ltd, Manchester, UK
Printed by Cromwell Press, Trowbridge, UK

Distributors

USA
AAPG Bookstore
PO Box 979
Tulsa
OK 74101-0979
USA
Orders: Tel. + 1 918 584-2555
Fax +1 918 560-2652
E-mail bookstore@aapg.org

India
Affiliated East-West Press Private Ltd
Marketing Division
G-1/16 Ansari Road, Darya Ganj
New Delhi 110 002
India
Orders: Tel. +91 11 2327-9113/2326-4180
Fax +91 11 2326-0538
E-mail affiliat@vsnl.com

Japan
Kanda Book Trading Company
Cityhouse Tama 204
Tsurumaki 1-3-10
Tama-shi, Tokyo 206-0034
Japan
Orders: Tel. +81 (0)423 57-7650
Fax +81 (0)423 57-7651
Email geokanda@ma.kcom.ne.jp

Contents

Alluvial fans: geomorphology, sedimentology, dynamics – introduction. A review of alluvial-fan research

ADRIAN M. HARVEY[1], ANNE E. MATHER[2] & MARTIN STOKES[3]

[1]*Department of Geography, University of Liverpool, PO Box 147, Liverpool L69 3BX, UK*
(e-mail: amharvey@liverpool.ac.uk.)
[2] *School of Geography, University of Plymouth, Drake Circus, Plymouth PL4 8AA, UK*
(e-mail: a.mather@plymouth.ac.uk.)
[3] *School of Earth, Ocean and Environmental Sciences, University of Plymouth, Drake Circus,*
Plymouth PL4 8AA, UK

This volume presents a series of papers on the geomorphology, sedimentology and dynamics of alluvial fans, selected from those presented at the 'Alluvial Fans' Conference held in Sorbas, SE Spain in June 2003. The conference was sponsored primarily by the British Geomorphological Research Group and the British Sedimentological Research Group, both organizations affiliated to the Geological Society of London.

It is some time since an international conference has been held that was exclusively devoted to the geomorphology and sedimentology of alluvial fans. The previous such conference was that organized by Terry Blair and John McPherson in 1995, and held in Death Valley, a classic setting for alluvial fans (Denny 1965; Blair & McPherson 1994*a*). Although many of the papers presented there have since been published, no dedicated volume on alluvial fans as a whole resulted from that meeting, so even longer has elapsed since there has been a specific publication devoted wholly to a series of papers on the geomorphology and sedimentology of alluvial fans (Rachocki & Church 1990).

South-east Spain was chosen as the venue for this conference, partly for logistic reasons and partly because it is a tectonically active dry region within which there is a wide range of Quaternary alluvial fans. These fans exhibit differing relationships between tectonic, climatic and base-level controls (Harvey 1990, 2002*a*, 2003; Mather & Stokes 2003; Mather *et al.* 2003), core themes in consideration of the dynamics of alluvial fans.

An emphasis within the previous alluvial fan literature has been on fans within the deserts of the American South-west (Bull 1977), with a focus especially on the alluvial fans of Death Valley (Denny 1965; Blair & McPherson 1994*a*). However, alluvial fans are not exclusive to drylands, nor to the particular combination of processes and morphology that characterize the Death Valley fans. Depositional processes may range from debris flows to sheet and channelized fluvial processes. Scales may range from small debris cones ($<$50 m in length: e.g. Wells & Harvey 1987; Brazier *et al.* 1988; Harvey & Wells 2003) to fluvially dominated megafans (up to 60 km in length: Gohain & Parkash 1990). Alluvial fans, although common in desert mountain regions (Harvey 1997), may occur in any climatic environment, in arctic (e.g. Boothroyd & Nummendal 1978), alpine (e.g. Kostaschuk *et al.* 1986), humid temperate (e.g. Kochel 1990) and even humid tropical environments (e.g. Kesel & Spicer 1985).

In all climatic environments alluvial fans may play an important buffering role in mountain geomorphic or sediment systems (Harvey 1996, 1997, 2002*b*). They trap the bulk of the coarse sediment delivered from the mountain catchment, and therefore affect the sediment dynamics downstream, either in relation to distal fluvial systems or to sedimentary basin environments. In doing so, fans preserve a sensitive sedimentary record of environmental change within the mountain sediment-source area, rather than a broad regional record as would, for example, pluvial lake sediments (Harvey *et al.* 1999*b*).

One of the aims of the 2003 conference was to bring together current research on the geomorphology, sedimentology and dynamics of alluvial fans, ranging from studies of modern processes through to studies of Quaternary fans and to those of ancient fan sequences within the geological record. We sought papers relating to a range of scales and of climatic and tectonic contexts. Some of the papers presented at the conference were 'in press' in other publications at the time (e.g. Harvey & Wells 2003) or are being published elsewhere (Garcia & Stokes in press; Saito & Oguchi 2005; Stokes *et al.* in press). In this volume we present a selection of the papers presented at the Sorbas meeting that spans a wide range of alluvial fan research. We group them into three main themes: those dealing with processes on fans; those dealing with the dynamics and morphology of Quaternary alluvial fans; and those dealing with the interpretation of

From: HARVEY, A.M., MATHER, A.E. & STOKES, M. (eds) 2005. *Alluvial Fans: Geomorphology, Sedimentology, Dynamics.*
Geological Society, London, Special Publications, **251**, 1–7. 0305–8719/05/$15 © The Geological Society of London 2005.

alluvial fan sedimentary sequences from the ancient rock record.

Processes on alluvial fans

In recent years there has been growing recognition of the range of sedimentary signals of fan depositional processes. The importance of sheet-flood processes as opposed to channelized fluvial processes has been recognized on desert alluvial fans (Blair & McPherson 1994a), and there is growing recognition of the distinctions within mass-flow processes between true debris-flow processes and those associated with hyperconcentrated flows (Wells & Harvey 1987; Blair & McPherson 1994a, 1998), aided by advances in the understanding of the rheology of mass flows in the lahar literature (e.g. Cronin *et al.* 1997; Lavigne & Suwa 2004) and in hydraulic engineering (e.g. Engelund & Wan 1984; Julian & Lan 1991; Rickenmann 1991; Wang *et al.* 1994). In this volume we include two papers on fan sedimentation processes, in markedly contrasted environments, by **Mather & Hartley** and by **Wilford *et al.*** in hyper-arid and humid environments, respectively.

Building on previous work (e.g. Bull 1977; Kostaschuk *et al.* 1986), it is realized that distinct fan surface gradients result from different depositional processes, but the concept of a specific 'slope gap' (Blair & McPherson 1994b) between alluvial fan and river gradients has since been demonstrated to have been flawed (Kim 1995; McCarthy & Candle 1995; Harvey 2002c; Saito & Oguchi 2005). To some extent Blair & McPherson's (1994a) classification of alluvial-fan styles, based on process combinations, reinforces the traditional concept of 'wet fans and dry fans' (Schumm 1977); however, the application of that concept to a climatic association with humid and arid climates, respectively, is clearly oversimplistic. On the contrary, arid environments tend to be more fluvially dominant than do humid regions (Baker 1977), and many studies have demonstrated an increase in fluvial activity on fans, resulting from climatic aridification (e.g. Harvey & Wells 1994), or an increase in hillslope sediment supply from debris-flow activity following a climatic change towards greater humidity (e.g. Gerson 1982; Bull 1991; Al-Farraj 1996).

There is increasing recognition that alluvial fans represent a continuum of depositional processes (Saito & Oguchi 2005) from small debris cones, characteristic of many mountain environments (e.g Brazier *et al.* 1988), especially in paraglacial zones (Ryder 1971), to large fluvially dominant fans, such as the Kosi megafan in India (Gohain & Parkash 1990). In this volume we include two papers on large fluvially dominated fans in contrasted environments

(by **Arzani** in Iran, and by **Gabris & Nagy** in Hungary).

Dynamics of Quaternary alluvial fans

Research on Quaternary alluvial fans tends to focus on the relationships between fan sediments and morphology, but has moved beyond earlier descriptive morphological studies (e.g. Denny 1965; Hooke 1967, 1968; Bull 1977) and basic morphometric studies (Bull 1977; Harvey 1984, 1997). More attention has been paid to the morphological style of alluvial fans insofar as this reflects fan setting, including the conventional tectonically active mountain-front setting and backfilled fans on tectonically stable mountain fronts (Bull 1978), or tributary-junction settings (Harvey 1997). Fan setting, in turn, affects the degree of confinement (Sorriso-Valvo *et al.* 1998; Harvey *et al.* 1999a) and accommodation space (Viseras *et al.* 2003). In addition, fan style reflects the geomorphic regime of the fan, from aggrading through to prograding telescopic fans and to fans undergoing dissection (Harvey 2002a). This is controlled fundamentally by the relationships between sediment supply to the fan and by flood power, in other words by the threshold of critical stream power (Bull 1979).

Perhaps the main emphasis over recent years in research on Quaternary alluvial fans has been to consider how the gross factors controlling alluvial-fan environments (catchment controls, tectonics and long-term geomorphic evolution, climatic controls, base level) interact to create fan morphological and sedimentary styles, and how fans respond to changes in the controlling factors. Studies of fan morphometry have developed to focus on these interactions (Silva *et al.* 1992; Calvache *et al.* 1997; Harvey *et al.* 1999a; Harvey 2002c; Viseras *et al.* 2003).

It has long been realized that of the controlling factors, catchment characteristics (including drainage basin area, relief and geology) control the supply of water and sediment to the fan, and therefore the process regime on the fan and the resulting fan morphology. Although emphasized in the geological literature (see below), tectonic controls may influence sediment production in the source area, and, together with gross topography, appear primarily to control fan location, fan setting and gross fan geometry (Harvey 2002a) rather than fan-sediment sequences. Tectonics appear to influence fan morphology and sedimentary sequences primarily through an influence on accommodation space (Silva *et al.* 1992; Viseras *et al.* 2003).

For Quaternary alluvial fans in a wide variety of environments, climatic factors appear to have an overwhelming control on fan morphology and sedimentary styles. This is not to say that there are fundamental differences between fans in different climatic environments; there are differences, of course, but the primary processes differ remarkably little between humid and arid environments, or between arctic and subtropical environments. In each case the fan morphology appears to be adjusted to the prevailing sediment supply and flood regime. If a climatic *change* alters flood power and/or sediment supply, the fan responds by a change in erosional or depositional regime, resulting in a change in the sedimentary environment. There are numerous examples of studies demonstrating where climatic shifts have resulted in changes to fan systems (e.g. Wells *et al.* 1987; Bull 1991; Harvey *et al.* 1999*b*; Harvey & Wells 2003). Studies that have evaluated the relative roles of tectonism and climate for changes in sedimentary and geomorphic sequences on Quaternary fans, demonstrate that climate is overwhelmingly the primary control (Frostick & Reid 1989; Ritter *et al.* 1995; Harvey 2004).

Most fans toe out at stable base levels, therefore base-level change is not a common cause of changes in fan behaviour on Quaternary fans. However, on fans that toe out to coasts or lakeshores and on fans subject to 'toe trimming' (Leeder & Mack 2001) by an axial river system, base-level change may be an effective mechanism for triggering incision into distal fan surfaces (Harvey 2004). However, base-level fall may not always trigger incision; this very much depends on the gradients involved. Indeed a base-level *rise* may trigger distal incision on coastal fans, if rising sea levels cause coastal erosion and fan profile foreshortening and steepening (Harvey *et al.* 1999*a*). Base-level change, such as a eustatic sea-level change or a change in the level of a pluvial lake, may be ultimately climatically controlled. The effectiveness of base-level change will depend on how it relates temporally to changes in sediment supply that are also climatically led. In cases where base-level change has been identified as an effective mechanism on Quaternary alluvial fans, the primary signal is a climatically led sediment signal, modified by the effects of base-level change (Bowman 1988; Harvey *et al.* 1999*b*; Harvey 2002*d*, 2004).

Within this volume we present several papers that deal with interacting controls on Quaternary alluvial fans in a range of mostly dry climate settings: in Andean mountain environments in Argentina (**Columbo**), and in deserts in the Middle East (**Al-Farraj & Harvey**), Chile (**Hartley *et al.***) and the American West (**Harvey**).

In the past, a full appreciation of the response of alluvial fans to environmental change or of their wider role within fluvial systems has been hampered by the lack of suitable dating methods, applicable to alluvial-fan sediments. While it is true that radiocarbon dating has been effectively applied to late Pleistocene or Holocene sequences of alluvial-fan development in humid regions (Brazier *et al.* 1988; Chiverrell *et al.* in press), the precise dating of dry-region alluvial fans over longer timescales has always been a problem. Traditional methods have involved the mapping and correlation of fan surfaces on the basis of relative age, expressed through desert pavement or soil development especially of calcreted surfaces (Lattman 1973; Harden 1982; Machette 1985; McFadden *et al.* 1989; Amit *et al.* 1993; Alonso-Zarza *et al.* 1998; Al-Farraj & Harvey 2000). In some cases fan sequences have been correlated with dated lake or coastal sediments (Harvey *et al.* 1999*a, b*). Three new dating techniques have potential in alluvial fan research. Cosmogenic dating, applied to depositional surfaces (Anderson *et al.* 1996) or to questions of sediment flux (Nichols *et al.* 2002), U/Th dating of pedogenic carbonates (Kelly *et al.* 2000; Candy *et al.* 2003, 2004), and applications of optically stimulated luminescence (OSL) dating to fluvial-terrace and alluvial-fan sediments have opened up new opportunities for alluvial-fan research. Two papers in this volume focus on recent applications of luminescence dating to alluvial-fan sediments (**Pope & Wilkinson**; **Robinson *et al.***).

Alluvial-fan sedimentary sequences

The challenge facing research on alluvial-fan sedimentary sequences is to develop interpretive models that are compatible with the findings of research on contemporary processes and on extant Quaternary alluvial fans. To some extent previous models may be seen as simplistic. The assumptions about overall coarsening-up sequences within alluvial-fan sediments (Steel 1974; Steel *et al.* 1977; Rust 1979) may be appropriate for distal-fan environments, when the proximal-fan sediments are being reworked by fanhead trenching, but are certainly not appropriate for proximal environments on aggrading fans (Harvey 1997). There we would expect an overall fining-up trend as the topography becomes progressively buried and each location effectively becomes more distal.

Another concept that clearly needs revision is the oversimplistic association of the wet/dry fans dichotomy with wet/dry climates, respectively (see above). This in part relates to the debate on the relative roles of climate, tectonism and base-level change in ancient alluvial-fan sediment sequences. Studies of Quaternary fans (see above) recognize that tectonics may play a major role in controlling the location and setting of alluvial fans, but fan sequences respond primarily to climatic controls. Within most of the previous geological literature tectonics are seen as the primary control of change in alluvial-fan sequences. We need to ask why there is this discrepancy. Are most of the geologists working on ancient sequences wrong? Is the climatic signal difficult to identify from the sediments alone, especially if some of the simplest climatic relationships are, in fact, oversimplifications? Perhaps this has something to do with scale. Most Quaternary fans are small in comparison with the scale of ancient systems described in the literature, therefore there is a question of preservation potential. Only the larger systems, especially those on the margins of sedimentary basins, and those developed over longer timescales are likely to be preserved in the geological record. It is known that responses to climatic change and to tectonic activity operate over different timescales (Harvey 2002*b*). Quaternary alluvial fans respond to climatic change over timescales of 10^2–10^4 years, whereas tectonic change operates over timescales well in excess of 10^4 years. Furthermore, during the global glacial–interglacial cycles of the Quaternary there were major changes in sediment input to fan environments in a wide range of climatic settings. Climatic changes during, say, the Triassic or the Cretaceous or the Neogene may have had less influence on geomorphic thresholds and therefore may have had a less overwhelming impact on sediment supply.

A second area of challenge in the understanding of ancient alluvial-fan sedimentary sequences lies in the development of an understanding of the three-dimensional (3D) geometry of alluvial-fan sedimentary bodies, and the interpretation of the 3D geometry in the context of models of sedimentary sequences in sedimentary basins. Of interest would be the application of sequence stratigraphy concepts to terrestrial environments, including alluvial fans, within which the salient events were not sea-level change, but climatic events. In such models (e.g. Weissmann *et al.* 2002) soil-covered fan surfaces could be seen as the sequence bounding surfaces. In this volume **Weissmann *et al*.** apply sequence stratigraphy concepts to the interpretation of the 3D geometry of Quaternary fans in the Central Valley of California. Although dealing with Quaternary fans, the methodology would be applicable to ancient fan sequences.

A final challenge for the application of concepts derived from modern fans to the interpretation of ancient sequences is to infer source-area characteristics from alluvial-fan sediments. A start has been made here by Mather *et al.* (2000) in applying morphometric relations derived from Quaternary fans to the assessment of drainage basin properties during the Pliocene in SE Spain. Pliocene fan geometry suggests that since emplacement of these sediments there have been major drainage reorganizations within the feeder catchments through river capture.

The influence of headwater river capture on downstream fluvial system development is itself an under-researched topic (Mather 2000).

Increasingly, there is more integration between studies of ancient fan sequences and studies of modern processes and of Quaternary fans. In each case the central theme of explanation lies in understanding the interactions between the tectonic, climatic and base-level controls. The final three papers in this volume illustrate this theme, addressing ancient fan sequences in different tectonic settings: **Nichols** on Tertiary fans of the Ebro Basin, **Wagreich & Strauss** on Tertiary fans in Austria and **Leleu *et al*.** on Cretaceous fans in Provence.

The future

Almost all the papers in this volume, whether dealing with ancient sequences, Quaternary fans or modern processes, address two fundamental questions. How do fans respond to the combinations of the controlling factors, and how is that response expressed in the morphology and the sediments? Challenges for the future lie partly in applied studies, for example hazards on alluvial fans, alluvial-fan sediments as reservoir rocks for water resources or in the ancient record, for oil. The major challenges, however, relate to the science itself. For Quaternary science: how far will more precise and more comprehensive dating allow integration of alluvial-fan research within wider basin-wide geomorphological and sedimentological models? For the study of ancient sequences: how far can the concepts derived from understanding modern fans be integrated more fully into basin sediment sequence models?

In addition to sponsorship from the British Geomorphological Research Group and the British Sedimentological Research Group, sponsorship for the conference was also received from the following organizations: Cortijo Urra Field Centre, Sorbas; Cuevas de Sorbas; Sorbas Municipality; and the Schools of Geography and of Earth, Ocean and Environmental Sciences, University of Plymouth. We would like to thank the staff of the publications section of the Geological Society for making this publication possible. We also thank the following who kindly acted as reviewers for the submitted papers: D. Bowman, P. Carling, M. Church, T. Elliott, P. Friend, A. Hartley, S. Jones, A. Lang, B. Luckman, M. Macklin, L. Mayer, G. Nichols, R. Robinson, P. Silva, C. Viseras and G. Weissmann.

References

AL-FARRAJ, A. 1996. *Late Pleistocene geomorphology in Wadi Al-Bih, northern U.A.E. and Oman: with special emphasis on wadi terraces and alluvial fans*. PhD Thesis, University of Liverpool.

AL-FARRAJ, A. & HARVEY, A.M. 2000. Desert pavement characteristics on wadi terrace and alluvial fan surfaces: Wadi Al-Bih, U.A.E. and Oman. *Geomorphology*, **35**, 279–297.

ALONSO-ZARZA, A.M., SILVA, P.G., GOY, J.L. & ZAZO, C. 1998. Fan-surface dynamics and biogenic calcrete development: Interactions during ultimate phases of fan evolution in the semiarid SE Spain (Murcia). *Geomorphology*, **24**, 147–167.

AMIT, R., GERSON, R. & YAALON, D.H. 1993, Stages and rate of the gravel shattering process by salts in desert Reg soils. *Geoderma*, **57**, 295–324.

ANDERSON, R.S., REPKA, J.L. & DICK, G.S. 1996. Explicit treatment of inheritance in dating depositional surfaces using in situ [10] Be and [26] Al. *Geology*, **24**, 47–51.

BAKER, V.R., 1977. Stream channel response to floods, with examples from central Texas. *Bulletin of the Geological Society of America*, **88**, 1057–1071.

BLAIR, T.C. & McPHERSON, J.G. 1994*a*. Alluvial fan processes and forms. *In*: ABRAHAMS, A.D. & PARSONS, A.J. (eds) *Geomorphology of Desert Environments*. Chapman & Hall, London, 354–402.

BLAIR, T.C. & McPHERSON, J.G. 1994*b*. Alluvial fans and their natural distinction from rivers based on morphology, hydraulic processes, sedimentary processes and facies assemblages. *Journal of Sedimentary Research*, **A64**, 450–489.

BLAIR, T.C. & McPHERSON, J.G. 1998. Recent debris-flow processes and resultant forms and facies of the Dolomite alluvial fan, Owens Valley, California. *Journal of Sedimentary Research*, **68**, 800–818.

BOOTHROYD, J.C. & NUMMENDAL, D. 1978. Proglacial braided outwash – a model for humid alluvial-fan deposits. *In*: MIALL, A.D. (ed.) *Fluvial Sedimentology*. Canadian Society for Petroleum Geologists, Memoir, **5**, 641–688.

BOWMAN, D. 1988. The declining but non-rejuvenating base-level – the Lisan Lake, the Dead sea, Israel. *Earth Surface Processes and Landforms*, **13**, 239–249.

BRAZIER, V., WHITTINGTON, G. & BALLANTYNE, C.K. 1988. Holocene debris cone evolution in Glen Etive, Western Grampian Highlands, Scotland. *Earth Surface Processes and Landforms*, **13**, 525–531.

BULL, W.B. 1977. The alluvial fan environment. *Progress in Physical Geography*, **1**, 222–270.

BULL, W.B. 1978. Geomorphic Tectonic Activity Classes of the South Front of the San Gabriel Mountains, California. United States Geological Survey, Contract Report 14-08-001-G-394, Office of Earthquakes, Volcanoes and Engineering, Menlo Park, CA.

BULL, W.B., 1979. Threshold of critical power in streams. *Bulletin of the Geological Society of America*, **90**, 453–464.

BULL, W.B. 1991. *Geomorphic Responses to Climatic Change*. Oxford University Press, Oxford.

CALVACHE, M., VISERAS, C. & FERNANDEZ, J. 1997. Controls on alluvial fan development – evidence from fan morphometry and sedimentology; Sierra Nevada, SE Spain. *Geomorphology*, **21**, 69–84.

CANDY, I., BLACK, S. & SELLWOOD, B.W. 2004. Quantifying time scales of pedogenic calcrete formation using U-series disequilibria. *Sedimentary Geology*, **170**, 177–187.

CANDY, I., BLACK, S., SELLWOOD, B.W. & ROWAN, J.S. 2003. Calcrete profile development in Quaternary alluvial sequences, southeast Spain; implications for using calcretes as a basis for landform chronologies. *Earth Surface Processes and Landforms*, **28**, 169–185.

CHIVERRELL, R.C., HARVEY, A.M. & FOSTER, G. Hillslope gullying in northwest England and southwest Scotland: response to human impact and/or climatic downturn. *Geomorphology*, in press.

CRONIN, J.J., NEALL, V.E., LECOINTRE, J.A. & PALMER, A.S. 1997. Changes in Whangachu River lahar characteristics during the 1995 eruption sequence, Rhuaepechu Volcano, New Zealand. *Journal of Volcanology & Geothermal Research*, **76**, 47–61.

DENNY, C.S. 1965. *Alluvial Fans in the Death Valley Region, California and Nevada*. United States Geological Survey, Professional Paper, **466**.

ENGELUND, F. & WAN, Z.H. 1984. Instability of hyperconcentrated flow. *Journal of Hydraulic Engineering, ASCE*, **110**, 219–233.

FROSTICK, L.E. & REID, I. 1989. Climatic versus tectonic controls of fan sequences: lessons from the Dead Sea, Israel. *Journal of the Geological Society, London*, **146**, 527–538.

GARCIA, A.F. & STOKES, M. Alluvial fan and pluvial lake response to the Pleistocene-Holocene transition in Jakes Valley, central Great Basin, USA. *Quaternary Research*, in press.

GERSON, R. 1982. Talus relics in deserts: A key to major climatic fluctuations. *Israel Journal of Earth Sciences*, **31**, 123–132.

GOHAIN, K. & PARKASH, B. 1990. Morphology of the Kosi megafan. *In*: RACHOCKI, A.H. & CHURCH, M. (eds) *Alluvial Fans: A Field Approach*. Wiley, Chichester, 151–178.

HARDEN, J.W. 1982. A quantitative index of soil development from field descriptions: examples from a chronosequence in central California. *Geoderma*, **28**, 1–28.

HARVEY, A.M. 1984. Aggradation and dissection sequences on Spanish alluvial fans: influence on morphological development. *Catena*, **11**, 289–304.

HARVEY, A.M. 1990. Factors influencing Quaternary alluvial fan development in southeast Spain. *In*: RACHOCKI, A.H. & CHURCH, M. (eds) *Alluvial Fans: A Field Approach*. Wiley, Chichester, 247–269.

HARVEY, A.M. 1996. The role of alluvial fans in the mountain fluvial systems of southeast Spain: implications of climatic change. *Earth Surface Processes and Landforms*, **21**, 543–553.

HARVEY, A.M. 1997. The role of alluvial fans in arid-zone fluvial systems. *In*: THOMAS, D.S.G. (ed.) *Arid Zone Geomorphology; Process, Form and Change in Drylands*, 2nd edn. Wiley, Chichester, 231–259.

HARVEY, A.M. 2002*a*. Factors influencing the geomorphology of dry-region alluvial fans. *In*: PEREZ-GONZALEZ, A., VEGAS, J. & MACHADO, M.J. (eds) *Aportaciones a la Geomorfologia de Espana en el Inicio del Tercer Milenio*. Instituto Geologico y Minero de Espana, Madrid, 59–75.

HARVEY, A.M. 2002*b*. Effective timescales of coupling in fluvial systems. *Geomorphology*, **44**, 175–201.

HARVEY, A.M. 2002*c*. The relationships between alluvial fans and fan channels within Mediterranean mountain fluvial systems. *In*: BULL, L.J. & KIRKBY, M.J. (eds), *Dryland Rivers: Hydrology and Geomorphology of Semi-arid Channels*. Wiley, Chichester, 205–226.

HARVEY, A.M. 2002*d*. The role of base-level change in the dissection of alluvial fans: case studies from southeast Spain and Nevada. *Geomorphology*, **45**, 67–87.

HARVEY, A.M. 2003. Uplift, dissection and landform evolution: The Quaternary. *In*: MATHER, A.E., MARTIN, J.M., HARVEY, A.M. & BRAGA, J.C. 2003. *A Field Guide to the Neogene Sedimentary Basins of the Almeria Province, South-east Spain*. Blackwell Science, Oxford, 225–322.

HARVEY, A.M. 2004. The response of dry-region alluvial fans to late Quaternary climatic change. *In*: ALSHARHAN, A.S., WOOD, W.W., GOUDIE, A.S., FOWLER, A. & ABDELLATIF, E.M. (eds) *Desertification in the Third Millenium*. Balkema, Rotterdam, 83–98.

HARVEY, A.M. & WELLS, S.G. 1994. Late Pleistocene and Holocene changes in hillslope sediment supply to alluvial fan systems: Zzyzx, California. *In*: MILLINGTON, A.C. & PYE, K. (eds) *Environmental Change in Drylands: Biogeographical and Geomorphological Perspectives*. Wiley, Chichester, 67–84.

HARVEY, A.M. & WELLS, S.G. 2003. Late Quaternary alluvial fan development, relations to climatic change, Soda Mountains, Mojave Desert, California. *In*: LANCASTER, N., ENZEL, Y. & WELLS, S.G. (eds) *Environmental Change in The Mojave Desert*. Geological Society of America, Special Paper, **368**, 207–230.

HARVEY, A.M., SILVA, P.G., MATHER, A.E., GOY, J.L., STOKES, M. & ZAZO, C. 1999*a*. The impact of Quaternary sea-level and climatic change on coastal alluvial fans in the Cabo de Gata ranges, southeast Spain. *Geomorphology*, **28**, 1–22.

HARVEY, A.M., WIGAND, P.E. & WELLS, S.G. 1999*b*. Response of alluvial fan systems to the late Pleistocene to Holocene climatic transition: contrasts between the margins of pluvial Lakes Lahontan and Mojave, Nevada and California, USA. *Catena*, **36**, 255–281.

HOOKE, R. LE B. 1967. Processes on arid region alluvial fans. *Journal of Geology*, **75**, 438–460.

HOOKE, R. LE B. 1968. Steady state relationships on arid-region alluvial fans in closed basins. *American Journal of Science*, **266**, 609–629.

JULIEN, P.Y. & LAN, Y.G. 1991. Rheology of hyperconcentrations. *Journal of Hydraulic Engineering, ASCE*, **117**, 346–353.

KELLY, M., BLACK, S. & ROWAN, J.S. 2000. A calcrete-based U/Th chronology for landform evolution in the Sorbas basin, southeast Spain. *Quaternary Science Reviews*, **9**, 995–1010.

KESEL, R.H. & SPICER, B.E. 1985. Geomorphic relationships and ages of soils on alluvial fans in the Rio General valley, Costa Rica. *Catena*, **12**, 149–166.

KIM, S.B. 1995. Discussion: Alluvial fans and their natural distinction from rivers based on morphology, hydraulic processes and facies assemblages. *Journal of Sedimentary Research*, **65A**, 706–708.

KOCHEL, R.C. 1990. Humid alluvial fans of the Appalachian Mountains. *In*: RACHOCKI, A.H. & CHURCH, M. (eds) *Alluvial Fans: A Field Approach*. Wiley, Chichester, 109–129.

KOSTASCHUK, R.A., MACDONALD, G.M. & PUTNAM, P.E. 1986. Depositional processes and alluvial fan–drainage basin morphometric relationships near Banff, Alberta, Canada. *Earth Surface Processes and Landforms*, **11**, 471–484.

LATTMAN, L.H. 1973. Calcium carbonate cementation of alluvial fans in southern Nevada. *Bulletin of the Geological Society of America*, **84**, 3013–3028.

LAVIGNE, F. & SUWA, H. 2004. Contrasts between debris flows, hyperconcentrated flows and stream flows at a channel of Mount Semeru, East Java, Indonesia. *Geomorphology*, **61**, 41–58.

LEEDER, M.R. & MACK, G.H. 2001. Lateral erosion ('toe-cutting') of alluvial fans by axial rivers: implications for basin analysis and architecture. *Journal of the Geological Society, London*, **158**, 885–893.

MACHETTE, M.N. 1985. Calcic soils of the southwestern United States. *In*: WEIDE, D.L. (ed.) *Soils and Quaternary Geology of the Southwestern United States*. Geological Society of America, Special Paper, **203**, 1–21.

MATHER, A.E. 2000. Impact of headwater river capture on alluvial system development: an example from SE Spain. *Journal of the Geological Society, London*, **157**, 957–966.

MATHER, A.E. & STOKES, M. (eds). 2003. Long-term landform evolution in southern Spain. *Geomorphology*, **50**, (Special Issue), 151–171.

MATHER, A.E., HARVEY, A.M. & STOKES, M. 2000. Quantifying long term catchment changes of alluvial fan systems. *Bulletin of the Geological Society of America*, **112**, 1825–1833.

MATHER, A.E., MARTIN, J.M., HARVEY, A.M. & BRAGA, J.C. 2003. *A Field Guide to the Neogene Sedimentary Basins of the Almeria Province, South-east Spain*. Blackwell Science, Oxford.

MCCARTHY, T.S. & CANDLE, A.B. 1995. Discussion: Alluvial fans and their natural distinction from rivers based on morphology, hydraulic processes and facies assemblages. *Journal of Sedimentary Research*, **65A**, 581–583.

MCFADDEN, L.D., RITTER, J.B. & WELLS, S.G. 1989. Use of multiparameter relative-age methods for age estimation and correlation of alluvial fan surfaces on a desert piedmont, Eastern Mojave Desert, California. *Quaternary Research*, **32**, 276–290.

NICHOLS, K.K., BIERMAN, P.R., HOOKE, R.L., CLAPP, E.M. & CAFFEE, M. 2002. Quantifying sediment transport on desert piedmonts using ^{10}Be and ^{26}Al. *Geomorphology*, **45**, 105–125.

RACHOCKI, A.H. & CHURCH, M. 1990. *Alluvial Fans: A Field Approach*. Wiley, Chichester.

Rickenmann, D. 1991. Hyperconcentrated flow and sediment transport on steep slopes. *Journal of Hydraulic Engineering, ASCE*, **117**, 1419–1439.

RITTER, J.B., MILLER, J.R., ENZEL, Y. & WELLS, S.G. 1995. Reconciling the roles of tectonism and climate in Quaternary alluvial fan evolution. *Geology*, **23**, 245–248.

RUST, B.R. 1979. Facies Models 2: Coarse alluvial deposits. *In*: Walker, R.G. (ed.) *Facies Models*. Geoscience Reprint Series 1. Kitchener, Ontario, Canada, 9–21.

RYDER, J.N. 1971. The stratigraphy and morphology of paraglacial alluvial fans in south central British

Columbia. *Canadian Journal of Earth Sciences*, **8**, 279–298.

SAITO, K. & OGUCHI, T. 2005. Slope of alluvial fans in humid regions of Japan, Taiwan and the Philippines. *Geomorphology*, **70**, 147–162.

SCHUMM, S.A. 1977. *The Fluvial System*, Wiley, New York.

SILVA, P., HARVEY, A.M., ZAZO, C. & GOY, J.L. 1992. Geomorphology, depositional style and morphometric relationships of Quaternary alluvial fans in the Guadalentin Depression (Murcia, Southeast Spain). *Zeitschrift für Geomorphologie Neue Folge*, **36**, 325–341.

STEEL, R.R. 1974. New Red Sandstone floodplain and piedmont sedimentation in the Hebridean Province, Scotland. *Journal of Sedimentary Petrology*, **44**, 336–357.

STEEL, R.J., MOEHLE, S., NILSON, H., ROE, S.L. & SPINNANGRE, A. 1977. Coarsening upwards cycles in the alluvium of Homelen Basin (Devonian), Norway: sedimentary response to tectonic events. *Bulletin of the Geological Society of America*, **88**, 1124–1134.

SORRISO-VALVO, M., ANTRONICO, L. & LE PERA, E. 1998. Controls on modern fan morphology in Calabria, Southern Italy. *Geomorphology*, **24**, 169–187.

STOKES, M., NASH, D.J. & HARVEY, A.M. Calcrete fossilisation of alluvial fans in SE Spain: groundwater vs pedogenic processes for calcrete development. *Geomorphology*, in press.

VISERAS, C., CALVACHE, M.L., SORIA, J.M. & FERNANDEZ, J. 2003. Differential features of alluvial fans controlled by tectonic or eustatic accommodation space. Examples from the Betic cordillera, Spain. *Geomorphology*, **50**, 181–202.

WANG, Z.W., LARSON, P. & XIANG, W. 1994. Rheological properties of sediment suspensions and their implications. *Journal of Hydraulics Research*, **32**, 495–516.

WEISSMAN, G.S., MOUNT, J.F. & FOGG, G.E. 2002. Glacially driven cycles in accumulation space and sequence stratigraphy of a stream-dominated alluvial fan, San Joachim Valley, California, U.S.A. *Journal of Sedimentary Research*, **72**, 270–281.

WELLS, S.G. & HARVEY, A.M. 1987. Sedimentologic and geomorphic variations in storm generated alluvial fans, Howgill Fells, northwest England. *Bulletin of the Geological Society of America*, **98**, 182–198.

WELLS, S.G., McFADDEN, L.D. & DOHRENWEND, J.C. 1987. Influence of late Quaternary climatic changes on geomorphic and pedogenic processes on a desert piedmont, Eastern Mojave Desert, California. *Quaternary Research*, **27**, 130–146.

Flow events on a hyper-arid alluvial fan: Quebrada Tambores, Salar de Atacama, northern Chile

ANNE E. MATHER[1] & ADRIAN HARTLEY[2]

[1]*School of Geography, University of Plymouth, Plymouth PL4 8AA, UK*
(e-mail: a.mather@plymouth.ac.uk)
[2]*Department of Geology & Petroleum Geology, University of Aberdeen,*
Aberdeen AB24 3UE, UK (e-mail: a.hartley@abdn.ac.uk)

Abstract: The Tambores alluvial fan is located within the hyper-arid Atacama Desert of northern Chile. We examine evidence of the range of flow processes operative in this environment from a combination of Pleistocene–Holocene fan deposits and a recent (2001) flood event (16 m^3 s^{-1}) in the fan feeder channel and upper alluvial-fan area. The field evidence suggests that peak flows recorded in the older deposits generated extensive sheetflood events dominated by antidune deposition in the upper fan area. These extreme, supercritical flows were generated by floods with sustained high sediment and water discharges and high stream power. Easily erodable alluvial source materials ensured high sediment discharge could be maintained within flood events. High stream power was ensured as a function of the tectonically exacerbated gradients within the source area. The 2001 event indicates the rapid rheological changes that can occur within an individual flood event, ranging from hyperconcentrated streamflow to mudflow. The flow deposits vary little in maximum clast size either between the varying flood events in the upper fan area, or down the fan gradient. This is due to a limited calibre of sediment being produced from the source area. The study highlights: (1) the range of flow rheologies that can be generated from a hyper-arid catchment both within and between flood events of varying magnitude and the associated difficulties in generating a reliable stratigraphy from the resultant deposits; (2) the high stream power and sediment discharge associated with major flood events and thus the nature of flood hazard in the catchment and on the fan; and (3) the limitations of sedimentological information such as maximum clast size as an indicator of peak flow characteristics in ancient deposits.

Hyper-arid regions by definition have limited water resources. In such environments the hydrological resources that do exist are commonly linked to alluvial fans (Ben David-Novak & Schick 1997; Houston 2001, 2002). Despite this lack of availability of water, floods provide one of the greatest hazards to populations in hyper-arid regions, as many population centres are built on alluvial fan bodies and thus in potentially hazardous locations, as the town of Antofagasta (coastal northern Chile) discovered in June 1991. More than 100 people were killed and extensive damage was caused by the 1991 flood event that followed 42 mm of rain which fell mainly within 3 h (Chong *et al.* 1991; Hauser 1997). Yet, studies of fluvial processes in hyper-arid areas are limited. The paucity of rainfall and lack of population usually mean that rainfall records are disparate. Ancient fluvial sediments in such environments can contribute to our understanding of the rare flood events that occur. However, the fluvial record is commonly biased towards the preservation of the deposits of larger magnitude floods at the expense of smaller ones. This is a direct consequence of the available stream power and thus erosive capability of the larger events. Hence, within ancient fluvial records, although there may be well-preserved evidence of the peak flood events, information on the

lower magnitude floods is typically poor, and determining the frequency of such events is thus difficult. Alluvial fans, however, are dominantly accretional sedimentary bodies that preserve a more complete record of flood events generated from the supplying fan catchment and feeder channel, unless the alluvial fan has become fully trenched and is thus bypassed by such events. Yet, despite this, studies of processes on hyper-arid alluvial fans are rare. Some examine hydrology (Ben David-Novak & Schick 1997; Houston 2002) or soils and weathering (Jones 1991; Amit *et al.* 1996; Berger & Cooke 1997). Research on sedimentology and process is mainly limited to studies within Death Valley in the USA (e.g. Blair 1999, 2000).

This paper aims to use the Pleistocene–Recent deposits of the Quebrada Tambores alluvial-fan system in the hyper-arid Atacama Desert of northern Chile to reconstruct the flow processes operative during fan construction and entrenchment. The paper examines the lowermost feeder channel and fan apex section of the alluvial-fan system. The older deposits record high-magnitude flood events from a relatively wetter (but still arid) period in the fan's history that was mostly responsible for the alluvial-fan construction. The recent event (February 2001) was generated from a low-magnitude flood and gives

From: HARVEY, A.M., MATHER, A.E. & STOKES, M. (eds) 2005. *Alluvial Fans: Geomorphology, Sedimentology, Dynamics.*
Geological Society, London, Special Publications, **251**, 9–29. 0305-8719/05/$15 © The Geological Society of London 2005.

an indication of the processes operative under the modern hyper-arid setting where incision and fanhead trenching appear to dominate. The older (Pleistocene and Holocene) depositional records give an indication of the nature of major floods that could be generated from rare extreme rainfall events under the present hyper-arid setting.

Geology and structure

The catchment area of Quebrada Tambores is dominated by Cretaceous and Tertiary age alluvial sediments of the Purilactis Group, overlying Paciencia Group and Neogene sediments (Fig. 1). These are predominantly conglomeratic material

within the Tambores catchment, apart from some lower sections of the Paciencia group in the lowermost elevations of the catchment area where finer muds, silts and evaporites are exposed. The area is significantly affected by still-active thrusting that exposes Neogene ignimbrites near the catchment mouth (Fig. 1) which are interbedded with a sequence of mid-Miocene–mid-Pliocene alluvial gravels, an as yet unnamed stratigraphic unit. Catchment slope and channel gradients are steepened by uplift along high-angle reverse faults that have been active throughout the Neogene. This uplift, in turn, has enabled (and is enabling) small quebradas to headcut across and breach hanging-wall drainage divides, facilitating expansion of the main quebrada by drainage capture (Fig. 2).

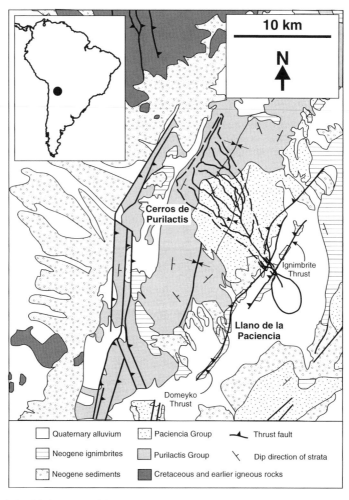

Fig. 1. Location and simplified geology of the study area with the catchment and fan area (shown in detail in Fig. 5) represented in grey outline.

(a)

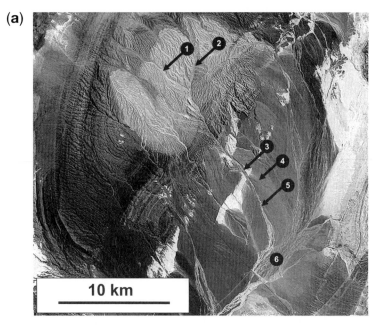

10 km

Fig. 2. General geomorphology of the study area. **a**) Satellite image showing the two main quebrada channels discussed in the text (1 and 2); the fan feeder channel (3); the relict older fan surface (4); fan intersection point (5); and axial fluvial system (6).

Climate

Over geological timescales the Atacama Desert has been dominated by a semi-arid climate from 8 to 3 Ma, punctuated by a period of greater aridity at about 6 Ma. Hyper-aridity became dominant at about 3 Ma (Hartley & Chong 2002). Where higher resolution records exist in the Quaternary it is apparent that fluctuations from hyper-arid to relatively wetter phases have occurred over varying timescales within the study area. A 106 ka salt core from the Atacama appears to indicate wetter periods within the Pleistocene at 75.7–60.7 and 53.4–15.5 ka, with the wettest period recorded at 26.7–16.5 ka. Many studies indicate a Late Pleistocene–early Holocene wetter period that started at >15.4 ka (Rech *et al.* 2002) or later (Bobst *et al.* 2001; Latorre *et al.* 2003). Within the highest resolution record of the Holocene, although timing of events varies according to location and methodology, there appears to be broad agreement that there was a period of greater rainfall in the early and mid-Holocene, and that the extreme hyper-aridity experienced today has dominated for the last 3 ka (Fig. 3). The palaeohydrology and vegetation records used to construct Figure 3 are also supplemented by sedimentological and archaeological evidence from the study area. Abundant debris-flow deposits, indicating more frequent storm events, are intercalated with archaeological remains dated at 6200–3100 C-14[C14] years BP (Grosjean *et al.* 1997).

The vegetation data derived from rodent middens would appear to indicate that some of the increased rainfall was related to increased seasonality with an increase in summer rainfall (particularly for the period 11.8–10.5 ka: Latorre *et al.* 2002). However, the overall volume changes in rainfall estimated from plant communities within the study area suggest annual precipitation for wetter periods was probably still within the arid realm, with rainfall increasing from the present 40–50 to 80–100 mm year^{-1} for the Late Glacial–Holocene wetter phase (Latorre *et al.* 2003). It has been proposed that the increases in seasonal wetness are associated with greater annual volumes of precipitation from the South American Summer Monsoon across the Andes, from the east, intensified by Walker Circulation (stronger easterlies) (Latorre *et al.* 2002; Rech *et al.* 2002).

The aridity of the modern climate of the Atacama Desert is influenced by four main factors (Houston & Hartley 2003). These are: (1) a pronounced rainshadow affect created by the high Andes, with annual rainfall dropping from over 300 mm year^{-1} at 5000 m above sea level (masl) to 20 mm year^{-1} at 2300 masl; (2) the descending stable air of the Hadley Circulation of the subtropical high-pressure belt; (3) the region's continentality from Atlantic moisture sources; and (4) the cold Peruvian ocean current that upwells along the Atacama coast and inhibits the uptake of moisture for any Pacific onshore winds. Within the study area used for this paper

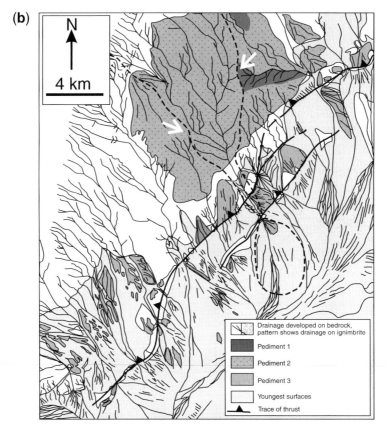

Fig. 2. b) A sketch map drawn from the satellite image. The Tambores catchment and fan are outlined by the dashed lines. White arrows indicate examples of stream piracy occurring within the catchment areas, with the modern drainage capturing older drainages preserved on pediment levels (pediment surface 1 is topographically highest and oldest, and 3 is the youngest and lowest).

rainfall ranges up to 80 mm year^{-1} (Fig. 4), but is generally much less. Seasonal precipitation is linked to the tail end of the South American Summer Monsoon, produced by continental heating over the Altiplano during the austral spring and summer. This is limited by the magnitude of the Bolivian High (Latorre *et al.* 2003). Modern interannual variation of rainfall figures (over 2–5 years) is attributed to the influence of the El Niño Southern Oscillation (Vuille 1999; Garreaud & Aceituno 2001), with highest rainfalls associated with La Niña periods when the Bolivian High is displaced southward and moisture-bearing easterly winds are enhanced.

Geomorphology of the fan catchment and depositional areas

The modern Tambores alluvial fan is located on the eastern side of the Andean Precordillera, in the Cordillera Domeyko, that spans a latitude in the order of 4° and runs parallel to the main Andean front.

The catchment area covers 52 km², with a basin relief of 889 m (Figs 5 & 6). The majority of the catchment area, above 3200 m, lies within a mixed 'Prepuna–Tolar' vegetation zone (Latorre *et al.* 2003), and the lower part (to approximately 2700 m) in the Prepuna vegetation zone. Within these vegetation zones most vegetation growth is constrained to surface washes. The geomorphology of the catchment area is dominated by a series of pediments that record earlier stages of drainage evolution in relation to the active geological structure (Fig. 2).

The fan feeder valley is 140 m wide at the top of the fan and contains two or three active channels of the order of 10–50 m wide. These channels are locally cut into the Pleistocene–Holocene deposits that have backfilled into the feeder channel. These deposits appear to grade to the top surface of the main alluvial fan into which the modern fan trench

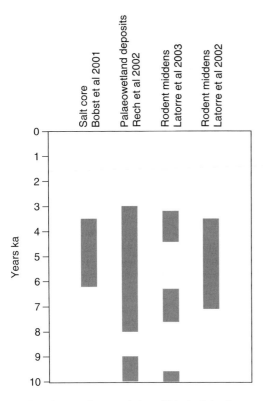

Fig. 3. Holocene climate variation within the Salar de Atacama region. Grey zones indicate period of relatively higher rainfall (but, in most cases, still arid). Sources of information are indicated along the top of the diagram.

Quaternary alluvial-fan sediment provenance and calibre

The lithology of the Tambores catchment is dominated by alluvial sediments that provide the main source material for the Quaternary alluvial-fan system. The alluvial-fan provenance is thus dominated by multigeneration conglomeratic material comprising mainly higher density igneous clasts together with lower density ignimbrite boulders sourced mainly from the lowest parts of the fan catchment (Fig. 1). These low-density blocks dot the surface of the alluvial-fan body and can be rafted considerable distances down the fan surface (Figs 5, 6a & 7a), with individual clast volumes of up to 33.2 m^3 located some 2–3 km from the nearest source area. Examinations of D_{max} (maximum clast size determined from measurements of the long axis) downfan indicate that whilst the ignimbrite boulders display a weak decrease in maximum size downfan, the higher density clasts do not (Fig. 6a).

Late Pleistocene–Holocene alluvial-fan sedimentology

The uppermost fan-surface deposits exposed within the upper (modern) fanhead trench, together with the terrace cuts evident in the equivalent surfaces within the lower parts of the fan feeder channel, allow an examination of the highest magnitude events which have affected the Tambores fan. The sections studied lie between A and A′ on Figure 5b. These deposits are dominated by gravel grade material that can be grouped into three main facies.

(1) *Bedded sheetform gravel with couplets*

The sheetform gravels (bed geometries with width: thickness >18) with couplets (dm-scale) dominate the sequence in the upper fan and lower part of the feeder channel. They comprise granule–pebble grade material that contains a matrix with mean sand: silt: clay percentages of 47: 44: 9 (Fig. 6b). The internal structure of the gravels is dominated by coarse-fine planar couplets (Fig. 7b). The couplets in the upper fan are dominantly subparallel to the fan surface, with the fan surface at 4°, and the couplets dipping downfan at approximately 5°–10°. The bases of the 1–2 m-thick beds in which the couplets are found are typically planar but may locally scour into the underlying deposits. Within the feeder channel the beds are seen to dip upstream at 10°–20° (Fig. 7c). Clast fabrics may be imbricated with the long axis transverse to the flow direction. Beds may be normally graded. Outsized ignimbrite boulders can be

is cut. The Pleistocene–Holocene deposits are inset within an older fan remnant for approximately 0.5–1 km, suggesting that the Pleistocene–Holocene deposits may have been restricted to a 200-m wide fanhead trench that had an intersection point approximately 0.5–1 km below the present fan apex.

The modern channels of the fan feeder valley merge at the fan apex into a single, 20 m-wide channel, which is incised to a depth of 4–5 m below the main fan surface at the fan apex. The modern intersection point lies 2.4 km below the fan apex. The alluvial fan deposits cover an area of 8 km^2 with surface slopes of 4° (upper fan) to 2° (lower fan) and are mainly located within a zone of 'absolute desert'.

Below the intersection point the fan grades over 3 km to an axial fluvial system that drains towards the SW (Fig. 2). The fluvial system comprises wide, shallow, poorly defined channels that drain eventually into a low-gradient playa area. Aeolian sand accumulation is common around vegetation in this zone.

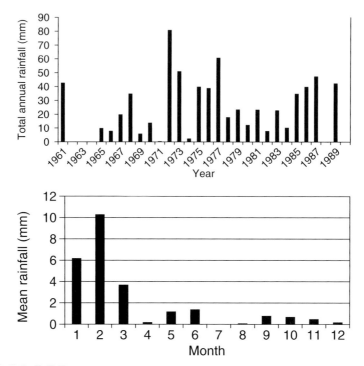

Fig. 4. Rainfall data (1961–1997, Ministerio de Obras, Chile) from San Pedro de Atacama, which is located in the lower elevations of the study area some 10 km to the south. (**a**) Total annual rainfall and (**b**) Mean monthly rainfall.

found lying on bedding surfaces, and buried by subsequent deposits. The overlying deposits show a disrupted sediment fabric around the larger boulders and may include scour features into the underlying beds (Fig. 7d).

(2) Sheetform massive gravel

These are rare within the succession and typically comprise poorly sorted–reverse graded beds less than 1 m in thickness, and often contain subvertical and outsize clasts (Fig. 7e). They are matrix- to clast-supported and tend to drape underlying beds, but may locally be weakly erosive into them.

(3) Channelform gravels

These are more common than (2) but less common than (1) above. The channelform gravels generally comprise shallow channels (width:depth of 5–10) typically 0.5 m deep. The channels are dominantly infilled with clast-supported pebbles and cobbles (D_{max}, excluding ignimbrite clasts, of 7 cm; D_{50} (median clast size) of 3.5 cm), which is persistently

coarser than the granule and pebble grades that comprise the bulk of the sheetform gravels (Fig. 7f).

Reconstruction of the flow conditions for the Late Pleistocene–Holocene alluvial fan

The sheetform gravels with couplets represent the deposits of supercritical flows from high-magnitude, high-sediment-load flood events from the catchment. The area studied for this paper is in the zone where catchment floods are at their peak and exit from an area of relative confinement (the fan feeder channel and fanhead trench) and expand downfan. It is in this zone that maximum transitions in flow behaviour will occur. The couplets have much in common with supercritical flows described from fan bodies elsewhere and interpreted as sheetflood deposits (Blair 1999, 2000). The lack of channelization and prevalence of sheetform geometries would suggest that the whole width of the feeder channel/fanhead trench was occupied by these flood events that expanded and shallowed across the fan surface. Imbrication would suggest that much of the deposition was by traction as bedload. The described couplets most probably relate to antidune development

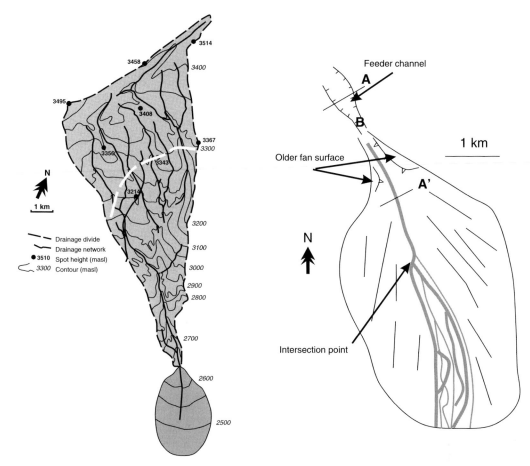

Fig. 5. Topography of the alluvial fan and catchment. (**a**) A general map showing the approximate northern limit to runoff generated by the February 2001 event (dashed white line), drawn from a 1:50 000 topographic map. (**b**) A sketch of the fan surface showing the key morphological elements discussed in the text. The Pleistocene–Holocene deposits were examined between A and A'. B indicates the position of the 2001 flood section survey. The sketch is drawn from aerial photographs and field data.

associated with the generation and destruction of supercritical standing waves during a sheetflood with high sustained water and sediment discharge. Such features have been described from the field (Blair 1987) and from laboratory flume experiments (Alexander *et al.* 2001). Upstream dipping strata are related to upstream growth and migration of water-surface waves and antidunes. Their preservation would suggest a non-violent, gentle termination to the wave train. Downstream-dipping coarse–fine couplets have been attributed to violently breaking rapid downslope washout (Blair 2000). In the latter case, as the turbulence from the washout abates, the finer material in temporary suspension returns to the bedload, creating the sharp based coarse–fine couplets (Blair 2000). The low angle would be consistent with long-wavelength bedforms (Barwis &

Hayes 1985). These sedimentary features suggest that the majority of the Late Pleistocene–Holocene deposits were generated from supercritical flows with a critical Froude number that could range from 0.5 to 1.8, although 0.84 is taken as the most common transition to antidunes in fine gravels (Carling & Shvidchenko 2002). The transition tends to occur at lower Froude numbers as the depth of water increases (Carling & Shvidchenko 2002). Thus, the prevalence of antidune features in the feeder channel and upper fanhead trench areas may be a reflection of deeper water depths here as a function of more laterally constrained flows, with flow depths decreasing downstream with less lateral constriction and higher transmission losses. The inferred less violent washouts within the feeder channel probably reflect more stable flow conditions

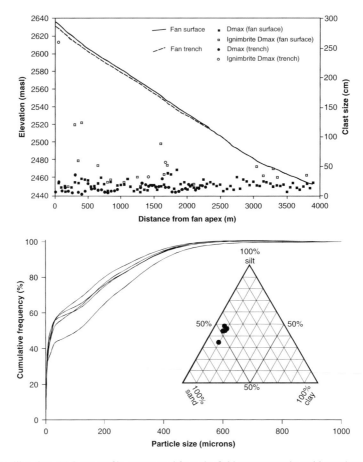

Fig. 6. Sediment calibre data. (**a**) Long profile constructed from the field survey together with maximum clast-size data (using long axis) derived from the fan surface and fanhead trench base; and (**b**) particle size analysis of the finer elements of the deposits selected from five beds in the upper fan area.

within the most constrained flow. The low amount of boulder-grade material reflects the limited availability of such clasts within the catchment area.

The sheetform massive gravels represent rare debris flows with lower water to sediment ratios than the sheetflood events. Their lack of abundance in the depositional sequence suggests they are less common events than the sheetflood deposits. The thickness (m-scale) of these deposits and lower water to sediment content may indicate that these punctuated debris-flow events may represent a combination of: (1) lower intensity rainfall events within the catchment when hillslope material was mobilized but less effective runoff generated; or (2) slope failures during large flood events.

The channelized gravels represent the highest water to sediment flows and may represent low-magnitude runoff events that simply rework the surface or the tail end of the larger flood events once the

main flood wave(s) have passed through the catchment.

Recent deposits: the February 2001 flood

The recent flood event examined was the result of a summer (February 2001) rainfall event. This rainfall event generated a 100–200 year return-period flood that peaked at 310 m³ s⁻¹ in the Rio Salado (Houston 2005), a tributary of the Rio Loa some 60–70 km north of Tambores. This rainstorm event occurred over a number of days from 25 to 28 February, with precipitation starting in the western Cordilleras on 22 February, and building south and west into the Precordillera by the 26. Maximum intensities were reported on 28 February at elevations of 3500 m (data from Houston 2005), but were half as much in the Precordillera where the Tambores catchment

Fig. 7. Sedimentology of the fanhead trench and fan feeder areas. (**a**) Example of an outsized ignimbrite clast in the upper fan. (**b**) Coarse–fine couplets associated with the upper fan area (flow from right to left). Pencil (circled) for scale. (**c**) Upstream dipping beds in lower part of the section (flow from left to right) in the feeder channel. (**d**) Outsize ignimbrite clast and associated scour (base indicated by dashed line). Lens cap (circled) for scale. (**e**) Debris-flow deposit with outsize and subvertical clasts. Lens cap (circled) for scale. (**f**) Channelform gravels (CH) cut into a sheetform bed with coarse–fine couplets (SC). Note also presence of outsize ignimbrite clast (top left). Lens cap (circled) for scale.

is located. Most rainfall appeared to fall on the western slopes of the Precordillera. Here the Tuina catchment, a similar catchment size (69.4 km^2) to Quebrada Tambores but located approximately 40 km to the north, generated an estimated peak flow of 39 m^3 s^{-1} from a catchment rainfall of 34 mm over the period 25–28 February and mobilized some 5 km^3 of sediment (Houston 2005).

The geomorphic and sedimentological effects of the 2001 flood event at the mouth of the Tambores catchment (i.e. the apex of the fanhead trench) were surveyed in the field by the authors in February 2002 using a combination of photography, GPS (Global Positioning System) and levelling survey. The site of the survey is indicated by 'B' in Figure 5b (S22° 49.015' and W68° 18.604'; WGS84). The flood event appears to have generated three distinct stages at the studied section. This same pattern was observed for the length of the flow event, which was constrained within the fanhead trench where it was artificially ponded in a quarry some 1.4 km below the fan apex.

Stage 1 – fluvial channel cutting

Evidence for Stage 1 is limited, but a comparison of photographs of the same channel reach (Fig. 8) suggests an erosive, streamflow was the first event to occur within the rainstorm. This cut through previously existing gravel bars, re-routing the main channel flow to the western wall of the main valley (Fig. 8). At the fan apex numerous ignimbrite boulders litter the fan surface and probably were transported by the Stage 1 flow as they are draped and coated by the Stage 2 deposits.

Stage 2 – mudflow

Stage 2 is represented by the deposits of a mudflow, typically less than 10 cm thick, which transported large amounts of cactus wood. This low-density wood tends to be restricted to the margins of the flow where the main trunks tend to be aligned parallel to the flow direction (Fig. 9a). The cactus wood is locally associated with clasts picked up from underlying deposits, which also tend to have a preferred alignment, with the long axis parallel to the flow direction. The mudflow is only preserved where it overtopped the Stage 1 flood channels that appear to have been the main conduit for the mudflow (Figs 8b & 9c). The mudflow deposits thus typically extend only a few metres laterally from the main active channel (Fig 8b). Most of the deposits have been removed by Stage 3 of the flood event (Fig. 9c). The deposits left by the flow bank around meanders and ramp up inverse gradients of as much as 10°

(Fig. 9b). Locally, a back-flow into the channel is recorded in the long axis of the cactus wood.

Stage 3 – fluvial channel flushing

Stage 3 deposits comprise a veneer of imbricated gravel deposits in the channel, and reworking of Stage 2 deposits and existing fan sediments. The flow was restricted to the main channel and some subsidiary channels on the north side of the valley. It was less laterally extensive than the Stage 2 deposit, being restricted to the channel. Flow depths peak at 40–50 cm within channel sections and average 20 cm where the flow occupies a single, wider channel further upstream. Observations suggest that two main quebrada tributaries within the catchment were active during the flood; however, the western Quebrada channel (1 on Fig. 2a) incises and truncates the eastern Quebrada (2 on Fig. 2a) flood deposits suggesting the flow duration was longer for the western tributary.

Stage 3 was most suitable for flood reconstruction as its peak limits could easily be recognized where it truncated the Stage 2 mudflow and the channel appears to have been little modified by deposition. Using techniques outlined in Clark (1996), Manning's roughness coefficient derived from channel slope (2°) was calculated at 0.0833. The peak discharge for the event could be calculated using Manning's equation at 16 m^3 s^{-1}. When maximum clast size data were incorporated following Clark (1996) the flow were estimated at 15 m^3 s^{-1}, with peak velocities of 1.5 m s^{-1}.

Reconstruction of the 2001 flood in Quebrada Tambores

The observations from the flood event suggest three pulses of flow alternating between streamflow and mudflow, as different subcatchments of the Tambores catchment were activated, a feature noted from flood flows in a variety of climate settings (e.g. the Howgills of NW England by Wells & Harvey 1987; and the Negev Desert of Israel by Greenbaum et al. 1998). The flood event was initiated with high water to sediment ratios that were dominantly erosional in the fan feeder and upper fan areas, incising new channels across older barforms (Fig. 8). This was followed by a lower water to sediment mudflow that overtopped the initial cut channel, asymmetrically overtopping banks on meander bends and ramping up reverse gradients, suggesting a viscous but high-velocity mudflow. The mudflow appeared to be sourced from a local catchment rich in muds and evaporates, and near to the fan apex. This event was then followed by a third pulse of activity dominated

Fig. 8. Images of the fan feeder channel (flow towards the viewer) showing the quebrada feeder channel (**a**) before the flood event in 1997 and (**b**) after the flood event in February 2002. Arrows indicate the same cliff face in both images. Note the channel avulsion that occurred in the 2001 event below this point.

by higher water to sediment ratios that removed the mudflow deposits from the channel, and was constrained to the channel. Within the feeder channel and upper fan the erosion and deposition from this event was limited and the flood appears to have waned rapidly.

Higher in the fan catchment no evidence of surface wash exists from the 2001 event. Although there is evidence of trashlines around vegetation, these are covered by aeolian material and motor vehicle tyre tracks that predate the 2001 flood (Fig. 10a), indicating a lack of surface activity

during the latter event and enabling a limit on the extent of runoff generation to be placed for the 2001 event (Fig. 5a). Areas unaffected by the flood display evidence of much larger flood events in recent times with 2 m flow depths locally (Fig. 10b).

Discussion

The late Pleistocene–Holocene sediments represent a period in the fan's history when the catchment was dominated by sedimentation from much larger flood

Fig. 9. Details of the February 2001 flood event photographed during the field survey of February 2002. (**a**) aligned cactus wood on the margins of the mudflow. Notebook for scale. (**b**) Mudflow that has ramped up a bank out of the main channel (where person is standing). (**c**) General view up the feeder channel (flow towards viewer) showing the truncation of the Stage 2 mudflow by the Stage 3 streamflow. The channel in the foreground is some 7 m across.

a

b

Fig. 10. (**a**) Upper part of the catchment unaffected by runoff in the 2001 event. Notice the aeolian material accumulated around the vegetation. The channel gradient is towards the viewer. (**b**) Evidence of previous, much higher, flood events (flow right to left) from quebrada sediments on ignimbrite boulder.

events than appear to occur under the current hyper-arid regime. These flood events had sustained high water and sediment discharges that generated antidune features in the lowermost feeder channel and upper fan areas. The preservation of the upstream dipping surfaces suggests no violent washout in the channel feeder area, perhaps indicating a rapid flow stage reduction as reported in proximal glacier outbursts floods in Iceland (Russell & Knudsen 2002) and the tropical Burdekin river in Australia

(Alexander & Fielding 1997). In both the latter cases the floods were of short duration, lasting only a few days, waned rapidly, and were associated with high water and sediment discharges. In Tambores we envisage that the sheetfloods lasted at least several hours to day(s) and were associated with sustained high sediment and water discharges that rapidly waned. These flows occupied the entirety of the modern feeder channel, spreading out onto the upper to midfan as supercritical sheetfloods with associated

Fig. 11. Sedimentology of a recent cut bank revealing evidence of past, historic flows showing the repeated intercalation of mudflow and hyperconcentrated fluvial flows. Note the mudflow at the top is the 2001 event.

standing waves that were unsteady and subject to frequent washouts in the more open upper fan. These floods appear to have been several orders of magnitude larger than those reported from the western Precordillera fans in February 2001, and were probably generated by heavy rainfall events that have not been witnessed in the area in recent times. As the sheetflood deposits construct most of the exposed fan sediments in the upper fan area it would suggest they relate to a wetter period in the regions history (Fig. 3), most probably corresponding to a period of increased intensity of the South American Summer Monsoon.

The 2001 flood event in Tambores was low magnitude (16 m^3 s^{-1}), and limited to the main feeder channel and fan apex. It was constrained to a channel that is incised by some 1–2 m into the older sediments by previous flood events, of which approximately 0.5 m belongs to this event. The calculated size of the flood in Tambores, together with the geomorphological evidence, would suggest that the Tambores catchment, situated on the eastern side of the Precordillera, escaped the main focus of the 2001 rainfall event reported by Houston (2005). The fact that this size and style of event has recurred in the recent past is evident in recent cut banks in recent deposits that show elements of both mudflows and streamflows hyperconcentrated in sediment (Fig. 11). There is, however, no record of the extreme events that appear to form the Pleistocene and Holocene elements of the fan body. Flows such as the February 2001 event appear to dominate the more recent flood history of the Tambores fan, although there is evidence of larger events occurring (Fig. 10b). These flows are pulsed with rapid changes in

flow rheology experienced during a single flood event. This reflects spatial variations in runoff experienced by the catchment as the storm event tracks across the catchment area and affects different lithologies such as the muds and evaporites of the lower catchment.

Comparison of the high- and low-magnitude events raises some interesting issues relating to the sedimentology. First, the D_{max} and D_{50} grain sizes between the events appear to be the same, in the same area of the fan. When downfan variations in D_{max} are examined (excluding the outsized, low-density ignimbrite clasts) there is no apparent change from proximal to distal locations in the older deposits (see the clast data on D_{max}, fan surface, above the intersection point in Fig. 6a), contrary to most facies models. This reflects the sediment availability within the catchment area, with the reworking of alluvial sediments that supply limited amounts of coarse calibre material to the modern alluvial system. Thus, the use of clast size in recreating flood conditions is more suitable for the lower magnitude events in this environment, and probably underestimates peak flows. Secondly, the 2001 Stage 2 mudflow was observed to overtop the bank of the channel and drape much older (Pleistocene and Holocene) deposits within the feeder channel. This indicates the difficulties in reconstructing stratigraphic sequences of events correctly in similar ancient deposits, particularly where exposed surfaces within sections cannot be distinguished by features related to desert pavement development. Lastly, the predominance of supercritical flow features indicates the potential hazard of extreme events

Fig. 12. Mantled slopes within the catchment area demonstrating the abundant potential sediment supply.

in arid catchments. Rainstorm events are infrequent and unpredictable, and, where conditions are appropriate, have the potential to generate rapid, high-magnitude flood hazards. In the eastern Atacama Desert the highest rainfalls appear to be associated with La Niña years. Where these correspond with seasonal variations brought about by an intensified South American Summer Monsoon it may be possible to recreate extreme flood events similar to those recorded within the ancient record. The extreme flow conditions and potential stream power of such events owe their origins to a number of factors within the study areas, which include tectonically enhanced stream gradients in the feeder channel and abundance of easily erodable multigeneration conglomerates. The current landscape indicates that, in the past, extensive badland landscapes have existed but that these are now mantled with abundant colluvial material (Fig. 12). This would suggest an unsteady state between sediment generation and removal, a factor not uncommon in hyperarid catchments where sediment being yielded from a catchment may have been generated under different (e.g. wetter) conditions (Clapp *et al.* 2000). This would indicate that in extreme flood events in such regions, hyperconcentration of sediment for sustained periods is possible, maximizing damage from such events.

Conclusions

Analysis of a modern low-magnitude and ancient high-magnitude flood event from a catchment in the

hyper-arid climate of the Atacama Desert indicates the range of flood events that can be generated from such catchments. Sedimentologically, the deposits indicate the problems of using clast size as an indicator of flood hydraulics, particularly in large-magnitude events, and the inaccuracies in basic facies models that predict sediment calibre changes downfan. These factors are significantly controlled by the lithology of the source area. The study also highlights the potential difficulties in establishing accurate stratigraphies within these environments where rapid variations in flow rheology can lead to the vertical superimposition of contemporaneous events over a wide age-range of older fan deposits.

The authors would like to thank the British Council and Royal Society for travel grants to aid this research (A. Mather) together with the logistical support of Prof. G. Chong, Universidad Catolica del Norte, and the survey assistance of Prof. J. Howell, University of Bergen. We also wish to thank P. Silva, G. Weissmann and A.M. Harvey for constructive comments on earlier versions of the manuscript.

References

ALEXANDER, J. & FIELDING, C. 1997. Gravel antidunes in the tropical Burdekin River, Queensland, Australia. *Sedimentology*, **44**, 327–337.

ALEXANDER, J., BRIDGE, J.S., CHEEL, R.J. & LECLAIR, S.F. 2001. Bedforms and associated sedimentary structures formed under supercritical water flows over aggrading sand beds. *Sedimentology*, **48**, 133–152.

AMIT, R., HARRISON, J.B.J., ENZEL, Y. & PORAT, N. 1996. Soils as a tool for estimating ages of Quaternary fault scarps in a hyper-arid environment – The southern Arava Valley, the Dead Sea Rift, Israel. *Catena*, **28**, 21–45.

BARWIS, J.H. & HAYES, M.O. 1985. Antidunes on modern and ancient washover fans. *Journal of Sedimentary Petrology*, **55,** 907–916.

BEN DAVID-NOVAK, H. & SCHICK, A.P. 1997. The response of Acacia tree populations on small alluvial fans to changes in the hydrological regime: Southern Negev desert, Israel. *Catena*, **29**, 341–351.

BERGER, I.A. & COOKE, R.U. 1997. The origin and distribution of salts on alluvial fans in the Atacama Desert, northern Chile. *Earth Surface Processes and Landforms*, **22**, 581–600.

BLAIR, T.C. 1987. Sedimentary processes, vertical stratification sequences and geomorphology of the Roaring River alluvial fan, Rocky Mountain National Park, Colorado. *Journal of Sedimentary Petrology*, **57**, 1–18.

BLAIR, T.C. 1999. Sedimentary processes and facies of the waterlaid Anvil Spring Canyon alluvial fan, Death Valley, California. *Sedimentology*, **46**, 913–940.

BLAIR, T.C. 2000. Sedimentology and progressive tectonic unconformities of the sheetflood-dominated Hell's Gate alluvial fan, Death Valley, California. *Sedimentary Geology*, **132**, 233–262.

BOBST, A.L., LOWENSTEIN, T.K., JORDAN, T.E., GODFREY, L.V., KU, T.L. & LUO, S.D. 2001. A 106 ka palaeoclimate record from the drill core of the Salar de Atacama, northern Chile. *Palaeogeography, Palaeoclimate and Palaeoecology*, **173**, 21–42.

CARLING, P.A. & SHVIDCHENKO, A.B. 2002. A consideration of the dune: antidune transition in fine gravel. *Sedimentology,* **49**, 1269–1282.

CHONG, G., PEREIRA, M., GONZÁLEZ, G. & WILKE, H.G. 1991. Los fenomenos de remocion en masa ocurridos en la region de Antofagasta en Junio de 1991. *Revista de la Facultad de Ingenieria y Ciencias Geologicas*, **7**, 6–13.

CLAPP, E.M., BIERMAN, P.R., SCHICK, A.P., LEKACH, J., ENZEL, Y. & CAFFEE, M. 2000. Sediment yield exceeds sediment production in arid region drainage basins. *Geology*, **28**, 995–998.

CLARK, A.O. 1996. Estimating probable maximum floods in the upper Santa Ana basin, Southern California, from stream boulder size. *Environmental and Engineering Geoscience,* **2**, 15–182.

GARREAUD, R.D. & ACEITUNO, P. 2001. Interannual rainfall variability over the South American Altiplano. *Journal of Climate*, **14**, 2779–2789.

GREENBAUM, N., MARGALIT, A., SCHICK, A.P., SHARON, D. & BAKER, V.R. 1998. A high magnitude storm and flood event in a hyper-arid catchment, Nahal Zin, Negev Desert, Israel. *Hydrological Processes,* **12**, 1–25.

GROSJEAN, M., NUNEZ, M., CARTAJENA, I. & MESSERLI, B. 1997. Mid-Holocene climate and culture change in the Atacama Desert, northern Chile. *Quaternary Research*, **48**, 239–246.

HARTLEY, A.J. & CHONG, G. 2002. Late Pliocene age for the Atacama Desert: Implications for the desertification of western South America. *Geology*, **30**, 43–46.

HAUSER, A.Y. 1997. *Los aluviones del 18 de Junio de 1991 en Antofagasta: un analisis critico, a 5 años del desastre.* Chile Servicio Nacional de Geologia y Mineria, Boletin No. 49.

HOUSTON, J. 2001. The year 2000 storm event in the Quebrada Chacarilla and calculation of recharge to the Pampa Tamarugal aquifer. *Revista Geologica de Chile*, **28**, 163–177.

HOUSTON, J. 2002. Groundwater recharge through an alluvial fan in the Atacama Desert, northern Chile: mechanisms, magnitudes and causes. *Hydrological Processes*, **16**, 3019–3035.

HOUSTON, J. 2005. The great Atacama flood of 2001 and implications for Andean hydrology. *Hydrological Processes,* in press.

HOUSTON, J. & HARTLEY, A.J. 2003. The central Andean west-slope rainshadow and it potential contribution to the origin of the hyper-aridity in the Atacama desert. *International Journal of Climatology*, **23**, 1453–1464.

JONES, C.E. 1991. Characteristics and origin of rock varnish from the hyper-arid coastal Deserts of Northern Peru. *Quaternary Research*, **35**, 116–129.

LATORRE, C., BETANCOURT, J.L., RYLANDER, K.A. & QUADE, J. 2002. Vegetation invasions into absolute desert: a 45 000 yr rodent midden record from the Calama-Salar de Atacama basins, northern Chile (Lat 22 degrees–24 degrees S). *Bulletin of the Geological Society of America*, **114**, 349–366.

LATORRE, C., BETANCOURT, J.L., RYLANDER, K.A., QUADE, J. & MATTHEI, O. 2003. A vegetation history from the arid prepuna of northern Chile (22–23 degrees S) over the last 13, 500 years. *Palaeogeography, Palaeoclimate and Palaeoecology*, **194**, 223–246.

RECH, J.A., QUADE, J. & BETANCOURT, J.L. 2002. Late Quaternary palaeohydrology of the central Atacama Desert (Chile (Lat 22 degrees – 24 degrees S), Chile. *Bulletin of the Geological Society of America*, **114**, 334–348.

RUSSELL, A.J. & KNUDSEN, Ó, 2002. The effects of glacier outburst flood flow dynamics on ice-contact deposits: November 1996 jökulhlaup, Skeióarásandur, Iceland. *In*: MARTINI, I.P., BAKER V.R. & GARZON, G. (eds) *Flood and Mefaflood Deposits: Recent and Ancient.* International Association of Sedimentologists Speical Publication, **32**, 67–83.

VUILLE, M. 1999. Atmospheric circulation over the Bolivian Altiplano during dry and wet periods and extreme phases of the Southern Oscillation. *International Journal of Climatology*, **19**, 1579–1600.

WELLS, S.G. & HARVEY, A.M. 1987. Sedimentologic and geomorphic variations in storm generated alluvial fans, Howgill Fells, northwest England. *Bulletin of the Geological Society of America*, **98**, 182–198.

Fans with forests: contemporary hydrogeomorphic processes on fans with forests in west central British Columbia, Canada

D.J. WILFORD[1], M.E. SAKALS[2], J.L. INNES[3] & R.C. SIDLE[4]

[1]BC Ministry of Forests, Bag 6000, Smithers, BC, Canada V0J 2N0
(e-mail: dave.wilford@gems3.gov.bc.ca)
[2]Bulkley Valley Centre for Natural Resources Research and Management, Box 4274, Smithers, BC, Canada V0J 2N0
[3]UBC Faculty of Forestry, 2424 Main Mall, Vancouver, BC, Canada V6T 1Z4
[4]Geohazards Division, Disaster Prevention Research Institute, Gokasho, Uji, Kyoto, 611–0011, Japan

Abstract: Alluvial and colluvial fans with forest cover are common in the valleys of west central British Columbia, Canada. Given the low population density of the region, most of these fans are uninhabited and the primary land use is forestry. The fans are desirable for timber harvesting due to the combination of the relatively easy access to their valley-bottom locations and their high-quality timber. However, they are also sites influenced by debris flows, debris floods and floods, and the interaction between conventional forest practices and these natural hydrogeomorphic processes has led to substantial financial costs and disturbance to forests and stream channels. Basic watershed morphometrics can be used to predict the dominant hydrogeomorphic process influencing forested fans. The hydrogeomorphically active zones of forested fans have characteristic site and stand features, and are referred to as the hydrogeomorphic riparian zone. Features within these zones can be used to determine the frequency and disturbance extent of hydrogeomorphic events. Appropriate management strategies can be developed to limit the effect of forestry activities on natural hydrogeomorphic processes.

Alluvial and colluvial fans are a common landform in the narrow, steep-sided valleys of the mountainous region of west central British Columbia. With the temperate climate, it is common for these fans to have forest stands composed of large trees that are attractive for forest harvesting. However, the large trees and gentle slopes of fans belie the fact that natural hazards are often present. Fans are often runout zones for debris flows and debris avalanches; in addition, they are subject to floods and debris floods. These events are collectively termed hydrogeomorphic processes. They are neither rare nor extreme (Innes 1985; Jakob & Jordan 2001), and it is not uncommon for contemporary hydrogeomorphic processes to be actively influencing at least a portion of the fan surface.

Hydrogeomorphic hazards on fans with forests have been described in detail by VanDine (1985) and Kellerhals & Church (1990), and studies of specific fans have led to the development of a method of hazard zonation (Thurber Consultants 1983). However, there is no body of research regarding forest practices on fans and there is not a hazard classification that can be used to facilitate forest management on fans. Thus, hydrogeomorphic hazards on fans are often not identified during forest development planning in British Columbia. Currently,

drainage structures on fans are built to accommodate 50 or 100 year flood events (Anon. 1995a) not hydrogeomorphic events, and the decision to retain riparian forests is based on the presence or absence of fish (Anon. 1995b) not on the protection that such forests provide. Not surprisingly, conventional forestry practices on fans have frequently been ineffective for their intended purposes (e.g. growing trees and providing safe access) and have exacerbated the impacts of natural hydrogeomorphic processes (e.g. broadcasting of sediment and channel avulsions) (Wilford et al. 2003).

This paper presents our findings regarding characteristics of forested fans and their watersheds with reference to contemporary hydrogeomorphic processes and hazard recognition for forest management planning.

Study area

The study area is in west central British Columbia, Canada, within the Western and Interior Systems of the Canadian Cordillera (Holland 1964). The inventoried fans lie across a broad geographic area between 53° 46′ and 55° 43′ N latitude, and 126° 00′ and 129° 10′ W longitude (Fig. 1). The Kitimat

From: HARVEY, A.M., MATHER, A.E. & STOKES, M. (eds) 2005. Alluvial Fans: Geomorphology, Sedimentology, Dynamics. Geological Society, London, Special Publications, **251**, 25–40. 0305–8719/05/$15 © The Geological Society of London 2005.

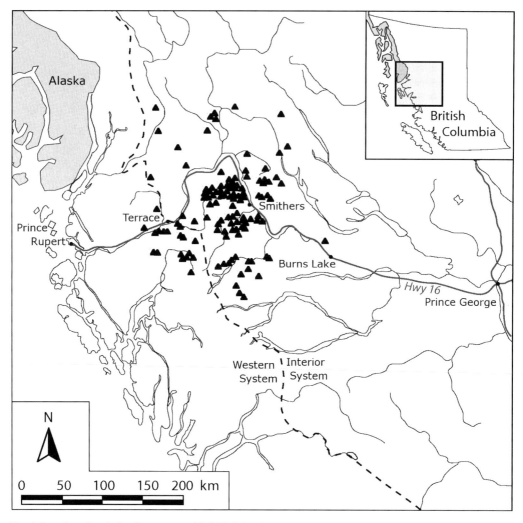

Fig. 1. Location of study fans in west central British Columbia, Canada.

Ranges are within the Coast Mountains of the Western System and consist of granitic mountains, characteristically round-topped and domed because they were overridden by large Pleistocene ice sheets. The Interior System includes the Skeena Mountains, Nass Basin, Hazelton Mountains and the Nechako Plateau. This system is underlain chiefly by volcanic and sedimentary rocks, and overall has less bedrock exposure and is less rugged than the Western System.

The study area was glaciated during the Fraser glaciation with ice retreat completed between 10 700 and 9300 years BP (Alley & Young 1978; Clague 1984). Much of the underlying bedrock is masked by extensive morainal deposits on hillslopes and glaciofluvial deposits in valley bottoms (Runka 1972). Fans are a post-glacial feature in the study area

reflecting paraglacial (Ryder 1971*a*, *b*; Church & Ryder 1972) and contemporary conditions.

Septer & Schwab (1995) compiled a record of major erosion events for this area that includes debris flows, debris floods and floods. Debris flows are relatively common in the smaller watersheds in the western portion of the study area, but also occur in the central and eastern portions. Debris floods occur throughout the study area. Floods are the dominant hydrogeomorphic process in the larger watersheds, but can also occur annually in debris-flood- and debris-flow-dominated watersheds.

The forests of the study area are described as coastal rainforests in the western portion, northern temperate in the central portion and sub-boreal in the eastern portion, reflecting the gradient from maritime to continental climates (Pojar *et al*. 1987; Banner *et al*. 1993; Mah *et al*. 1996). All study fans were forested, although some have been logged to varying degrees. Tree species present on the study fans were *Picea sitchensis, Picea glauca, Picea engelmannii, Picea glauca x engelmannii, Thuja plicata, Abies lasciocarpa, Abies amabilis, Tsuga heterophylla, Pinus contorta* var. *latifolia, Pinus albicaulis, Alnus rubra, Alnus incana, Betula papyrifera, Populus trichocarpa,* and *Populus tremuloides* (Banner *et al*. 1993). Logging operations have not been conducted in most watersheds above the fans, and where present were only to a very limited extent.

Methods

Fans were selected according to several criteria, with provision of a reasonable cross-section of hydrogeomorphic processes and geographic distribution across the study area being important. To gain an understanding of natural processes it was essential that the hydrogeomorphic processes not be influenced by human land use. Only watersheds with no, or very limited, human land-use activities (e.g. logging, mining, road building) were therefore selected. Fans with human land use were included only if the use did not obscure identification of the hydrogeomorphic processes.

Sediment deposits were used to differentiate debris flow, debris flood and flood fans. Debris flows are homogeneous (single phase) sediment–water mixtures that have non-Newtonian flow, sediment concentrations between 70 and 90% by weight, and result in the formation of marginal levées and terminal lobes (Costa 1984, 1988; VanDine 1985; Smith 1986; Pierson & Costa 1987; Wells & Harvey 1987; Hungr *et al*. 2001). Debris floods, also referred to as hyperconcentrated flows, occur when essentially all material on the streambed surface is mobilized but the mixing is not complete (there is a rapid increase in solids concentration toward the bed). Debris floods have Newtonian–non-Newtonian flow, sediment concentrations between 40 and 70% by weight, and sediment deposits are bars, fans, sheets and splays (Costa 1988; Hungr *et al*. 2001). Floods in gravel, cobble or boulder channels rarely mobilize the entire bed. Floods have Newtonian flow, sediment concentrations between 1 and 40% by weight, and sediment deposits are bars, fans, sheets and splays (characteristically not as extensive or thick as debris floods) (Wells & Harvey 1987; Costa 1988; Hungr *et al*. 2001).

Where evidence of more than one process was found, the dominant process was determined based on the following order: debris flow, debris flood and flood. The order was based on practical forest management considerations. One of the major forest management issues on fans is damage to drainage structures (Wilford *et al*. 2003). The common approach to designing drainage structures in British Columbia is to use calculated peak water flows (50 or 100 year design floods) (Anon. 1995*a*). However, debris flows can produce peak discharges up to 40 times greater than anticipated flood flows, and debris floods can have peak discharges of double the flood flows (Hungr *et al*. 2001).

The stream-channel gradient at the fan apex was measured and channel conditions described (e.g., whether debris jams near the fan apex were influencing channel gradient). Forest stands were described on the study fans. A forest stand (or cohort) is a group of trees of the same age that establish after a disturbance, such as on sediment from a hydrogeomorphic event (Oliver & Larson 1996). Aerial photographs and forest inventory maps were used to identify cohorts wider than 20 m, and cohorts less than 20 m wide were identified in the field. The positions of the different stands were described (e.g. whether they were linked to the stream channel and could have established following a hydrogeomorphic event). In the field we determined the cause of stand establishment (e.g. forest fires, hydrogeomorphic event, windthrow, etc.). Timber volumes were determined using variable plot cruising (Watts 1983; Helms 1998).

In addition to the cohort descriptions, 10 site features were used to describe the interactions between hydrogeomorphic events and forest stands on fans. We use the term 'hydrogeomorphic riparian zone' to identify this forested zone of interaction. *Recent sediment* splays are apparent due to the lack, or limited amount, of organic matter and minor vegetation on the forest floor (Fig. 2). When trees are buried by sediment their basal flare is frequently absent and the trees appear to emerge from the ground with a uniform cylindrical shape (*buried trees*) (Fig. 3). If the depth of burial is sufficient to kill a tree, subsequent rotting of the tree can lead to the formation of a cylindrical hole in the ground, termed a *tree hole* (these features are only present where sites are not subsequently influenced by sediment deposition) (Fig. 4). If the depth of burial is not sufficient to kill a tree, adventitious roots are established just below the new soil surface. *Exposed adventitious roots* can often be observed if there has been subsequent erosion of the deposited sediment (Fig. 5). Erosion of sediment can expose webs or mats of tree roots that are providing a degree of *soil reinforcement* (Fig. 6). *Scars on trees* are a result of abrasion from boulders and logs transported by hydrogeomorphic events (Fig. 7). Vegetation along a stream channel can enhance the deposition of sediment by reducing stream velocities (*levée enhancement*). This vegetation can also trap small woody debris, forming *woody dykes* (Fig. 8). A *log step* is formed when the stems of toppled trees store sediment on their upslope side (Fig. 9). *Log retaining walls* are formed when multiple tree stems transported by hydrogeomorphic events lodge against standing trees and trap sediment on their upslope side (Fig. 10). Immediately after debris-flow events, large boulders may be wedged between trees 1–2 m or more above the ground surface, and mud and abrasion on the trees provide an indication of the depth of the flow (Fig. 11).

When logging or road construction had been undertaken on a fan, practices were described (e.g. roads were surveyed, drainage structures measured and their condition

Fig. 2. A recent deposit of sediment in the hydrogeomorphic riparian zone with limited organic accumulation or moss cover.

described). Impacts to roads, drainage structures, planted areas and stream channels were described, as were the effects these features had on natural hydrogeomorphic processes (e.g. whether there was greater or lesser disturbance due to the nature of the anthropogenic feature). Impacts were classed as major if damage was significant, for example drainage structures required replacement, roads were washed out, streams were impacted to the point where there was no or very poor fish habitat, and plantations were buried in sediment or eroded. Impacts were classed as limited if there was localized erosion on roads, and limited impacts to drainage structures, plantations, forest sites and fish habitat.

Boundaries for the watersheds of fans were established using a digital elevation model (DEM) and Geographic Information Systems (GIS). The DEM has a cell size of 25 $\times$ 25 m and 90% of the vertical data are accurate to within 10 m of their true elevation. The lowest point in a watershed was the apex of the fan (i.e. fans were not included in the watersheds). We determined basic watershed morphometric measures: area, relief and length. The Melton ratio, namely watershed relief divided by the square root of watershed area, was calculated for each watershed (Melton 1957; Patton & Baker 1976; Jackson *et al.* 1987).

Results and Discussion

Of the many fans in the study area that met the selection criteria, 65 were selected for study. Ten fans were in a natural forested state. Fifty-five fans had a range of forestry activities from single road crossings to complete clearcutting. The dates of forestry activities ranged from recent to 50 years old, although the majority of forestry activities had occurred in the past 20 years.

Debris flow was the dominant hydrogeomorphic process on 13 fans, debris flood on 36 fans and flood on 16 fans. Basic morphometric measurements of the fans and their watersheds are presented in Table 1. The Melton ratio has been used to differentiate flood- and debris-flow watersheds (Jackson *et al.* 1987; Bovis & Jakob 1999). This differentiation was not expected in the study area due to the range in climatological and geological conditions. However, by plotting the Melton ratio against the watershed length, Wilford *et al.* (2004) were able differentiate the three hydrogeomorphic processes for this study area (Table 2). It is possible that the geological differences were masked by glacial deposits, as suggested by Runka (1972). Climatological differences could be the reason that more debris-flow fans were identified in the western maritime portion of the study area;

Fig. 3. A buried tree with the characteristic lack of a basal or butt flare. The tree appears to emerge from the ground with a uniform cylindrical shape.

however, debris-flow fans were also identified in the central and eastern continental portions.

The size of fans ranged from 0.05 to 6 km². In some settings, determination of fan size can be an important geomorphic measurement related to bedrock geology and weathering processes (Hooke & Rohrer 1977). However, determination of fan area was not an integral measurement in this project. Delineation of the distal boundaries of forested fans from aerial photographs can be difficult because forest cover often obscures the landform boundary where there are subtle changes in slope, particularly along less geomorphically active margins. In many cases detailed fieldwork would be required to locate boundaries. Most of the fans were located in narrow valleys with main valley streams actively truncating the deposits, reducing their size through the downstream transport of a portion of delivered material.

Watersheds ranged in size from 0.25 to 99 km². Debris flows were generated in the smallest watersheds, with an average size of 1.3 km² and a range of 0.25–4.1 km². This is within the size range identified in the literature (VanDine 1985). Debris floods were

generated in larger watersheds, with an average of 7 km² and a range of 0.7–31 km². Floods were generated in the largest watersheds, with an average of 34 km² and a range of 1.4–99 km².

In general, the study fans have a classic conical shape. Some fans were confined laterally by valley walls or bedrock knolls for some distance from the apex and had a shape more akin to a steep flood plain. Elongated or oblong fans were associated with channel entrenchment. Entrenchment in one case was due to debris-flow levées that were at least as old as the forest stand (in excess of 250 years).

The degree of stream channel confinement on study fans ranged from unconfined for the length of the fan to entrenched for a major portion of the fan. Stream channel gradient at the apex ranged from 1° to 22°. Flood fans had an average stream channel gradient at the apex of 3.0°, with a range of 1.0–10.5°. In two cases the gradient at the apex of flood fans was greater than the 4° limit reported by Jackson *et al.* (1987). In one case the channel was entrenched with a gradient of 10.5° and did not reflect the gradient of the fan at the apex, which was 5.5°. In both cases it is likely that the contemporary hydrogeomorphic process (i.e. flood) did not reflect the mode of origin of the fan (i.e. debris flood). Debris-flood fans had an average stream channel gradient at the apex of 7.7°, with a range of 1.75–18°. In seven cases the gradient at the apex of debris-flood fans was ≤4°. This was most probably due to sediment wedges in the channels that locally reduced the channel gradient, and was most apparent in one channel where a large log jam just below the apex resulted in a significant sediment accumulation with a gradient of 1.75°. Debris-flow fans had an average stream channel gradient at the apex of 12.4°, with a range of 7°–22°.

Fan-surface cross-sections ranged from smooth to dissected. Characteristically, the larger fans had relatively smooth, convex cross-sections while the smaller fans had dissected cross-sections with a generally convex shape. Field identification of forested fans, particularly small fans, can be problematic without aerial photographs. Coalescing small forested fans along lower hillslopes can often be mistaken by forest practitioners as a series of streams on a colluvial footslope. This misidentification can limit the appropriate assessment of hazards because individuals may not be looking for evidence of hydrogeomorphic activity.

Elevated and hydrogeomorphically inactive fan surfaces were observed on 13 of the 65 study fans. It is possible that these are early Holocene or paraglacial fan surfaces (Ryder 1971*a*, *b*); however, the surfaces were not dated. These older surfaces were from 2 to 10 m above the hydrogeomorphically active fan surface and ranged from a major portion of the present fan landform to minor remnants along

Fig. 4. Tree holes are the result of rotting following deep burial of tree stems. These features are only found in areas that are not subject to subsequent sediment accumulation. This tree hole is 2 m deep.

Fig. 5. Adventitious roots emerge from the stem of a buried tree just below the new soil surface. In this case the soil was subsequently eroded and the adventitious roots have been exposed.

Fig. 6. The root web in a forest stand provides reinforcement to the soil mass, resulting in a degree of resistance to the formation of new channels. In this case the bark and wood of the roots has been scoured, indicating a period of bedload movement over and through the root web prior to the undermining of sediment beneath the roots.

the adjacent hillslopes. Forest ecosystems and stand volume differences between the elevated and active fans surfaces on nine fans were in stark contrast (Table 3 and Fig. 12). Timber volumes were from three to nine times higher on the lower, more active fan surfaces than the elevated fan surfaces. The differences were most probably due to factors that reduce forest site productivity: coarse-textured soils on the elevated surfaces coupled with entrenchment of the stream relative to the elevated surface that resulted in a much lower water-table. In addition, the ecosystem disturbance by hydrogeomorphic processes (including the periodic broadcasting of sediments and inundation of active zones on fans) most probably acted to enhance site productivity on the lower sites.

Forest cover has been used to identify where contemporary hydrogeomorphic events have occurred on fans (e.g. Kellerhalls & Church 1990), with a focus primarily on high-power or forest-stand clear-ing events. A forest stand, consisting of a group of trees of the same age that established after a disturbance (such as on the sediment from a geomorphic event), is termed a cohort (Oliver & Larson 1996); its age can provide a reasonably accurate date of the occurrence of the event (Schweingruber 1996) (Fig. 13). We found cohorts as strips of uniform forest stands, linked to the stream channel. Based on dating of tree scars and young cohorts, it appears as though trees establish on the sediment from stand-clearing events within 3–5 years. We also observed that cohorts of shade-tolerant tree species establish on sediment spread under a forest canopy by lower power events that do not remove the original forest stand.

Sixty-one of the study fans had forest cover on at least part of the area influenced by hydrogeomorphic events. On these fans we documented the occurrence of 10 site features (Table 4). Buried trees, log steps, recent sediment splays and levée enhancers were common features (over 84% occurrence) that highlight the sediment storage role of forests on fans. Similar observations on sediment storage in forests were made by Gomi *et al.* (2004). Tree scars were observed in almost 80% of the study fans, an occurrence level that supports the use of dendroecology for dating hydrogeomorphic events (Schweingruber 1996; Strunk 1997). Woody dykes were observed in approximately half of the fans. These features, along with levée enhancers, tend to maintain water in the stream channel, but also tend to block the re-entry of water once it has left the channel. By containing the stream within its channel, stream power is maintained and sediments may be potentially transported further downfan than would be observed on non-forested fans. Exposed adventitious roots were observed in a quarter of the study fans, but notably, adventitious roots were present on all tree species in our study area, not just deciduous trees (i.e. *Populus trichocarpa* and *Alnus rubra*) as generally considered. Exposed root webs as a result of sediment scour were observed in a quarter of the study fans. In most cases the bark and wood of the roots had been scoured, which indicated a period of bedload movement over and through the root web prior to the undermining of sediment beneath the roots. From these observations it appears that root webs offer a degree of resistance to the formation of new channels. Entrenchment under some root mats was more than 2 m. Channel avulsions on forested fans are also hindered through the presence of downed logs, tree stems and shrubs, which act to disperse flow, reduce velocities by increasing hydraulic roughness and thus lead to deposition rather than scour. Log retaining walls were found not to be common when all hydrogeomorphic processes are considered (13% occurrence), but were observed in 30% of the debris-flow fans. The most impressive log retaining walls had vertical heights of

Fig. 7. A scar on a tree stem resulting from abrasion by boulders and logs transported by debris flow. Scars can provide the approximate date of an erosion event (e.g. during the dormant, early growth or late growth season of a year). The scar on this tree predates the fine textured sediment deposited on the tree by a more recent debris flow.

Fig. 8. A woody dyke formed by the accumulation of small woody debris against streamside vegetation. To a degree, these structures serve to keep water in the stream channel, maintaining stream power and limiting channel avulsions.

Fig. 9. A log step created by the recent deposition of sediment behind a log. These features can be over 1 m in height and create a micro-scale stepped profile on a fan surface.

Fig. 10. Log retaining walls are formed when multiple tree stems transported by hydrogeomorphic events lodge against standing trees and trap sediment on their upslope side. These structures can be over 4 m in height. This log retaining wall on a forested debris flow fan is storing a considerable volume of sediment. These structures result in localized areas of reduced gradient above the log structures, but the net result of sediment storage is an increase in the overall slope gradient of a fan surface.

Fig. 11. Immediately after debris-flow events large boulders may be wedged between trees 1–2 m or more above the ground surface, and mud and abrasion on the bark provide an indication of the depth of the flow.

Table 1. *Basic watershed and fan morphometrics by hydrogeomorphic process*

	Hydrogeomorphic process		
	Flood	Debris flood	Debris flow
Watershed			
Mean area	34 km^2	7.0 km^2	1.3 km^2
Range	1.4–99 km^2	0.7–31 km^2	0.25–4.1 km^2
±1 standard deviation	3–65 km^2	0.3–13.6 km^2	0.2–2.4 km^2
Mean relief	1.1 km	1.2 km	0.9 km
Range	0.4–2.1 km	0.5–1.7 km	0.5–1.4 km
±1 standard deviation	0.5–1.7 km	0.9–1.5 km	0.7–1.3 km
Mean Melton ratio	0.23	0.57	0.95
Range	0.08–0.49	0.26–1.13	0.66–1.21
± 1 standard deviation	0.13–0.33	0.31–0.83	0.76–1.14
Channel gradient at apex			
Mean	3.0°	7.7°	12.4°
Range	1.0°–10.5°	1.75°–18.0°	7.0°–22.0°
± 1 standard deviation	1.0°–5.4°	3.5°–11.9°	7.8°–17.0°

over 2 m and stored considerable volumes of sediment. During investigations of fans in other British Columbia locations we observed log walls up to 4 m in height. Log retaining walls and log steps result in localized areas of reduced gradient above the log structures, but the net result of sediment storage is an increase in the overall slope gradient of the fan surface (Irasawa *et al.* 1991). These log structures form a stepped or segmented fan profile at the micro-scale that is of a different origin than the macro-scale

Table 2. *Differentiating hydrogeomorphic processes using the Melton ratio and watershed length (from Wilford* et al. *2004)*

Hydrogeomorphic process	Watershed attribute	Class limits
Floods	Melton ratio	<0.30
Debris floods	Melton ratio and watershed length	Melton: 0.30–0.6 When Melton >0.6, length ≥2.7 km
Debris flows	Melton ratio and watershed length	Melton >0.6 and length <2.7 km

Table 3. *Field indicators of hydrogeomorphic activity on the 61 study fans with forests in the areas influenced by hydrogeomorphic processes*

Feature	Frequency of occurrence %
Buried trees	97
Log steps	93
Recent sediment splays	88
Levée enhancers	84
Scars from events on trees	79
Woody dykes	49
Exposed adventitious roots	26
Soil reinforcement	24
Log retaining walls	13
Tree holes	8

Table 4. *Comparison of forest stands on elevated and active fan areas*

Fan	Forest cover			
	Elevated fan surface		Active fan surface	
	Average tree height (m)	Stand volume (m³ ha⁻¹)	Average tree height (m)	Stand volume (m³ ha⁻¹)
Gosnell 6	11	76	26	703
Gosnell 7	10	183	21	810
CP095-2	19	260	26	707
Sibola	12	116	29	770

segmented profile found on arid fans (i.e. as a result of channel entrenchment and fan elongation). Tree holes are subtle features that were observed in approximately 10% of the fans. Some of these holes were 2 m deep and, in general, present safety hazards for the unobservant.

It is apparent that forests on fans influence hydrogeomorphic process through sediment storage, channel confinement and resistance to channel avulsions. The forests in this zone can be considered to be riparian forests (i.e. influenced by streams and influencing streams). Riparian forests provide many physical and biological functions, including nutrient inputs, bank stabilization, channel hydraulic structures through large woody debris inputs, shade and deposition zones for sediment from upland sources (Beschta *et al.* 1987; Bisson *et al.* 1987; Bjornn & Reiser 1991; Murphy & Meehan 1991; Naiman *et al.* 2000). A function that has been inadequately described in the literature is the role of riparian

Fig. 12. An example of the contrast between (**a**) the high-volume forest stand on the more hydrogeomorphically active fan surface and (**b**) the low-volume forest stand on the adjacent elevated fan surface.

forests with regard to hydrogeomorphic processes beyond or outside the stream channel. Our results present a strong case for the recognition of this role. To strengthen the concept of this role a new term is suggested – the hydrogeomorphic riparian zone. This zone of contemporary influence (i.e. within the last 200 years) is generally apparent through characteristic site features and forest composition (e.g. cohorts established on sediment from hydrogeomorphic events).

A point of interest is the role that large woody debris plays in the stream channels on forested fans. In addition to the structure afforded by large clasts transported by previous mass-wasting events, large woody debris provides hydraulic structure to stream channels in forested areas (Keller & Swanson 1979; Thomson 1991). While entrenchment was observed on some of our forested fans it is possible that the degree of entrenchment may be less than on non-forested fans due to the presence of large woody debris. This woody debris is not only on the surface but is also incorporated into the sediment of the sub-surface of the fan and can be subsequently exposed by degrading stream channels (Fig. 14)

Sato (1991) described a fan in Japan that was prone to overbank flows. To protect farmland and residential areas, 'flood-control forests' were planted along the watercourse in 1905. Over time some of the forests were removed and homes built. In 1990 a highly destructive flood with mobile logs and sand caused severe damage, primarily in

Fig. 13. A 35 year-old cohort of *Picea sitchensis* and *Tusga heterophylla* growing on the sediment of a high-power debris flow. The event removed the previous forest stand on this site.

Fig. 14. Erosion of sediment on this logging trail has exposed buried logs and stumps. One of the buried logs is acting as a hydraulic structure, limiting the amount of erosion. A buried stump has been exhumed by erosion in the centre foreground.

reaches without flood-control forests. Sato notes that the flood-control forests along the river prevented overflows of drifted logs and led to reduced water flows and sediment broadcasting. It is also apparent that avulsions did not occur in the forested reaches.

Forestry activities on fans can influence natural hydrogeomorphic processes in several ways (Wilford *et al.* 2003). Removal of streamside forests in zones where the forest was storing sediment and maintaining the stream channel resulted in an apparent increase in sediment transport across the fan surface. The long-term effect of riparian logging could be the reduction in large woody debris recruitment, and subsequent channel entrenchment. We observed this on fans that had been logged only 15 years previously, and it is expected that these streams will continue to exhibit strong entrenchment and channel widening before attaining a new quasi-equilibrium. Wilford *et al.* (2003) documented that roads constructed on fans can become new stream channels, particularly if they climb in elevation toward the stream channel. A common observation was that drainage structures designed to pass water can become obstructed with sediment and woody debris, leading to channel avulsions. To avoid aggravating the impacts of natural hydrogeomorphic events, it is essential that forest stand and site features associated with hydrogeomorphic activity be recognized and appropriate plans be developed. Documentation of practices that did not cause aggravation of natural events lead us to suggest a series of 'best forest management practices' for fans based on the assessment of hydrogeomorphic hazards. Riparian forests should not be harvested where the forest and its associated debris are interacting with hydrogeomorphic events. Roads should drop in elevation across a fan to the stream crossing, or provision made to ensure that roads do not become new stream channels (frequent cross drains and rolling grades). Drainage structures must be designed to pass water, sediment and debris associated with hydrogeomorphic events, or to withstand these events without leading to channel avulsions (materials pass over the structure and continue down the channel).

While impacts to fan surfaces as a result of forestry activities were observed on all three types of fans, we were initially surprised to find that the frequency of major impacts was greatest on flood fans (69% of the 16 study flood fans with forestry activities), intermediate on debris-flood fans (50% of the 30 study debris-flood fans with forestry activities) and lowest on debris-flow fans (11% of the nine study debris-flow fans with forestry activities). This situation can be explained in several ways. Flood fans have the lowest gradients and, hence, appear to have a low potential for problems, so potential

hazards are not explored and no extra precautionary measures are taken. Conversely, debris-flow fans are steep and contemporary deposits are more obvious (e.g. levées and lobes), raising concerns that lead to road locations lower on fan surfaces and the establishment of forested reserves. The oldest logging in the study area occurred on flood fans, and was carried out when environmental hazards were generally not considered. The most recent logging has been on debris-flow fans, and foresters are now more aware of hydrogeomorphic hazards. However, substantial impacts are still being created through inappropriate forest management. It is essential for foresters to recognize the hydrogeomorphic hazards present on fans.

Through our investigation of fans in the three climatological/forest types we have found that the indicators of hydrogeomorphic hazard and the role of forests within the hydrogeomorphic riparian zone on fans are consistent. We have also found that the interactions of forest practices with hydrogeomorphic hazards are dependent on process, but are also not affected by climatic regime or forest type. This consistency is beneficial to our aims of further understanding and increasing the awareness of hydrogeomorphic hazards on fans.

Conclusions

Fans with forests have received only limited attention from the scientific community. This is probably because the hydrogeomorphologic features are frequently obscured, making study more difficult than on arid, non-forested fans. As a result there is limited scientific basis for current forest management on fans. This study was undertaken in a region with considerable geomorphic and climatic variation. For this area, it has been demonstrated that hydrogeomorphic processes can be predicted through the use of basic watershed attributes. Site features were identified that are considered to be universal with regards to the role forests play in hydrogeomorphically active areas of fans. Such site features within these hydrogeomorphic riparian zones provide valuable information regarding contemporary hydrogeomorphic activity: power of events, disturbance extent and frequency of events. Forest harvesting and road construction should only be undertaken on fans when the benefits and risks have been fully evaluated, and appropriate plans have been developed.

We are grateful to the British Columbia Ministry of Forests, Forest Renewal BC, Forestry Innovation Investment BC, Forest Investment Account and the University of British Columbia Faculty of Forestry for providing financial support. M. Church and an anonymous reviewer provided valuable comments.

References

ALLEY, N.F. & YOUNG, G.K. 1978. *Environmental Significance of Geomorphic Processes in the Northern Skeena Mountains and Southern Stikine Plateau*. British Columbia Ministry of Environment, Resource Analysis Branch Bulletin, **3**.

ANON. 1995a. *Forest Road Engineering Guidebook*. British Columbia Ministry of Forests and British Columbia Ministry of Environment Forest Practices Code Guidebook.

ANON. 1995b. *Riparian Management Area Guidebook*. British Columbia Ministry of Forests and British Columbia Ministry of Environment Forest Practices Code Guidebook.

BANNER, A., MACKENZIE, W., HAEUSSLER, S., THOMSON, S., POJAR, J. & TROWBRIDGE, R. 1993. *A Field Guide to Site Identification and Interpretation for the Prince Rupert Forest Region*. British Columbia Ministry of Forests, Land Management Handbook, **26**.

BESCHTA, R.L., BILBY, R.E., BROWN, G.W., HOLTBY, L.B. & HOFSTRA, T.D. 1987. Stream temperature and aquatic habitat: fisheries and forestry interactions. *In:* SALO, E.O. & CUNDY, T.W. (eds) *Streamside Management: Forestry and Fishery Interactions*. University of Washington Institute of Forest Resources, 191–232.

BISSON, P.A., BILBY, R.E. *ET AL.* 1987. Large woody debris in forested streams in the Pacific northwest: past, present, and future. *In:* SALO, E.O. & CUNDY, T.W. (eds) *Streamside Management: Forestry and Fishery Interactions*. University of Washington Institute of Forest Resources, 143–190.

BJORNN, T.C. & REISER, D.W. 1991. Habitat requirements of salmonids in streams. *In:* MEEHAN, W.R. (ed.) *Influences of Forest and Rangel and Management on Salmonid Fishes and Their Habitats*. American Fisheries Society, Special Publications, **19**, 83–138.

BOVIS, M.J. & JAKOB, M. 1999. The role of debris supply conditions in predicting debris flow activity. *Earth Surface Processes and Landforms*, **24**, 1039–1054.

CHURCH, M. & RYDER, J.M. 1972. Paraglacial sedimentation: A consideration of fluvial processes conditioned by glaciation. *Bulletin of the Geological Society of America*, **83**, 3059–3072.

CLAGUE, J.J. 1984. *Quaternary Geology and Geomorphology, Smithers–Terrace–Prince Rupert area, British Columbia*. Geological Survey of Canada, Memoir, **413**.

COSTA, J.E. 1984. Physical geomorphology of debris flows. *In:* COSTA, J.E. & FLEISHER, J.P. (eds) *Developments and Applications of Geomorphology*. Springer, New York, 268–317.

COSTA, J.E. 1988. Rheologic, geomorphic, and sedimentologic differentiation of water floods, hyperconcentrated flows, and debris flows. *In:* BAKER, V.R., KOCHEL, R.C. & PATTON, P.C. (eds) *Flood Geomorphology*. John Wiley, New York, 113–122.

GOMI, T., SIDLE, R.C. & SWANSTON, D.N. 2004. Hydrogeomorphology of sediment transport in headwater streams, Maybeso Experimental Forest, southeast Alaska. *Hydrological Processes*, **18**, 667–683.

HELMS, J.A. (ed.). 1998. *The Dictionary of Forestry*. Society of American Foresters, Bethesda, MD.

HOLLAND, S.S. 1964. *Landforms of British Columbia, A Physiographic Outline*. British Columbia Department of Mines and Petroleum Resources Bulletin, **48**.

HOOKE, R.LE B. & ROHRER, W.L. 1977. Relative erodability of source-area rock types, as determined from second-order variations in alluvial-fan size. *Bulletin of the Geological Society of America*, **88**, 1177–1182.

HUNGR, O., EVANS, S.G., BOVIS, M.J. & HUTCHINSON, J.N. 2001. A review of the classification of landslides of the flow type. *Environmental and Engineering Geoscience*, **7**, 221–238.

INNES, J.L. 1985. Magnitude-frequency relations of debris flows in northwest Europe. *Geografiska Annaler*, **67A**, 23–32.

IRASAWA, M., ISHIKAWA, Y., FUKUMOTO, A. & MIZUYAMA, T. 1991. Control of debris flows by forested zones. *In: Proceedings of the Japan–United States Workshop on Snow Avalanche, Landslide, Debris Flow Prediction and Control, September 30–October 2, 1991, Tsukuba, Japan*. Science and Technology Agency of the Japanese Government, 543–550.

JACKSON, L.E., KOSTASCHUK, R.A. & MACDONALD, G.M. 1987. Identification of debris flow hazard on alluvial fans in the Canadian Rocky Mountains. *In:* COSTA, J.E. & WIECZOREK, G.F. (eds) *Debris Flows/Avalanches: Process, Recognition, and Mitigation*. Geological Society of America Reviews in Engineering Geology, **VII**, 115–124.

JAKOB, M. & JORDAN, P. 2001. Design flood estimates in mountain streams – the need for a geomorphic approach. *Canadian Journal of Civil Engineering*, **28**, 425–239.

KELLER, E.A. & SWANSON, F.J. 1979. Effects of large organic material on channel form and fluvial processes. *Earth Surface Processes and Landforms*, **4**, 361–380.

KELLERHALS, R. & CHURCH, M. 1990. Hazard management on fans, with examples from British Columbia. *In:* RACHOCHI, A.H. & CHURCH, M. (eds) *Alluvial Fans: A Field Approach*. John Wiley, New York, 335–354.

MAH, S., THOMSON, S. & DEMARCHI, D. 1996. An ecological framework for resource management in British Columbia. *Environmental Monitoring and Assessment*, **39**, 119–125.

MELTON, M.A. 1957. *An Analysis of the Relation Among Elements of Climate, Surface Properties and Geomorphology*. Office of Naval Research Department of Geology, Columbia University, New York, Technical Report, **11**.

MURPHY, W.L. & MEEHAN, W.R. 1991. Stream Ecosystems. *In:* MEEHAN, W.R. (ed.) *Influences of Forest and Rangeland Management on Salmonid Fishes and Their Habitats*. American Fisheries Society, Special Publications, **19**, 17–46.

NAIMAN, R.J., BILBY, R.E. & BISSON, P.A. 2000. Riparian ecology and management in the Pacific coastal rain forest. *BioScience*, **50**, 996–1011.

OLIVER, C.D. & LARSON, B.C. 1996. *Forest Stand Dynamics*. John Wiley, New York.

PATTON, P.C. & BAKER, V.R. 1976. Morphometry and floods in small drainage basins subject to diverse hydrogeomorphic controls. *Water Resources Research*, **12**, 941–952.

PIERSON, T.C. & COSTA, J.E. 1987. A rheologic classification of subaerial sediment–water flows. *In:* COSTA, J.E.

& WIECZOREK, G.F. (eds) *Debris Flows/Avalanches: Process, Recognition, and Mitigation. Geological Society of America Reviews in Engineering Geology*, **VII**, 1–12.

POJAR, J., KLINKA, K. & MEIDINGER, D.V. 1987. Biogeoclimatic ecosystem classification in British Columbia. *Forest Ecology and Management*, **22**, 119–154.

RUNKA, G.G. 1972. Soil Resources of the Smithers–Hazelton Area. British Columbia Department of Agriculture, Soil Survey Division, Kelowna, BC.

RYDER, J.M. 1971*a*. The stratigraphy and morphology of paraglacial alluvial fans in south-central British Columbia. *Canadian Journal of Earth Science*, **8**, 279–98.

RYDER, J.M. 1971*b*. Some aspects of the morphometry of paraglacial alluvial fans in south-central British Columbia. *Canadian Journal of Earth Science*, **8**, 1252–1264.

SATO, T. 1991. Flood disaster with drifted logs and sand in Aso Volcano. *In: Proceedings of the Japan–United States Workshop on Snow Avalanche, Landslide, Debris Flow Prediction and Control, September 30– October 2, 1991, Tsukuba, Japan*. Science and Technology Agency of the Japanese Government, 497–506.

SCHWEINGRUBER, F.H. 1996. *Tree Rings and Environment Dendroecology*. Paul Haupt, Berne.

SEPTER, D. & SCHWAB J.W. 1995. *Rainstorm and Flood Damage: Northwest British Columbia 1891–1991*. British Columbia Ministry of Forests Land Management Handbook, **31**.

SMITH, G.A. 1986. Coarse-grained nonmarine volcaniclas-tic sediment: terminology and depositional process. *Bulletin of the Geological Society of America*, **97**, 1–10.

STRUNK, H. 1997. Dating of geomorphological processes using dendro-geomorphological methods. *Catena*, **31**, 137–151.

THOMSON, B. 1991. *Annotated Bibliography of Large Organic Debris (LOD) With Regards to Stream Channels and Fish Habitat*. British Columbia Ministry of Environment Technical Report, **32**.

THURBER CONSULTANTS. 1983. *Debris Torrent and Flooding Hazards, Highway 99, Howe Sound*. Report to the British Columbia Ministry of Transportation and Highways.

VANDINE, D.F. 1985. Debris flows and debris torrents in the southern Canadian Cordillera. *Canadian Geotechnical Journal*, **22**, 44–68.

WATTS, S.B. (ed.). 1983. *Forestry Handbook for British Columbia*, 4th edn. Forestry Undergraduate Society, Faculty of Forestry, University of British Columbia, Vancouver, BC.

WELLS, S.G. & HARVEY, A.M. 1987. Sedimentologic and geomorphic variations in storm-generated alluvial fans, Howgill Fells, northwest England. *Bulletin of the Geological Society of America*, **98**, 182–198.

WILFORD, D.J., SAKALS, M.E. & INNES, J.L. 2003. Forestry on fans: a problem analysis. *Forestry Chronicle*, **79**, 291–295.

WILFORD, D.J., SAKALS, M.E., INNES, J.L., SIDLE, R.C. & BERGERUD, W.A. 2004. Recognition of debris flow, debris flood and flood hazard through watershed morphometrics. *Landslides*, **1**, 61–66.

The fluvial megafan of Abarkoh Basin (Central Iran): an example of flash-flood sedimentation in arid lands

NASSER ARZANI

Geology Department, University of Payame-Nour, P.O. Box 81465–617, Kohandej Road, Esfahan, Iran (e-mail: arzan2@yahoo.com)

Abstract: The Abarkoh Basin is situated in central part of the NW–SE-trending Gavkhoni–Sirjan depression of Central Iran. The studied fluvial megafan represents a Quaternary fan, which is located in the western part of this intermontane, extensional basin. This fan covers an area of 940 km^2 and with a gentle slope (about 0.5°) over a 45 km-length terminates in a playa lake in the centre of the Abarkoh Basin. Its catchment is the Abadeh Basin (>2000 km^2), which is a NW–SE-trending intermontane basin and mainly floored with the Quaternary alluvial-fan sediments. The studied megafan was built by stream-dominated processes, as revealed by the presence of thick-bedded, clast-supported gravels–sandy gravels (gravelly–sandy conglomerates) in the proximal-fan areas that grade into thick marls with sinuous, lenticular channel-filled, grain-supported, imbricated, pebbly conglomerates in the proximal–medial fan regions. Thick calcareous marls and lenticular, single channel-fill sandstone–conglomerate deposits characterize the distal part of the fan and pass laterally into sediments of the playa lake fringe. The aim of this paper is to present an example of a fluvial megafan in an arid–semi-arid setting, to highlight the flash-flood, sheetflood–channelized sedimentation in megafans, and to stress their importance in water resources in arid lands. Fluvial fans are frequently identified in the rock record. The fluvial megafan of the Abarkoh Basin represents one of only a few described, large-scale, stream-dominated, Quaternary megafans in arid–semi-arid settings.

A fluvial megafan has been defined as a large (10^3–10^5 km^2), fan-shaped (in plan view) mass of sediment deposited by the migration of a permanent or intermittent channelized stream with a point source from the outlet of a large mountainous drainage network (Gohain & Parkash 1990; Horton & DeCelles 2001). These types of subaerial fans are distinct from typical sediment-gravity flow-dominated and stream-dominated alluvial fans in term of their size (alluvial fans rarely exceed 250 km^2), lower slope, presence of floodplain areas and absence of sediment-gravity flows (Singh *et al.* 1993; Stanistreet & McCarthy 1993). All documented modern and ancient fluvial megafans are examples from perennial or semi-permanent channelized rivers and mostly the fan and/or its catchment are located in areas with humid climate (e.g. Gohain & Parkash 1990; Singh *et al.* 1993; DeCelles & Cavazza 1999; Horton & DeCelles 2001; Shukla *et al.* 2001). In this paper I apply the definition of a fluvial megafan to a stream-dominated alluvial fan with a large (>30 km) radial length, and with slope gradients from 0.2° to 1°, deposited by periodic flash-flood sedimentation in arid–semi-arid settings (see Blair & McPherson 1994*a, b* for a discussion of alluvial-fan classification). Fluvial megafan sediments record the history of sediment dispersal and deposition in sedimentary basins adjacent to mountain belts. An understanding of the geomorphology and facies distribution of Quaternary fluvial megafans, and their response to flash flooding in

arid–semi-arid settings, is important in constructing facies models to interpret geological records and to improve our understanding of the large-scale geomorphology of ancient mountain belts.

Fluvial megafans are frequently identified in the rock record, and are important mineral, water and petroleum reservoirs (see Galloway & Hobday 1996 for a review). The Kosi fan in Nepal and India (Gohain & Parkash 1990; Singh *et al.* 1993; Goodbred 2003) and the Okavango fan of Botswana (McCarthy *et al.* 1991; Stanistreet & McCarthy 1993) are documented modern examples of the fluvial-dominated megafans. However, the small number of modern examples of stream-dominated megafans that are documented means that our understanding of these important types of large alluvial fans, especially in arid–semi-arid climates, is limited (see Collinson 1996; Miall 1996; Blair 1999*a, b*). The purpose of this paper is to present a case study of a large (>900 km^2), gently sloping (about 0.5°), Quaternary fluvial fan supplied by periodic flash floods, formed under an arid–semi-arid climate. The geomorphology and sedimentary facies analysis of a megafan in the west of the Abarkoh Basin, which comprises the medial part of the Gavkhoni–Sirjan depression of Central Iran, is described (Figs 1 & 2). The Abarkoh megafan is a good example of a Quaternary fan because of its large drainage basin (Abadeh Basin >2000 km^2), and well-exposed Quaternary conglomerates (sinuous meandering channels and wide sheets) and marls of the medial fan (35 km downstream from the apex).

From: HARVEY, A.M., MATHER, A.E. & STOKES, M. (eds) 2005. *Alluvial Fans: Geomorphology, Sedimentology, Dynamics.* Geological Society, London, Special Publications, **251**, 41–59. 0305-8719/05/$15 © The Geological Society of London 2005.

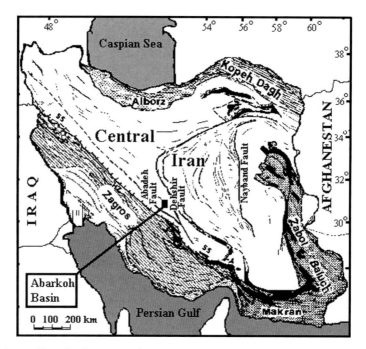

Fig. 1. Geological map of Iran showing seven major tectono-sedimentary units. The Abarkoh Basin is located in Central Iran, which is the unit between the Zagros and Sanandaj–Sirjan (SS) belts in the west, Alborz and Kopeh Dagh in the north, Zabol-Baluch in the east and Makran unit in the south (adapted and modified from Berberian & King 1981; see also Heydari *et al.* 2000 for details).

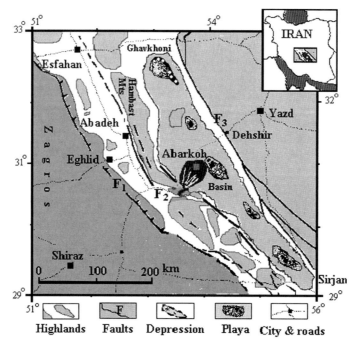

Fig. 2. Location and structural map of the Abarkoh and Abadeh basins. The Ghavkhoni–Abarkoh–Sirjan depression is shown to the west of the Dehshir fault in Central Iran (F1, F2 and F3 are the Zagros, Abadeh and Dehshir faults, respectively).

In many arid countries life would unsustainable without a water supply and the importance of alluvial fans as aquifers has been well recognized by ancient (e.g. >2500 years in Central Iran) people that settled in these regions. The morphology as well as the facies distribution of the alluvial fan controls the distribution of recharge, lateral flow and discharge zones in the fan aquifers (Houston 2000). This case study is from an arid land located in a desert margin, where more than 850 shallow (<100 m)–deep (up to 400 m) irrigation water wells with 103 Qanat systems (>270 km of underground tunnels) tap groundwater from the studied megafan. The Qanat (Kareze) is an ancient groundwater tapping system. It includes the wells (up to 100 m deep) drilled into the water-table, in the proximal–medial-fan sediments and a gently sloping underground tunnel (up to several kilometres) with a series of vertical shafts (Wulff 1986; Arzani 2003).

Geological and tectonic setting

Iran is divided into seven geological provinces, each with a distinct sedimentary and tectonic history (Fig. 1). The studied area is in the middle part of a NW–SE-trending extensional basin, the Gavkhoni–Sirjan depression of Central Iran (Figs 1 & 2). This depression is more than 600 km long and has been part of a larger graben (over 1800 km) that runs from NW to SE across Iran (Taraz *et al.* 1980; Berberian & King 1981). The width of this basin is about 100 km in the studied area, and is bounded by the Dehshir highlands and faults in the east and the Hambast Mountains and faults in the west (Fig. 2). The latest marine transgression in Central Iran occurred through this depression, and deposited Oligocene–Miocene reefal limestones and green marls of the Qom Formation (Berberian & King 1981; Arzani 2003). During later continental sedimentation, they formed the basement, especially in the northern and central parts of the Abarkoh Basin, on which the Quaternary alluvial-fan and playa-lake sediments accumulated (Amidi *et al.* 1983; Arzani 2003).

The studied fan sediments form horizontal terraces that have no fossils and generally have been referred to as Quaternary in age (Taraz 1974; Amidi *et al.* 1983). The Quaternary basin-fill sediments change from thick conglomerates near the depression margins to thick marls towards the basin centre (Arzani 2003). The Oligocene–Miocene basement topography and their early Quaternary erosion also controlled the thickness of the alluvial sediments. Gravity and magnetic surveys shows that the Abarkoh playa is filled with more than 500 m of sediment (Tabatabei 1994). This is also confirmed by the recently drilled (rotary and cable) irrigation water wells on the periphery of the playa.

The drainage basin of the studied megafan is to the west and comprises the Abadeh Basin. It is a NW–SE-trending depression that is easterly tilted and is nearly parallel to the Abarkoh Basin. The former is bounded by the Zagros highlands on the west and Hambast Mountains in the east (Figs 1 & 2). These highlands provide Quaternary alluvial-fan sediments to both the Abadeh and Abarkoh basins.

The Zagros range comprises Cretaceous limestones and grades into the metamorphic complex of the Eghlid towards the south (Taraz 1974). Both limestones and metamorphic rocks are dense, folded and fractured, and prone to disintegration into boulders and pebbles under the arid–semi-arid conditions of the east Abadeh and Eghlid regions. The Zagros thrust fault runs through the eastern edge of these highlands.

The bedrock geology of the Hambast Mountains includes the Permian (limestones) and Triassic (limestones, marls and dolomites) rock units (see Taraz *et al.* 1980; Heydari *et al.* 2000 for details). The apex of the studied megafan is in one of the transverse valleys that connect the drainage basin of the Abadeh depression to the Abarkoh Basin. The Abadeh fault on the western edge and two major faults in the east bound the Hambast range. These faults are nearly parallel to the Zagros fault in the west of the Abadeh depression, and the Dehshir fault in the east of the Abarkoh depression (Figs 1 & 2). Differential displacement along these major faults formed the higher level Abadeh and lower level Abarkoh depressions, and provided sufficient sediment supply into the drainage basin and accommodation space in the depositional site of the studied megafan.

Methods

Recently incised longitudinal channels and exposed terraces of the Abarkoh megafan were visited, sampled and compared with about 1400 sediment samples from 94 recently drilled (rotary and cable) water wells. The sampling depth interval in drilled wells was 3 m, the maximum clast size in the conglomerate and sandstone beds was determined in the exposed terraces by measuring the *a*-axis of the 10 largest clasts within a 1 m^2 exposure. Sand/silt/clay ratios in wet, muddy drilled samples were measured in the field using plastic bags between two fingers. This was checked using wet-sieve analysis of four selected samples. Clast composition was determined in the field and checked in 12 thin sections in the laboratory. Carbonate content of the marl samples was measured using weak hydrochloric acid digestion on 10 selected samples from medial and distal fan areas.

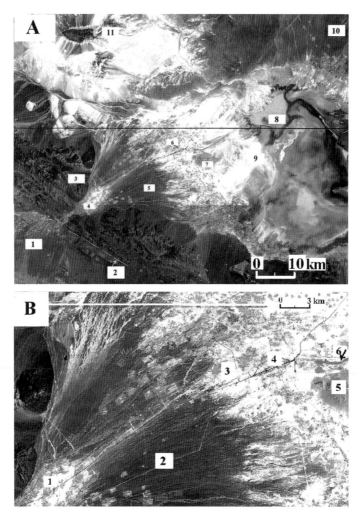

Fig. 3. Satellite images (TM, Landsat 4, 16 January 1988) of the Abarkoh megafan. (**A**) The general view showing: (1) the SE part of the Abadeh Basin (catchment); (2) Abadeh fault; (3) Hambast Mountains; (4) proximal and lower feeder channel areas; (5) lower proximal fan areas; (6) Abarkoh City (midfan areas); (7) Oligocene–Miocene outcrops; (8) playa; (9) mud–salt flats in playa fringe; (10) Dehshir highlands; and (11) Oligocene–Miocene outcrops. (**B**) Proximal–medial and upper distal fan areas showing: (1) proximal and lower feeder channel areas; (2) proximal fan regions; (3) gravelly meander belts of the medial fan areas; (4) Abarkoh City; (5) Oligocene–Miocene outcrops; and (6) wide gravelly meanders at the lower parts of the medial fan areas.

This has been complemented with data from 700 shallow, hand-drilled irrigation water wells and 103 Qanats. The visited Qanats were mainly in the upper–midfan areas, but the wells were scattered all over the fan and near the playa fringe. Geophysical data (electrical resistivity method using the Schlumberger system) were also used (Nico 1981; Meyangi 1985). Rate of precipitation, flood velocity and volume was estimated in place (during field sessions, 1999–2003) and compared with the available data (Bagheri 1995). An available geological map (1/250 000 of Abadeh, Iranian Geological Survey, Amidi *et al.* 1983), topographic maps (1/25 000–1/250 000 Iranian Cartographic Centre) and satellite images (TM, Landsat 4 and ETM Landsat 7) have been used to characterize the catchment, fan drainage and the geology of the area.

Geomorphology and hydrology

The Abarkoh megafan has an area of 940 km^2 with a radial length of 45 km, a maximum width of 40 km and a gentle slope of 0°25′ (0.006). It typifies one of

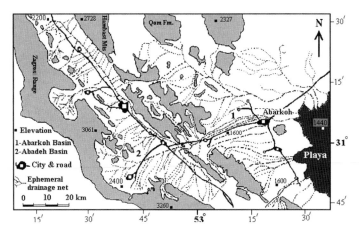

Fig. 4. Simplified map of the Abadeh and Abarkoh basins. Roads, drainage nets, highlands (grey areas) and playa are shown (drawn based on the geological map of Abadeh, 1/250 000 sheet, Geological Survey of Iran, Amidi *et al.* 1983, topographic maps of Abadeh and Abarkoh, 1/25 000–1/250 000, Iranian Cartographic Centre and satellite images, TM, Landsat 4 and ETM Landsat 7, sheet numbers 162R38 and 162R39).

the large Quaternary fans in Central Iran. This fan originates from a valley on the western highlands and terminates in a playa lake in the centre to eastern part of the Abarkoh Basin (Fig. 3).

The Abadeh Basin, with an area of more than 2000 km² (more than 20 km wide and 100 long), is the catchment of the Abarkoh megafan. It is a depression nearly parallel to the Abarkoh Basin and is bounded by the Zagros highlands on the west and Hambast Mountains in the east (Figs 2 & 4). The Zagros highlands are as much as 1500 m above the Abadeh depression. The interbasin Hambast Mountains are as much as 800 and 1600 m above the Abadeh Basin floor and Abarkoh playa, respectively. Differential subsidence along the Zagros thrust fault in the west and Abadeh fault in the east resulted in an easterly tilted basin floor. Following this latter structure, the axial drainage system of the Abadeh Basin is in the east and along the foothills of the Hambast Mountains (Fig. 4). The drainage net of this basin receives precipitation mainly from the Zagros highland in the west. The catchment area is covered by the stream-dominated to debris-flow Quaternary alluvial fans. These fans are relatively small (1–5 km in radial length) and mostly steep (>2° in slope), and provide a ready supply of gravel–mud-grade material. The catchment has a mature drainage network characterized by a nearly straight fifth-order (cf. Strahler 1964) feeder channel that leads southward and then eastward to the fan apex (Figs 4 & 5). At the apex of the studied megafan, the feeder channel is approximately 700 m wide, 2.5 km long and cross-cuts the Hambast Mountain range. It connects the

higher level (2500–1850 m above sea level (masl)) drainage basin of the Abadeh to the lower level (1850–1450 masl) Abarkoh Basin to the east (Figs 4–6).

The present main active channel (up to 150 m wide and 2 m deep) is along the northern margin of the studied megafan. The other channels (up to 80 m wide and 1 m deep) radiate from the apex in a W–E direction towards the playa lake fringe (Figs 3 & 5). The main channels change into more distributary channels in the playa fringe and, other than the main channel that continues towards the playa lake, the other channels terminate in mud flats or between the sand dune fields (Fig. 3).

Quaternary conglomeratic terraces (up to 4 m thick) are exposed 20–35 km downstream from the fan apex and part of Abarkoh City has been built upon them (Fig. 7). They are interbedded with yellowish marls. In plan view, the terraces either form sinuous-shaped, channel-filled, gravelly meanders or gravelly–sandy sheets and are mainly directed downstream (Fig. 8). The geometry of the channels can be directly measured from the 0.5–2 km length of exposed meanders to the south of Abarkoh City. The channel (bankfull) width and depth are 3–45 m and 0.4–2 m, respectively. In the seven well-exposed channels, the meander wavelengths are 85–290 m and the sinuosity of the channel thalwegs is between 1.2 and 1.8 (moderate–high). The wide gravelly–sandy sheets are 120–800 m wide and 20–70 cm thick deposits.

The climate of the Abarkoh is arid, with average rainfall ranging from 29 to 132 mm (Bagheri 1995). The catchment area of the studied megafan is arid–

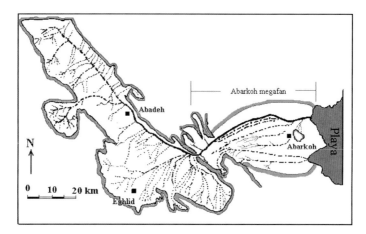

Fig. 5. Drainage nets in the Abadeh Basin and the catchments of the Abarkoh megafan (based on the available geological and topographic maps, and satellite images, see Fig. 3 for details). The fifth-order (cf. Strahler 1964) main channel (solid lines) of the Abadeh Basin drains into the Abarkoh Basin through a valley about 700 m wide in the Hambast Mountains (see Fig. 3). The recent main channel flows in the northern margin of the Abarkoh megafan. The piedmont plotting method follows Blair & McPherson (1994*a*).

semi-arid, with average rainfall ranging between 60 mm in the Abadeh Basin and 300 mm in the headwaters of some rivers in the Zagros highlands (Bagheri 1995). The precipitation is mostly as storm rainfall (in late autumn and early spring) and less frequently as snow on the Zagros hills. The summer maximum daily temperature (for >15 days) is up to 42 °C in the Abadeh plain, and up to 50 °C in the Abarkoh plain and near the playa borders. Other than the cultivated lands mainly near the villages and cities, the vegetation cover is barren–very poor in the rest (*c.* 80%) of the basin floor, and restricted to scattered small bushes and sparse grasses. There are farm lands and villages located in the proximal–distal part of the studied fan. The old city of Abarkoh, with about 40 000 population and 80 km² in area, is located in the medial part, 35 km downfan from the apex (Figs 2 & 3). Catastrophic floods have destroyed roads and part of this city and several villages in the last 100 years.

The large catchment area, the Abadeh Basin, produces flash floods with large volumes of water being focused towards the feeder channel. As an example, the latest catastrophic flood, which occurred on 7 December 2003, was only after 47 mm of rainfall that precipitated in 15 h. This was enough to carry a flash flood with an instantaneous discharge of 92.3 m³ s⁻¹ and a total water volume of 3.45 ×10⁶ m³ through the apex of the studied megafan (estimated using Gauckler–Manning equation, see Williams 1988 for details). The flood followed the major drainage along the northern margin of the megafan and also on other W–E paths (Figs 3 & 5). The flash flood in shallow channels

transformed into wide water sheets, 5–8 km downstream from the fan apex, and added to the torrential rainfall on the proximal fan areas and conducted downfan by sheetfloods. The water spread laterally into zones about 2 km in width and 5 km in length (towards the east) of very shallow flow (very shallow ill-defined channels). About 10–12 km downstream from the fan apex, and through the braided distributary channels of the surface of the fan, the flow entered the main channels. The central channel terminated downfan in cultivated lands and formed a pond near the Abarkoh City (Fig. 7). This flood is comparable to the event of September 1994, with an estimated instantaneous discharge of 188 m³ s⁻¹ and total water volume of 4.2×10³ m³ (Bagheri 1995).

Sedimentary facies and distribution

Proximal-fan facies

Description. The proximal-fan area is dominated by facies 1, grey thick-bedded, clast-supported gravels (conglomerates) to sandy gravels (sandy conglomerates). This facies is exposed in the apex and along parts of the main incised channels. It is also recorded in the subsurface from the drilled-wells in the proximal parts of the studied megafan (Figs 6 & 9–11).

In the apex, facies 1 represents the Quaternary basin-fill sediments and the recent deposits of the feeder channel. The total thickness of the basin-fill

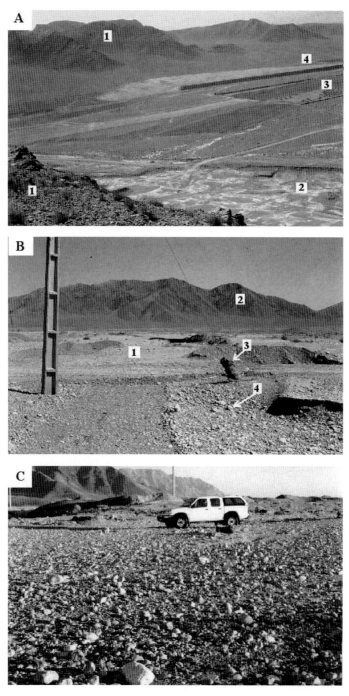

Fig. 6. The feeder channel of the Abarkoh megafan. (**A**) 500 m downfan from the apex and (1) the Permian–Triassic highland at the side of the valley, (2) new braided channel, (3) cultivated land and (4) the changed flood path at sides of the artificial embankments. The width of the valley is about 600 m on the left. (**B**) The feeder channel pavement with coarse gravels (1) and the Permian–Triassic highlands of the Hambast Mountains at the side of the feeder valley (2), the remnants of the flood deposits (rubbish, arrowed) on the electric cable (3), which shows the depth (>40 cm) of the last flood and the rounded pebbles in the channel floor (4). (**C**) Part of the feeder channel, 700 m downstream from the fan apex, showing the gravelly pavement.

Fig. 7. Conglomeratic terraces about 35 km downstream from the Abarkoh fan apex. The upfan area is to the left. (**A**) Up to 80 cm-thick conglomeratic terraces (1), interlayered with yellowish marls (2), on which the Abarkoh City has been built (3, old mosque towers) and a recently drilled irrigation water well (4). The cultivated land is in the muddy sediment (5) of a flood pond. (**B**) A close view of the conglomerate (1) and marl (2) layers in (A). (**C**) A close view of the conglomerate grains (clast-supported and imbricated) in the area shown in (**B**). The car key (circled), for scale, is 7 cm long.

Fig. 8. Conglomerate and marl exposures near the Abarkoh City. (**A**) Part of sinuous-shaped (4–10 m wide) well-cemented conglomerate (1) on thick marls (2). The upfan is towards the right. (**B**) Gravelly–sandy meander (1) deposited on marl (2). The conglomeratic meander belt has a pinch-out and a lens shape. The sandy–muddy beds (3) thin and pinch out toward the centre of the meander belt. The conglomeratic sheet (4) and the Oligocene–Miocene (5) exposures are also shown (looking downfan). The rucksack (circled), for scale, is 40 cm high.

sediments is not known. However, a hand-dug well, 500 m downstream from the apex, shows that the total thickness of the conglomerates is more than 50 m. The feeder channel (up to 700 m wide and 2.5 km long) is floored with the coarse gravelly (pebble–cobble) sediments (Fig. 6). The cobbles are scattered and comprise up to 15% of the grains in 1 m^2, and the rest are well-rounded pebbles (up to 70%) and granules (about 15%). The cobbles and pebbles are dominantly light grey Cretaceous limestones, and, less frequently, yellowish upper Triassic dolomites. Dark grey–black chert and whitish marble pebbles are also occasionally present. The mean size of the 10 largest clast sizes in 4 m^2 of exposure in the channel floor is about 30 cm. The

pebbles are rounded, equant (spheroids) to bladed in shape and show imbrication, with a–b planes transversely aligned and mainly dipping upslope (towards the west). In the feeder channel, coarse grains form elongate and isolated coarse gravel bars (up to 6 m wide and more than 15 m long). They appear between small, erosional gullies (1–12 m wide and 10–40 cm deep) in the main channel. The finer gravels in the gullies and the cobble–pebble bars show a braided-stream appearance (Fig. 6). Pebble–cobble gravel lags are also present above the erosional scours cut into the deposits of the main channel.

Along the incised channels the best exposures of facies 1 are 10–15 km downstream from the apex.

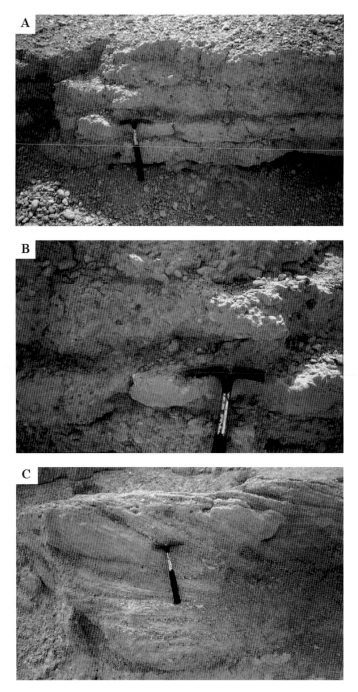

Fig. 9. Exposed terraces of the interlayered conglomerates and sandstones of the studied proximal fan (about 15 km downstream from the apex, north of the village of Faragheh and 32 km NW of Abarkoh City). (**A**) Planar stratified conglomerate beds, with grain-supported, well-rounded and imbricated gravels. (**B**) Closer view of the conglomerate beds. (**C**) Trough cross-bedding in sandy-granular beds. The geological hammer, for scale, is 35 cm long.

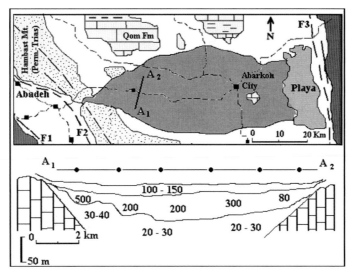

Fig. 10. Simplified geological map (above) and geophysical (geoelectric, resistivity method) cross-section (below) along the line A1–A2 in the proximal region of the Abarkoh megafan. The arrangement of the electrodes was according to the Schlumberger method with measured distance (A–B spacing in each station along the A1–A2 line) of 300 m. The upper (100–150 ohm m^{-1}) and lower (80–500 ohm m^{-1}) layers have been interpreted as gravelly deposits and conglomerates, and the basal (20–40 ohm m^{-1}) part is considered as marl (adapted and modified from Nico 1981; Meyangi 1985).

They are interlayered, pebbly gravels and coarse sands that are locally exposed as terraces (up to 2 m thick) along the sides of a 60–150 m wide river. The pebbly gravels form planar beds 7–45 cm thick. They are interstratified with the planar beds (5–80 cm thick) of coarse sands and fine granules with scattered pebbles (Fig. 9). Both the pebbly and sandy beds are clast-supported. The bladed cobbles of the overlying bed also show imbrication mainly dipping toward the west. The mean size of the 10 largest clasts in 1 m^2 of exposure is 8 cm. The lateral continuity of the beds is uncertain, as the exposures are locally covered. It seems that the coarse gravel and sandy beds form rhythmic planar couplets.

Facies 1 is recorded in the subsurface and 8–15 km downfan from the apex (in a radial distance) and where the width of the fan is about 6 and 20 km, respectively (Figs 3, 10 & 11). In this area more than 16 hand-dug, and five drilled, water wells and more than 10 Qanats are located. The geophysical data and the vertical shafts of the Qanats/wells show that the total thickness of the light–dark grey conglomerates–sandstones of the basin-fill sediments is up to 100 m. The size of the crushed pebbles from the recent drilled wells is up to 4 mm, of which up to 92% are from the Cretaceous limestones and the rest from the Permian–Triassic exposures. The hardness of the bed to be drilled indicates that the beds are well cemented with carbonate

(drusy sparite) cement. Between the cemented horizons, very coarse sandy–fine gravelly samples, which are not cemented, are also present. It is difficult to distinguish the types of conglomerate and sand beds because of the sampling interval (3 m). However, it does indicate the presence of interlayered conglomerates and less frequent sandstone beds.

Interpretation. Facies 1 is interpreted as sheetflood and traction-current deposits to very wide bedload-dominated fluvial channel deposits (Table 1). Sheetfloods transport sediment both in suspension and as bedload, and under conditions of high bed-shear stress only the coarsest clasts are deposited, leading to clast-supported gravels. At the lower current velocities, finer clasts and sand infiltrate into the open clast framework (Blair & McPherson 1994a; Miall 1996; Blair 1999b; Jones 2002). During the Quaternary the studied proximal deposits aggraded and/or prograded downfan from the feeder channel at the fan apex. Sediment-charged flash floods from the large catchment transformed into sheetfloods after a short distance from the feeder channel due to a lack of lateral confinement on the fan surface. Similar sheetflood, traction-current gravels, deposited in proximal areas, have been described for the late Cretaceous–Eocene arid-climate fan system in northern Chile (Hartley 1993). They are also comparable to the waterlaid Anvil

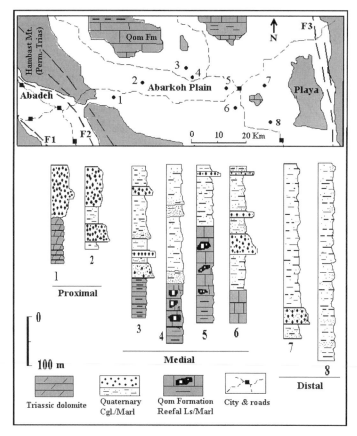

Fig. 11. Simplified geological map (above) and sedimentological logs (below) of the selected drilled water wells in the Abarkoh Basin. Well numbers 1–8 are located in the proximal–distal parts of the Abarkoh megafan. F1, F2 and F3 are the Zagros, Abadeh and Dehshir faults, respectively (modified from Arzani 2003).

Spring Canyon alluvial fan of Death Valley in California (Blair 1999*b*).

The imbricated, coarse-gravelly sediments in the apex of the Abarkoh fan indicate the deposition from a feeding channel system capable of transporting clasts up to 30 cm in diameter. In the large catchment area episodic thundershowers result in periodic high-magnitude runoff, creating flash floods (cf. Schick 1988). The Abadeh drainage basin focuses runoff through the main fifth-order channel and gravelly sediments within the catchment are transported by these floods (Figs 4 & 5). The limited vegetation in such an arid environment is also an important factor in producing flash floods. The abundance of Cretaceous limestone pebble–cobbles relative to Permo-Triassic clasts reflects the shape of the drainage basin with more rainfall and larger alluvial fans in the west of Abadeh Basin and along the Zagros highlands. However, reworking the Quaternary alluvial-fan deposits of the Abadeh Basin provides a ready supply of gravel–mud-grade sediments, fluxed by episodic flood events to the Abarkoh megafan. The isolated coarse gravel bars between the erosional gullies within the main feeder channel form braided distributary channels, with pebble–cobble gravel lags present above erosional scours. These represent a secondary winnowing process that affects the gravelly deposits of the main channel, and results from dissection and down-cutting during the waning flood stage or by other low-magnitude floods. These deposits could be termed 'compound or multistorey channels' (cf. Graf 1988; Miall 1996).

Medial–distal-fan facies

Description. The medial–distal fan area is dominated by facies 2, grey lenticular gravelly–coarse-grained sandy deposits, that is interbedded with facies 3, yellow–brownish calcareous silty marls.

Table 1. *Summary of the lithofacies in Abarkoh megafan*

Location	Facies	Description	Interpretation
Proximal	1	Grey, thick-bedded, clast-supported gravels (conglomerates) to sandy gravels (sandy conglomerates) form rhythmic planar couplets	Sheetflood and traction current to very wide bedload-dominated fluvial channel deposits
Medial–distal	2	Grey, lenticular gravelly–coarse-grained sandy deposit that is interbedded with facies 3. In plan view, facies 2 includes either narrow, sinuous-shaped gravelly meanders (up to 45 m in width and >2 km in length) or wide sheets (up to 800 m in width) of interlayered conglomerate and coarse sandstone beds. In cross-section, facies 2 shows lenticular beds with an erosional base that laterally pinches out towards both sides and with overall lensoid bodies up to 2 m thick. Individual lenticular beds are 5–60 cm thick and composed of gravelly–coarse sandy deposits. The gravelly deposits are grain-supported, poorly sorted–sorted, carbonate-cemented conglomerate that are interlayered with the coarse sandstone beds	Multistorey, single-channel fills of the medial–distal fan areas. The thin, sand–mud interlayers of the margins of the channels show the lateral accretion beds that indicate lateral migration of these channels
	3	Yellow–brownish, calcareous silty marls	Non-channelized, fluvial overbank deposits
Distal (playa fringe)	4	Brown, calcareous and gypsiferous marls, travertine and aeolian sands	Interaction between playa lake, aeolian and the terminal fluvial deposits

Towards the playa margin facies 2 is finer and is interbedded with facies 4, brown calcareous and gypsiferous marls, travertines and aeolian sands. The medial fan facies 2 and 3 are exposed 20–35 km from the fan apex and near Abarkoh City (Figs 7 & 8). In plan view, facies 2 includes either narrow, sinuous-shaped, gravelly meanders (up to 45 m in width and >2 km in length) or wide sheets (up to 800 m in width) of interlayered conglomerate and coarse sandstone beds. This facies overlies widespread yellow, thick marls of the facies 3. In cross-section, facies 2 shows lenticular beds with an erosional base that laterally pinches out towards both sides and with overall lensoid bodies up to 2 m thick (Figs 8, 12 & 13). Individual lenticular beds are 5–60 cm thick and are composed of gravelly–coarse sandy deposits. The gravelly deposits are grain-supported, poorly sorted–sorted, carbonate-cemented conglomerates that are interlayered with the coarse sandstone beds. The rounded, equant (spheroid) to bladed pebbles of the conglomerates occur within a coarse sandy matrix. The mean size of the 10 largest clasts in a 1 m² wall exposure is 5 cm. The pebbles are mainly light grey Cretaceous limestones (up to 70%), yellowish Triassic dolomites (up to 15%), dark grey

Permian limestones and cherts (up to 10%) with a minor fraction of metamorphic rock fragments. Sedimentary structures in facies 2 include subhorizontal stratification with imbricated clasts, moderately developed wedge-planar and wedge–trough beds of sandy pebble–gravel in backsets sloping 10°–30° towards the west (Figs 8 & 12). The interlayered thick sandy beds also display trough cross-stratification. In the sides of the meanders, and where the gravelly beds pinch out, there are thin, fine sand–mud layers that pinch out in the opposite direction and toward the meander axis (Fig. 8).

The small, minor channels that cross-cut the main meanders are mostly located in the meander belt, and are 7–120 cm wide and up to 45 cm deep (Fig. 13). The distributary channels are filled with interlayered very fine pebble–granule conglomerate and thin sandstone layers. The contacts of the sand and gravel layers are sharp, and show multiple phases of erosion.

The wide gravelly–sandy sheets are 120–800 m in width and 30–70 cm thick. They are interbedded with yellowish marls. The fine-pebble–granule beds are poorly sorted and show sharp planar–undulatory contacts with the marls. They either grade to sandy–muddy beds or are abruptly overlain by

Fig. 12. A sinuous meander channel belt in the medial fan SW of Abarkoh City. (**A**) The conglomeratic channel fills (1) deposited on marl (2). (**B**) The conglomeratic channel fills (1) deposited on nodular marl (2). Upstream is to the right.

another fine-gravel bed. The sediments are weakly carbonate cemented, and are exposed between the main meander channels and/or in separate bodies within the marls (Fig. 8).

Facies 3, the yellowish marls, form the dominant facies in the medial–distal fan areas. They are well exposed near Abarkoh City and are also present in the subsurface as thick beds alternating with conglomerate–sandstone beds (Figs 7, 8 & 14). Massive (mottled), nodular–laminated mud–silt beds, which include desiccation and traces of rootlets, are present. The ratio between the fine sand–silt and the clay–carbonate content of the marls, measured in the field and checked by the selected samples in the laboratory, varies between 5/95 and 20/80%. They are very calcareous and the carbonate content varies between 15 and 60%, and is mostly >40%. The high carbonate content of the marls results in the partial dissolution and formation of cavernous porosity in underground and surface exposures. Within the marls, scattered fine-medium pebble–granules of well-rounded, equant–bladed light grey Cretaceous limestones and, rarely, yellowish Triassic dolomite grains are present.

At the distal end of the studied megafan and towards the playa fringe, the silty marls of the facies 3 grade into gypsiferous marls and travertine and aeolian deposits of the facies 4. Facies 3 is also widespread where the main fluvial channels change into a more distributary (terminal) system near the playa fringe and deposit fluvial fine sediment in mud flats and/or within the aeolian interdune areas (Fig. 15).

The distribution of the conglomerate–sandstone beds interbedded with the marls is very variable in the subsurface and in the vertical shafts of Qanats, and in drilled and hand-dug wells in the medial-fan

Fig. 13. A gravelly meander belt (A) and its small chute channel (B) south of Abarkoh City. (**A**) Gravelly meander belt (1) and basal marls (2), location of B (3), an irrigation water well (4) and looking south. (**B**) Small channels cross-cutting the main meanders. (1) Main channel deposits, (2) marls, (3) side of the main channel and (4) minor channel fills. The latter is filled with thin, basal sands and mud layers. These are later filled with interlayered very-fine-pebble–granule conglomerate and sandstone thin layers.

area (Fig. 11). It is difficult to correlate distinct conglomerate beds in the closely spaced wells. The sinuous shape of the gravelly deposits can be inferred from the variability of the facies in adjacent Qanat shafts or new wells drilled adjacent to the old ones. The distal extent of the facies 2 and 3 is evident by the presence of thick (up to 400 m), silty–clayey, calcareous marls and/or their interbedded thick (>40 m) channel-filled conglomerates–sandstones near the playa fringe (wells 7 and 8 in Fig. 11).

Interpretation Facies 2 and 3 are interpreted as deposition in bedload-dominated fluvial channels and floodplain deposits, respectively (Table 1). Facies 2 represent multistorey, single-channel fills of the

medial–distal-fan areas. The geometry of the channel fills represents incised meanders and/or wide shallow channels within the marl substrate. The thin sand–mud interlayers of the margins of the channels display lateral accretion that indicates lateral migration of these channels. It has been widely documented that channel patterns in alluvial settings are primarily dependent on water discharge, sediment load and basin slope (Miall 1996; Field 2001). The channel cross-section geometry also partly reflects the nature of the substrate into which the channel has been cut (Miall 1996; Jones 2002). The incised channels are the conduits through which floods are transferred across the upper fan from the catchment towards the medial fan, where they form

Fig. 14. A sink hole (A) and details of the sediment exposure (B) in its sides, north of Abarkoh City. (A) Sinkhole is in the basin-filled alluvial sediments deposited on Oligocene–Miocene reefal limestones (see Fig. 11, well 4). (B) The selected part of the sinkhole in (A) showing the side wall with interlayered gravel–sand (1), mud (2) and sand (3) beds deposited with an erosive base on the thick marl (4) beds. The camera bag, for scale, is 15 cm wide.

the meanders and shallow, wide channels, and finally terminate toward the playa fringe (lake). The sinuous-shaped gravel meanders, which display multistorey sediment fills, and their related flood-plain deposits are characteristic of the medial–distal fan areas. However, although high-sinuosity channels can occur in almost any fluvial setting, it has been pointed out that high-sinuosity, gravel-bed rivers are not common (Miall 1996).

The small channels that cross-cut the main meanders are interpreted as crevasse and/or chute channels. The presence of these types of channels and their association with the meanders has been widely documented (e.g. Collinson 1996; Miall 1996). However, the small channels are not well preserved and there is no evidence for widespread crevasse

splay sedimention in this area. It is possible that these deposits were not preserved and/or have been reworked, as the present outcrops mainly form cultivated terraces and occur within the city of Abarkoh.

Facies 3 are the widespread facies in medial–distal fan areas and represent non-channelized, fluvial overbank deposits. They are highly affected by post-depositional alteration. Mottled, nodular marls with roolet traces are interpreted as palaeosols in floodplain areas. Laminated interbedded sand, silt and mud represent distal floodplain deposits from clastic sources. Massive muddy facies represent deposition from broad flow zones on the alluvial plain and from standing water during low-stage channel abandonment and in floodplain ponds (e.g. Graf 1988; Collinson 1996). The calcareous nature of the marls and their scattered

Fig. 15. Distal sediments of the Abarkoh megafan and its interaction with the playa fringe sediments. Fluvial fine (1, mud) sediment deposited within the aeolian interdune areas (2) and desiccated into flakes (1), and later covered with aeolian sediments. The geological hammer, for scale, is 35 cm long.

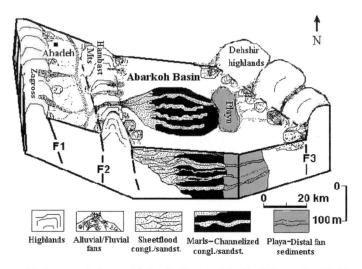

Fig. 16. Schematic model of geomorphology and facies distribution of the Abarkoh megafan. F1, F2 and F3 are the Zagros, Abadeh and Dehshir faults, respectively. The total thickness of the alluvial basin-fill sediments is more than 400 m downfan (see Fig. 11, well 8).

Cretaceous carbonate clasts indicate the main lithology of the catchment area, which is mostly carbonate in composition. The other source of the carbonate is pedogenesis. The presence of carbonate clasts within the marls represents the arid climate of the Abadeh and Abarkoh basins (cf. Miall 1996).

Facies 4 is characteristic of the playa fringe, and represents interaction between playa lake, aeolian and the terminal fluvial deposits. Distributary channel patterns are characteristic of terminal fans, and reflect both loss of stream power and spatially/temporally fluctuating discharge. However, even at present, the major channel of the studied megafan enters the playa lake and the term 'terminal fan' is not suitable, as it reflects the absence of a terminal base level (Collinson 1996; Jones 2004).

Depositional model

The general facies and depositional model of the studied fan is shown in Figure 16, where it is referred to as a fluvial megafan, formed by flash-flood sedimentation in an arid climate. A fluvial megafan is a type of alluvial fan with its general geomorphic definition but relatively large size, lower slope, presence of floodplain areas and absence of gravity-flow deposits (Singh *et al.* 1993; Stanistreet & McCarthy 1993). Blair & McPherson (1994*a, b*) recommended limiting the term 'alluvial fan' to steep systems with a slope range of 1.5°–25°, in which debris flows and sheetfloods are common. This has been criticized by others, who distinguished alluvial fans based on their radial sediment dispersal and their cone shape from fluvial systems with linear-elongated patterns (e.g. Galloway & Hobday 1996; Miall 1996). They stated that alluvial fans are a type of fluvial depositional system in which their geomorphic character is much more important than their fluvial style. Stanistreet & McCarthy (1993) proposed a triangular classification scheme in which the three main processes of sediment-gravity flows, braiding and meandering were considered. This can be compared to the classification of Collinson (1996), who recognized the three main classes: gravity-flow, fluvial and terminal fans. However, fluvial megafans are principally dominated by stream processes but are not to be confused with the stream-dominated alluvial fans. The latter are smaller in size (up to $250\,km^2$) and generally steeper in slope (e.g. Evans 1991; DeCelles & Cavazza 1999; Shukla *et al.* 2001).

Documented modern examples of the fluvial-dominated megafans are the giant Kosi fan in Nepal and India (Gohain & Parkash 1990; Singh *et al.* 1993; Shukla *et al.* 2001; Goodbred 2003) and the Okavango fan of Botswana (McCarthy *et al.* 1991; Stanistreet & McCarthy 1993). The Kosi fan of the Himalayan foreland basin in Nepal and India has a radial length of 160 km (apex to toe), an area of $1000–10\,000$ km^2 and a slope of 0.05°–0.18° (Gohain & Parkash 1990). The main fluvial channel of this fan has migrated across the surface and reworked the fan sediments. Gravelly–sandy to dominantly sandy braided channels with various channel bars are common in proximal areas, and low- to high-sinuosity meandering channels are typical of the distal parts of the Kosi fan (Singh *et al.* 1993). The other example of a megafan is the Okavango fan of Botswana. It has a radial length of 150 km, an area of $18\,000$ km^2, low gradient and is highly vegetated – which makes it the largest documented subaerial fan. The proximal–medial deposits are mainly confined to meandering and anastomosed channel complexes, and show radial, lensoid, sandy channel-fill facies (McCarthy *et al.* 1991).

Fluvial megafans are genetically linked to relatively large drainage networks with a distinct drainage outlet. Horton & DeCelles (2001), based on the geomorphology of the modern fluvial megafans of the central Andes and the Himalayas, suggested a minimum required drainage area of about $10^4\,km^2$ to generate a fluvial megafan. The smaller catchment area of the Abarkoh megafan indicates the importance of sediment availability and flash flooding in arid–semi-arid settings. However, as is true for all type of subaerial fans, the development of a megafan is influenced by tectonic, climatic, lithological and geomorphological parameters in a mountain belt (e.g. Collinson 1996; Miall 1996; Blair 1999*a, b;* Ritter *et al.* 2001).

Conclusions

Flash floods and sheetfloods–channelized flows, under an arid climate and in an intermontane extensional basin, formed the general geomorphology and facies distribution of the Abarkoh megafan. In its large catchment area, the Abadeh Basin, episodic thundershowers, in an arid–semi-arid climate, resulted in periodic high-magnitude runoff and created flash floods towards the feeder channel at the fan apex. These floods transported gravel–mud-grade sediments, provided by a ready source in the mature drainage network and basin-floor Quaternary alluvial-fan sediments of the Abadeh Basin. Sediment-charged flash floods transformed into sheetfloods after a short distance from the feeder channel by the loss of lateral confinement on the fan surface. The floods transformed downstream in turn into channelized flows in the medial–distal fan regions. Thick interlayered gravelly and sandy sediments deposited upfan and aggraded and/or prograded downfan from the feeder channel at the fan apex. These deposits passed downfan into thick marls with sinuous, lenticular channel-filled gravels and sands in the medial–distal fan areas. The studied megafan is a Quaternary example of a fluvial megafan formed by stream-dominated processes in an arid climate in which debris flows, which would be expected in such climatic settings, were not important.

The author would like to thank J. Khajeddin from the Isfahan University of Technology, for his help in satellite image processing and interpretation. Helpful comments on the manuscript were provided by S.J. Jones (University of Durham), and editors A.M. Harvey and M. Stokes. The University of Payame-Nour kindly granted research support.

References

AMIDI, S.M., TARAZ, H., AGHNABATI, A. & NABAVI, S.M. 1983. *Geological Map of Abadeh.* 1/250 000 scale, No. G.9 Geological Survey of Iran, Tehran.

ARZANI, N. 2003. The tragedy of ancient qanats in Kavir borders and arid lands, a case study from Abarkoh Plain, Central Iran. *Iranian International Journal of Science,* **4,** 73–86.

BAGHERI, M.E. 1995. *Water Resources of the Abarkoh Plain.* Department of Water Resources, Yazd Province, Report, **1/120.**

BERBERIAN, M. & KING, G.C.P. 1981. Towards a paleo-geography and tectonic evolution of Iran. *Canadian Journal of Earth Sciences*, **18**, 210–265.

BLAIR, T.C. 1999*a*. Sedimentology of the debris-flow-dominated water Spring Canyon alluvial fan, Death Valley, California. *Sedimentology*, **46**, 941–967.

BLAIR, T.C. 1999*b*. Sedimentary processes and facies of the waterlaid Anvil Spring Canyon alluvial fan, Death Valley, California. *Sedimentology*, **46**, 913–940.

BLAIR, T.C. & McPHERSON, J.G. 1994*a*. Alluvial fans and their distinction from rivers based on morphology, hydraulic processes, sedimentary processes, and facies. *Journal of Sedimentary Research*, **64**, 451–490.

BLAIR, T.C. & McPHERSON, J.G. 1994*b*. Alluvial fan processes and forms. *In*: ABRAHAMDS, A.D. & PARSONS, A. (eds) *Geomorphology of Desert Environments*. Chapman & Hall, London, 354–402.

COLLINSON, J.D. 1996. Alluvial sediments. *In*: READING, H.G. (ed.) *Sedimentary Environments: Processes, Facies and Stratigraphy*. Blackwell Science, Oxford, 37–83.

DeCELLES, P.G. & CAVAZZA, W. 1999. A comparison of fluvial megafans in the Cordilleran (Upper Cretaceous) and modern Himalayan foreland basin systems. *Bulletin of the Geological Society of America*, **11**, 1315–1334.

EVANS, J.E. 1991. Facies relationship, alluvial architecture, and paleohydrology of a Paleogene, humid-tropical alluvial-fan system: Chumstick Formation, Washington State, U.S.A. *Journal of Sedimentary Petrology*, **61**, 732–755.

FIELD, J. 2001. Channel avulsion on alluvial fan in southern Arizona. *Geomorphology*, **37**, 93–104.

GOHAIN, K. & PARKASH, B. 1990. Morphology of the Kosi Megafan. *In*: RACHOCKI, A.H. & CHURCH, M. (eds) *Alluvial Fans: A Field Approach*. Wiley, Chichester, 151–178.

GALLOWAY, W.E. & HOBDAY, D.K. 1996. *Terrigenous Clastic Depositional Systems. Applications to Fossil Fuel and Groundwater Resources*. Springer, Berlin.

GOODBRED, S.L. 2003. Response of the Ganges dispersal system to climate change: a source-to-sink view since the last interstade. *Sedimentary Geology*, **162**, 83–104.

GRAF, W.L. 1988. Definition of flood plains along arid-region rivers. *In*: BAKER, V.R., KOCHEL, R.C. & PATTON, P.C. (eds) *Flood Geomorphology*, John Wiley & Sons, New York, 231–241.

HARTLEY, A.J., 1993. Sedimentological response of an alluvial system to source area tectonism: the Seilao Member of the Late Cretaceous to Eocene Purilactis Formation of northern Chile. *In*: MARZO, M. & PUIGDEFABREGAS, C. (eds) *Alluvial Sedimentation*. International Association of Sedimentologists, Special Publications, **17**, 489–500.

HEYDARI, E., HASSANZADEH, J. & WADE, W.J. 2000. Geochemistry of central Tethyan Upper Permian and Lower Triassic strata, Abadeh region, Iran. *Sedimentary Geology*, **137**, 85–99.

HORTON, B.K. & DECELLES, P.G. 2001. Modern and ancient fluvial megafans in the foreland basin systems of the central Andes, southern Bolivia: implications for drainage network evolution in fold-thrust belts. *Basin Research*, **132**, 43–63.

HOUSTON, J. 2000. Groundwater recharge through an alluvial fan in the Atacama Desert, northern Chile: mechanisms, magnitude and causes. *Hydrological Processes*, **16** (15), 3019–3035.

JONES, S.J. 2002. Transverse rivers draining the Spanish Pyrenees: large scale pattern of sediment erosion and deposition. *In*: JONES, S.J. & FROSTICK, L.E. (eds) *Sediment Flux to Basin: Causes, Controls and Consequences*. Geological Society, London, Special Publications, **191**, 171–185.

JONES, S.J. 2004. Tectonic controls on drainage evolution and development of terminal alluvial fans, southern Pyrenees, Spain. *Terra Nova*, **16**, 121–127.

McCARTHY, T.S., STANISTREET, I.G. & CAIRNCROSS, B.C. 1991. The sedimentary dynamics of active fluvial channels on the Okavango fan, Botswana. *Sedimentology*, **38**, 471–478.

MIALL, A.D. 1996. *The Geology of Fluvial Deposits, Sedimentary Facies, Basin Analysis and Petroleum Geology*. Springer, Berlin.

MEYANGI, Y. 1985. *Geoelectric Survey in the Abarkoh Plain*. Department of Water Resources, Tehran, Report, **5/86**.

NICO, M.M. 1981. *Water Resources of the Abarkoh Plain*. Department of Water Resources, Fars Province, Report, **1361**.

RITTER, J.B., MILLER, J.R. & HUESK-WULFORST, J. 2001. Environmental controls on the evolution of alluvial fans in Buena Vista Valley, North Central Nevada, during late Quaternary time. *Geomorphology*, **36**, 63–87.

SCHICK, A.P., 1988. Hydrologic aspect of floods in extreme arid environments. *In*: BAKER,V.R., KOCHEL, R.C. & PATTON, P.C. (eds) *Flood Geomorphology*, John Wiley & Sons, New York, 189–203.

SHUKLA, U.K., SINGH, I.B., SHARMA, M. & SHARMA, S. 2001. A model of alluvial megafan sedimentation: Ganga Megafan. *Sedimentary Geology*, **144**, 243–262.

SINGH, H., PARKASH, B. & GOHAIN, K. 1993. Facies analysis of the Kosi megafan deposits. *Sedimentary Geology*, **85**, 87–113.

STANISTREET, I.G. & McCARTHY, T.S. 1993. The Okavango fan and the classification of subaerial fan systems. *Sedimentary Geology*, **85**, 115–133.

STRAHLER, A.N. 1964. Quantitative geomorphology of drainage basins and channel networks. In: Chen, V.A. (ed.) *Handbook of Applied Hydrology*, 40–74.

TABATABEI, H., 1994. *Gravity And Magnetic Surveying of the Central Part of Abarkoh Basin*. National Iranian Oil Company, internal report.

TARAZ, H., 1974. *Geology of the Surmaq-Deh Bid Area, Abadeh Region, Central Iran*. Geological Survey of Iran, Tehran, Report **37/148.**

TARAZ, H., GOLSHANI, F. *ET AL.* 1980. *The Permian and the Lower Triassic System in Abadeh Region, Central Iran*. Memoirs of the Faculty of Science, Kyoto, Japan, **47**, 62–133.

WILLIAMS, G.P. 1988. Paleofluvial estimates from dimensions of former channels and meanders. *In*: BAKER, V.R., KNOCHEL, R.C. & PATTON, P.C. (eds) *Flood Geomorphology*, John Wiley & Sons, New York, 321–334.

WULFF, H.E. 1986. The qanats of Iran. *Scientific American*, **218**, 94–106.

Climate and tectonically controlled river style changes on the Sajó–Hernád alluvial fan (Hungary)

GYULA GÁBRIS & BALÁZS NAGY

Eötvös Loránd University of Budapest, Department of Physical Geography, H–1117 Budapest, Pázmány sétány 1/c, Hungary (e-mail:gabris@ludens.elte.hu)

Abstract: Based on geomorphological field investigations, sediment analysis, radiocarbon and palynological data, changes in fluvial style have been recognized on one of the most important low-angle fluvial-dominant alluvial fans on the margin of the Great Hungarian Plain (Hungary). Late Pleistocene and Holocene climatic and tectonic controls are reflected partly by meandering and anastomosed channel pattern changes, and partly by erosional step features on the cone that mark erosional and accumulational phases. This work has led to the surface mapping of a 'horizontal stratigraphy' as part of a larger research project in the Tisza region.

Low-angle fluvial-dominated fans have received only a little attention in the literature. Our work is devoted to an alluvial fan of such a type. The Sajó and Hernád rivers originate from the Carpathian Mountains and have built a complex alluvial mountain-front fan (Harvey 1988) on the margin of the central plain of the Carpathian Basin. This is one of the most important alluvial fans of the Great Hungarian Plain and reaches far into the lowland (Figs 1 & 2). The contour lines of the topographical map suggest a single alluvial fan with its apex near the town of Miskolc and a radial extent of 25–35 km. In comparison to most fans described in the literature, the extent is great and the gradient is gentle (Rannie 1990): the average inclination is only 0.0007 from the apex to the Tisza valley. Geomorphological research allows the Sajó–Hernád alluvial fan to be divided into an old plain, characterized by a branching channel system (anastomosing pattern) with levées and sand ridges, three other levels with meandering channels and point bars, and, finally, the young active river course on the fan's northern margin. Thus, the morphological structure is complex but not as diverse as, for example, the Kosi megafan in India (Gohain & Parkash 1990). Fluvial dynamics were controlled mainly by the climatic events of the late Quaternary. Erosional steps separate the above-mentioned surfaces of the fan into different levels, but there are no terraces *stricto sensu* as in the case of the Tokachi Plain in Japan (Ono 1990).

The extent of the fan (Fig. 3) was determined by geological investigations (Franyó 1966), which proved the distal end to be on the other side of the Tisza valley and also demonstrated the internal structure of the fan. The thickness of the fluvial sequences is considerable in contrast to the low gradient of the fan surface. In temperate regions studies have demonstrated the evolution of such fans to be dominated by climatically led changes in river behaviour. In the case of the Sajó–Hernád fan, alternating coarser and finer strata were interpreted also

in the context of climatic geomorphology, but the sedimentation was different between the valley of the Sajó River and the fan. In the glacial stages gravelly and sandy material was deposited in the valley as a terrace body, and only fine sediments were transported and accumulated on the fan. In contrast, during the interglacial periods, the river cut into the gravelly terraces, and the coarser material was remobilized, transported and finally deposited on the fan (Gábris 1970). The fan tends to show successive changes in channel pattern morphology following the climatic changes of the Late Pleistocene and the Holocene. Rhythmic tectonic sinking of the foreland has broken the climate-driven evolution of the alluvial fan.

The surface of the alluvial fan is relatively young: most of the deposits date from the end of the Pleistocene and have been partially reworked during the Holocene. The major fan was traversed and cut over by the Tisza River at the end of Pleistocene period. The exact age of this event is uncertain.

Some years ago the Pleistocene–Holocene boundary was generally accepted as the time of the flow direction change of the Tisza (Borsy & Félegyházi 1983; Borsy 1990). However, some new results support an earlier appearance of this great river, and suggest that this river was not the Tisza (the main river of the Plain) but the Bodrog and its tributaries coming from the NE Carpathians (Gábris 1998). This question is mainly of regional interest and is not important for this paper.

The presented geomorphological map (Fig. 4) clearly shows the geographical distribution of the varied meandering and braided river patterns, as well as the erosional steps, formed during episodes of incision. This map was drawn using detailed topographic maps (1:10 000), aerial photographs and field observations. Distinct regions are characterized by different river patterns resulting from different fluvial mechanisms (deposition and lateral erosion).

From: HARVEY, A.M., MATHER, A.E. & STOKES, M. (eds) 2005. *Alluvial Fans: Geomorphology, Sedimentology, Dynamics.* Geological Society, London, Special Publications, **251**, 61–67. 0305–8719/05/$15 © The Geological Society of London 2005.

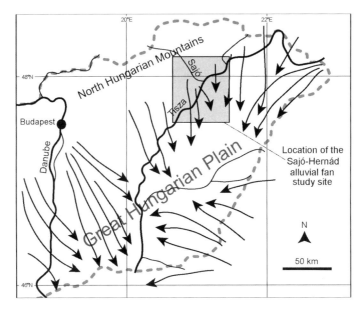

Fig. 1. Location of the Sajó–Hernád alluvial fan on the margin of the Great Hungarian Plain. The arrows show the direction of alluvial fan build-up in the Pleistocene.

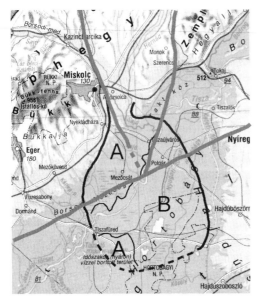

Fig. 2. The Sajó–Hernád alluvial fan. Area A – in this area the sediments of the alluvial fan are on the surface. Area B – in this area the sediments of the alluvial fan are under the surface. The map also shows the main fault lines of this region.

These are separated by erosional step features, which mark deepening phases. In this horizontal stratigraphy six distinct phases of river pattern change and four incisions have been detected. Figure 5 shows the relative chronology of these river style variations.

Methods of chronology and reconstruction of fluvial evolution

Precise dating of these changes is relatively weak. We have five new boreholes for pollen analysis, but others are known from the region's relevant literature, the results of which have been also used here. In the oldest meanders (P-1) the pollen content is not sufficient for qualitative analysis and dating: although rare pollen grains indicate a cold steppe vegetation in this part of the Great Plain (Nagy & Félegyházi 2001), which is supported by malacological analyses (Krolopp & Sümegi 1995). Three other boreholes have incomplete pollen diagrams: the lower part of the meander lag contains very little pollen. These three boreholes (P-2, P-3 and P-4) were located in the same generation of meanders, but in different regions. All of these incomplete pollen diagrams indicate the age of meander formation to be older than the Late Glacial. The borehole taken from the youngest channel generation (P-5 and P-6) have provided a pollen diagram suitable for accurate analyses.

An Upper Tisza Project was initiated by University of Newcastle in co-operation with Hungarian partners, including Eötvös University of Budapest. On the northern part of the investigated area, more precisely around the town of Polgár, a number of radiocarbon dates (Beta Analytic Laboratories, Florida) contribute to the age determination of the meanders (Davis & Passmore 1998). Field research, radiocarbon dating (Centrum voor Isotopen

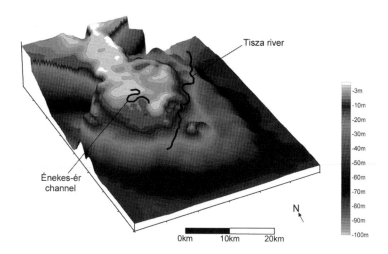

Fig. 3. Elevation model of the gravel surface of the studied alluvial fan.

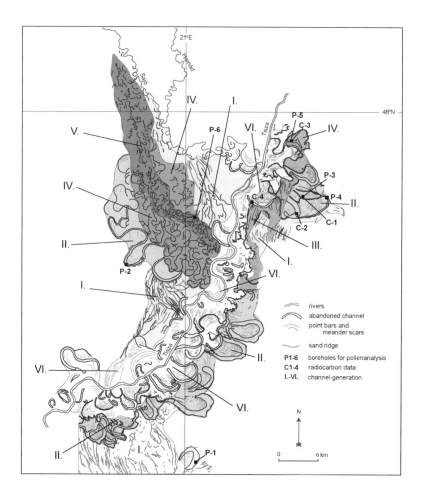

Fig. 4. Geomorphological map of studied region.

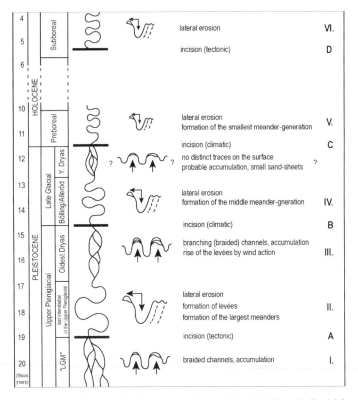

Fig. 5. Diagram of the reconstructed fluvial style changes on and around the Sajó–Hernád alluvial fan.

Onderzoek, Groningen, the Netherlands) and pollen analysis were also carried out in co-operation with the Vrije Universiteit, Amsterdam (Vandenberghe *et al.* 2003) and the University of Debrecen (Hungary).

Description of the phases of the river style changes

The large and oldest surface of the Sajó alluvial fan, which continues far to the SW, is characterized by loess-covered linear sand ridges, interpreted as levées of a braided river system (phase I). (The different phases characterized by braided and meandering patterns are indicated by Roman numerals and the incision episodes are indicated by letters, both on the map (Fig. 4), the diagram (Fig. 5) and in the text.) The wind modified the sandy areas, and deflation landforms occur locally (Gábris *et al.* 2002). Based on a radiocarbon date (18 010 ± 90 C[14] BP; charcoal, G-A 16093) and on a pollen diagram (Gietema 2000), its age appears to relate to the coolest period of the Upper Pleniglacial between 19 and 23 ka cal BP (calibrated years before persent) (18–21 ka uncalibrated C[14] BP), hence to the Last Glacial Maximum.

A very small sandy patch with dunes in the old centre of Polgár town (NE edge of the area) presumably represents this episode.

The termination of this phase is clearly marked by a distinct incision phase (A). An erosional step developed between the preceding surface and the lower plain characterized by a meandering river pattern. This incision period was probably caused by tectonic subsidence at the end of the growth of the alluvial fan.

The meanders of greatest size (phase II) occur in two geographically separated regions (Fig. 4), on the Sajó alluvial fan and the eastern border of the Tisza valley. These two meander generations seem to be similar in age, but different in measure and origin. In the Tisza valley the meanders are larger and indicate the main river of the Great Hungarian Plain (Nagy & Félegyházi 2001), and on the alluvial fan the meanders of this generation are smaller in correlation of the smaller discharge of the Tisza's tributaries. We have three pollen diagrams from this meander generation. The first diagram from the alluvial fan (P-2) suggests an age older than Late Glacial in spite of the lack of pollen in the bottom of the core (Fig. 6). The two other boreholes were drilled in a large and narrow palaeochannel of the Tisza valley. The

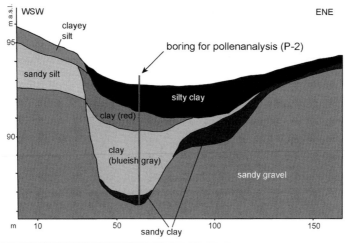

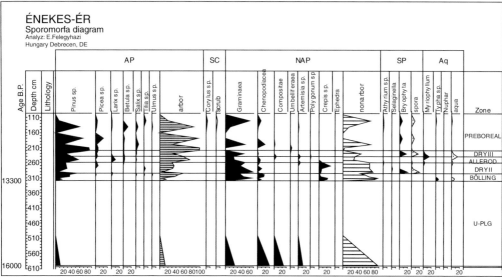

Fig. 6. Cross-section (**a**) and pollen diagram (**b**) of the Énekes-ér meander.

meander lag contains very little pollen in the lower part of the sediment core but what is there indicates an extremely cold period. We had the opportunity to draw a pollen diagram based on the data from the higher part of the core. In one of these meanders (C-1) we have an uncalibrated or conventional radiocarbon date: $20\,470 \pm 310$ C^{14} BP; charcoal, Beta-105027 (Passmore *et al.* submitted) from the floor of the meander. On the basis of the radiocarbon dating and the pollen content of the abandoned meander filling we suggest a hypothesis: that the meander was formed during the increased water discharge just after the Last Glacial Maximum when periglacial circumstances dominated in the Carpathian Mountains of the water catchment area of the great river

(Gábris 1995). Sedimentation in the oxbow lake has continued until recent times.

The next phase (III) is characterized by the second braided period. The remains of this morphology occur only in two small areas in the Tisza valley on the NE edge of the studied area (Fig. 4).

On the alluvial fan a large region represents the fourth period (phase IV) characterized by abandoned medium-sized meanders, but these surfaces lie below the preceding level of the meandering rivers and this position indicates a downcutting period (phase B) in the fluvial development. A core site was located on the alluvial fan of the Sajó River but only some Pinus sp. pollen were found at the base of the borehole that are not sufficient to indicate the age and the environment of the period of its formation. Even though, according to data of the latest borehole (P-5), the channel's sediment contains pollen from the early period of Younger Dryas (Magyari 2002). Therefore, meander formation may relate to the Alleröd.

During the transition between the cool, dry Younger Dryas and the warmer, more humid Preboreal the rivers cut down again (Vandenberghe *et al.* 1994) into the surface of the alluvial fan (phase C). As a result, in this level, rivers widened their meanders and formed the next channel generation (phase V). This smallest meander generation of the Sajó River extends over the axis of the alluvial fan. The boundary between this surface and the level of the preceding meander generation is marked by a distinct step. Based on the pollen

diagram P-6 (Fig. 7), the Sajó River moved to the NE margin of the alluvial fan (to the current location) and the infilling process of the abandoned meanders started in the sub-boreal.

In the late Holocene the Tisza River cut again into the surface (phase D), and this is indicated by well-developed erosional steps on both sides along the studied reach. On this lowest surface of the Tisza valley there are overdeveloped young meanders (phase VI). One of the abandoned curves has a radiocarbon age: 4070 ± 100 C^{14} BP (charcoal, Beta-105024). This part of the Tisza River is also characterized by meanders artificially cut off in the last 150 years, and clearly the result of modern anthropogenic effects.

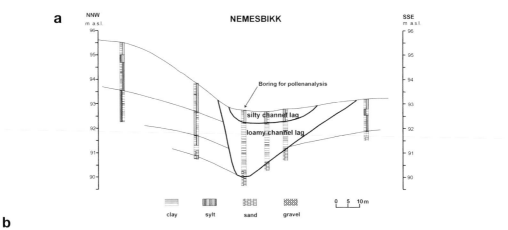

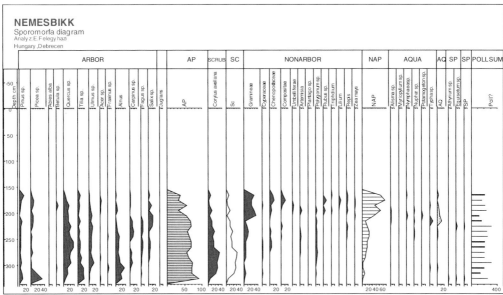

Fig. 7. Cross-section (**a**) and pollen diagram (**b**) of the Nemesbikk meander.

Conclusions

A geomorphological map was drawn following the detailed topographic maps (1:10 000), aerial photographs and fieldwork observations in order to determine the geographical distribution of the different channel pattern types on the Sajó–Hernád alluvial fan. Based on the regions characterized by different (meandering and braided) river styles and separated mainly by erosional steps – which mark deepening phases – a 'horizontal stratigraphy' has been established. Six distinct phases of river pattern change and four incisions were detected in this stratigraphy. These phases generally match the results of research in Poland and the Netherlands (Vandenberghe *et al.* 1994). Figure 5 shows the relative chronology of the river styles variations – in some cases dated by palynological and radiocarbon data.

During 1999–2000 our investigation was supported by the FKFP 0180/99 project of the Office for Higher Education of Hungary and during 2001–2003 by the OTKA T034979 project of the National Scientific Research Fund. We appreciate the facilities and the assistance in preparation of pollen samples in the Laboratories of the Vrije Universiteit Amsterdam. We are grateful to Dr. E. Félegyházi (University of Debrecen, Department of Physical Geography and Geoinformatics) for the pollen analysis of the borehole materials originating from different palaeo-meanders.

References

BORSY, Z. 1990. Evolution of the alluvial fans of the Alföld. *In:* RACHOCKI, A.H. & CHURCH, M. (eds) *Alluvial Fans: A Field Approach.* Wiley, Chichester, 229–248.

BORSY, Z. & FÉLEGYHÁZI, E. 1983. Evolution of the network of water courses in the northeastern part of the Great Hungarian Plain from the end of the Pleistocene to our days. *Quaternary Studies in Poland,* **4,** 115–124.

DAVIS, B.A.S. & PASSMORE, D.G. 1998. *Upper Tisza Project: Radiocarbon Analyses of Holocene Alluvial and Lacustrine Sediments.* Interim Report on Current Analyses to the Excavation and Fieldwork Committee, University of Newcastle.

FRANYÓ, F. 1966. Der Schwemmfächer der Flüsse Sajó und Hernád im Spiegel der Geologischen Ereignisse des Quartärs. *Földrajzi Értesítő,* **15,** 153–178.

GÁBRIS, GY. 1970. Method for a research of recent riverbed-changes on the example of the alluvial cone of the Sajó. *Földrajzi Közlemények,* **18,** 294–303 (in Hungarian).

GÁBRIS, GY. 1995. River activity as a function of changing palaeoenvironmental conditions during the Lateglacial–Holocene period in Hungary. *In:* FRENZEL, B. (ed.) *European River Activity and Climatic Change During the Lateglacial and Early Holocene European Palaeoclimate and Man,* Volume **9.** G. Fischer, Stuttgart, 205–212.

GÁBRIS, GY. 1998. Late Glacial and Post Glacial development of drainage network and the palaeohydrology in the Great Hungarian Plain. *In:* BASSA, L. & KERTÉSZ, Á. (eds) *Windows on Hungarian Geography.* Akadémiai kiadó, Budapest, 23–36.

GÁBRIS, GY., HORVÁTH, E., NOVOTHNY, Á. & UJHÁZY, K. 2002. History of environmental changes from the Last Glacial period in Hungary. *Praehistoria,* **3,** 9–22.

GIETEMA, S. 2000. *Pollenanalyse van verschillende riviergeneraties van de Tisza, Hongarije.* Vrije Universiteit Amsterdam.

GOHAIN, K. & PARKASH, B. 1990. Morphology of the Kosi Megafan. *In:* RACHOCKI, A.H. & CHURCH, M. (eds) *Alluvial Fans: A Field Approach.* Wiley, Chichester, 151–178.

HARVEY, A. 1988. Controls of alluvial fans development: the alluvial fans of the Sierra de Carrascoy, Murcia, Spain. *Catena,* **13,** Suppl., 123–137.

KROLOPP, E. & SÜMEGI, P. 1995. Palaeoecological reconstruction of the Late Pleistocene, Based on Loess Malacofauna in Hungary. *GeoJournal,* **36,** 213–222.

MAGYARI, E. 2002. *Climatic versus human modification of the Late Quaternary vegetation in Eastern Hungary.* PhD Thesis, University of Debrecen, Hungary.

NAGY, B. and FÉLEGYHÁZI, E. 2001. Investigation of the Late Pleistocene channel-system of the Sajó–Hernád alluvial fan. *Acta Geographica Debrecina,* **35,** 221–232 (in Hungarian).

ONO, Y. 1990. Alluvial fans in Japan and South Korea. *In:* RACHOCKI, A.H. & CHURCH, M. (eds) *Alluvial Fans: A Field Approach.* Wiley, Chichester, 91–107.

PASSMORE, D.G., SHIEL, R.S., DAVIS, B.A.S., MAGYARI, E., MOORES, A. & KUTI, L. Quaternary palaeochannels of the Upper Tisza valley, Northeast Hungary. *Journal of Quaternary Science,* submitted.

RANNIE, W.F. 1990. The Portage La Prairie 'Floodplain Fan'. *In:* RACHOCKI, A.H. & CHURCH, M. (eds) *Alluvial fans: A field approach.* Wiley, Chichester, 179–193.

VANDENBERGHE, J., KASSE, C., BOHNCKE, S. & KOZARSKI, S. 1994. Climate-related river activity at the Wechselian–Holocene transition: a comparative study of the Wartha and Maas rivers. *Terra Nova,* **6,** 476–485.

VANDENBERGHE, J., KASSE, K., GABRIS, GY., BOHNCKE, S. & VAN HUISSTEDEN, K. 2003. Fluvial style changes during the last 35.000 years in the Tisza valley. *In: XVI. INQUA Congress, 23–30 July 2003, Reno, Nevada, USA, Abstracts.*

Quaternary telescopic-like alluvial fans, Andean Ranges, Argentina

F. COLOMBO

Departament d'Estratigrafia, Paleontologia i Geociències Marines,
Facultat de Geologia, Universitat de Barcelona, E-08028 Barcelona, Spain
(e-mail: colombo@ub.edu)

Abstract: The largest rivers that drain the Argentine Andean Ranges are characterized by incised valleys in high mountains and by a variety of Quaternary terraces. The terraces display a fan geometry with the apex located upstream of a tributary junction. Their convex-up morphology suggests that these terraces are related to a series of alluvial fans developed where the tributaries join the main river. The succession of alluvial aggradation and degradation is controlled by local base-level variation conditioned by temporary lake development in the main river valley. All these factors give rise to the inset segmented (terraced) morphology of the fan surfaces, yielding a telescopic-like relationship. The variation in the morphology and number of terraces suggests that they are not controlled by a general/regional base level. Neither tectonic activity nor significant climatic changes account for the alluvial fans at the confluences of the tributaries and the main river. Significant variations in rainfall or thunderstorms induced by the El Niño Southern Oscillation (ENSO) could explain the genesis of these telescopic-like alluvial fans.

The main Argentine rivers that drain the Andean Ranges (between approximately 24°S and 33°S) have incised valleys and tributaries displaying a variety of terraces composed of gravel deposits. This study concerns the terraces developed in the valleys of the main rivers, such as the Mendoza River in the province of Mendoza, the San Juan River in the province of San Juan and the Rosario River (Quebrada del Toro) in the province of Salta. This area exceeds 1000 km in length.

The terrace levels are not continuous along each valley and correlation is not possible between adjacent valleys. The terraces are usually displayed at the junction of a tributary with the main river. The terraces show characteristic fan morphology with a radial palaeocurrent distributions. Moreover, their steep longitudinal slope and cross-sectional convex geometry suggest that the terraces are individual alluvial fans generated by intense sedimentary activity at the confluences of the main rivers and the tributaries (Colombo *et al.* 1996, 2000).

These alluvial fans (tributary alluvial fans, TAF) are characterized by a variety of terrace levels. The younger are found in front of the older ones at progressively lower levels. Thus, each major alluvial fan shows a segmented geometry made up of several morphological breaks on their upper surfaces. We propose to term them telescopic-like alluvial fans (Bowman 1978; Janocko 2001) because of their geometrical characteristics. There are a number of studies of segmented alluvial fans (Blissenbach 1954; Bull 1964, 1968, 1979; Harvey 1984, 1987*a*; Harvey *et al.* 1999, 2003); some of these relate to tectonic activity (Bull & McFadden 1977; Wallace 1978; Harvey 1987*b*; Silva *et al.* 1993; Ferrill *et al.* 1996; Shaoping & Guizhi 1999; Stokes & Mather 2000), whereas others deal with climatic variations (Bull 977, 1991; Harvey 1990, 1996). This work is focused on the generation of telescopic-like alluvial fans and on their sedimentological significance. The morphology of TAF is studied by theodolite surveys and their sedimentology by means of facies analysis. A search for fossil remains to obtain ^{14}C data was also undertaken.

General distribution

Along the Argentine sector of the Andean Ranges, telescopic-like alluvial fans, located in the main river valleys, are common geomorphological features and are well developed in the Quebrada del Toro, situated at approximately 24°S in the province of Salta (NW Argentina), and in the Mendoza river located at approximately 33°S in the province of Mendoza. Both the Jáchal (30°S) and the San Juan rivers (31° 30´S) provide some good examples (Fig. 1) in the San Juan province.

The surfaces of the telescopic-like fans, which are related to different terrace levels (Colombo *et al.* 1996, 2000), are termed T_0 at the level of the current terrace of the river to T_n, which is the highest observed level. This labelling scheme mimics the increasing age of the higher terrace levels; the current river (T_0) reflects the lowest topographic level, whereas the highest (T_n) corresponds to the older level.

From: HARVEY, A.M., MATHER, A.E. & STOKES, M. (eds) 2005. *Alluvial Fans: Geomorphology, Sedimentology, Dynamics.* Geological Society, London, Special Publications, **251**, 69–84. 0305–8719/05/$15 © The Geological Society of London 2005.

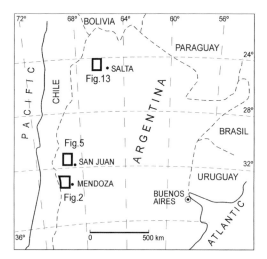

Fig. 1. Location of the areas along the Argentine Andes where the Quaternary telescopic-like alluvial fans are displayed.

Mendoza River area

This is located in the central part of the Argentine Andean Ranges, and consists of the approximately W–E-orientated large valley of the Mendoza River, which drains the high summits and reaches the town of Mendoza (Fig. 2). This area extends from approximately 32°35'S to 33°S and from 69°07'W to 69°57'W, and displays diverse levels of TAF deposits.

In this region the Precordillera structure is made up of large eastward-propagating thrust sheets. The main faults, and the biggest folds and valleys, display a N–S orientation of their major axes. There are large N–S-oriented valleys and mountains, such as the Uspallata valley, where extensive TAFs have their source area. One important TAF exists at the mouth of the Invernaditas Quebrada (Fig. 3), with five terrace levels (T_0–T_4). It seems that its dimensions (1800×1200 m) reflect the significance of its extensive drainage basin. Other areas (Fig. 4) that display large TAFs are Potrerillos (P), Guido (G) and Polvaredas (PV) located along the Mendoza River valley.

San Juan River area

This is the large regional river that predates the Andean structures and cuts them at an orthogonal angle in some places (Fig. 5). In the main valley there are several terraces corresponding to the alluvial fans at the confluences of the tributaries and the San Juan River (Figs 6 & 7). In the region there are other examples of telescopic-like fans in the Castaño Viejo area.

Sasso River. This tributary river (Fig. 8) has a source area made up of shales and greywackes, yielding large amounts of clastic sediments. The slope of the Sasso River varies between 5.3 and 3.4% (with a median of 4.35%) on reaching the San Juan River, which displays a reduced width in the confluence area (approximately 200 m). There are five levels of terraces from youngest to oldest (T_0–T_4) with an altitude difference of 171 m.

Sassito River area. This tributary river (S in Fig. 5) has a N–S source area developed in the predominantly Carboniferous shales, producing a large supply of clastic materials. There are diverse levels of terraces (T_0–T_4), with slopes varying between 13.3 (lowest levels) and 6.5% (highest levels), with a mean of 8.3%. The main facies of these alluvial terraces consist of relatively well-stratified, poorly sorted gravels with a poorly sorted sandy matrix and some randomly distributed out-sized clasts.

Albarracín River area. This river (A in Fig. 5), which is one of the largest tributaries of the San Juan River, displays five levels of well-differentiated alluvial terraces. These terraces (Fig. 9) show a radial pattern starting from an area where the palaeocurrents are focused upstream of the Albarracín River, reaching the right bank of the San Juan River. Based on topographical measurements of each terrace, the cross-section shows convex-up surfaces (Fig.10). This, together with the fact that each terrace is incised by another level, suggests that these terraces are unstable and that the Albarracín River cuts downward to reach its equilibrium profile. The longitudinal section shows that the younger alluvial terrace surfaces have a slope that is larger than those of the older terraces. This indicates that the episodic incision is controlled by the level of the San Juan River, which shows a difference of almost 1 m below the current terrace of the Albarracín fan (T_0) in the area of its confluence (Fig. 11). The current terraces corresponded to different incision episodes as a result of diverse events, and to the sedimentary activity of the lateral fan that could generate alluvial segments further and further from the apical area. The most recent segment and the present riverbed were incised by the San Juan River. Nearby is the La Laja area (L in Fig. 5) where older terraces display higher slopes than the recent (T_0) ones owing to local tectonic activity (Colombo *et al.* 2000).

Jáchal River area. This large regional river (T in Fig. 5), which has an antecedent character, displays a number of terraces (Fig. 12) along its incised valley through the Precordillera. There are more than five terrace levels (T_0–T_4) that are related to alluvial terraces, made up of coarse-grained clastics and

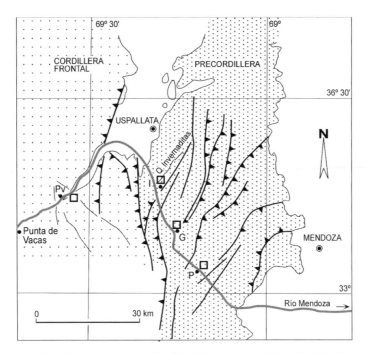

Fig. 2. Mendoza River region. The main telescopic-like alluvial fans are in the Polvaredas (PV), Invernaditas (I), Guido (G) and Potrerillos (P) areas marked as insets. The Frontal Cordillera and Precordillera are represented by different patterns where the main faults and overthrusts are displayed.

Fig. 3. The Invernaditas telescopic-like alluvial fan where some levels of alluvial terraces (T_0–T_4) are displayed. The T_1 terrace (arrow) has a thickness of 1.8 m.

interfingered with fine-grained shallow lacustrine materials (Busquets *et al.* 2002).

Quebrada del Toro area

This corresponds to the Rosario River valley located to the west and NW of the town of Salta (Fig. 13),

with an approximate NNW–SSE orientation and characterized from north to south (Fig. 14) by the Quebrada del Tastil and the Quebrada del Toro (24° 22′ to 24°54′S and 65° 39′ – 65° 57′W) large valleys.

Diverse TAFs exist along the main valley (Igarzabal 1991; Igarzabal & Medina 1991) where the tributaries join the main river (Fig. 15). Five different terrace levels (T_0–T_4) varying in size and

Fig. 4. Alluvial terraces (T_0–T_4) depicted at the Guido (G in Fig. 2) telescopic-like alluvial fan. Railway bridge for scale.

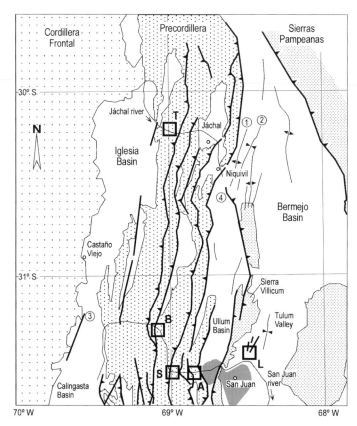

Fig. 5. Locations of studied examples sited mainly in the Precordillera. The Tambolar (B), Sasso and Sassito (S), Albarracín (A) and La Laja (L) areas in the San Juan River valley. The Totoralitos (T) area is in the Jáchal River valley. (1) Anticline axis; (2) syncline axis; (3) main fault; and (4) overthrust. The grey pattern corresponds to the San Juan alluvial fan.

Fig. 6. Alluvial terraces (T_0–T_6) in the telescopic-like alluvial fan at the junction of a tributary and the San Juan River valley. The total thickness of the terraces is 16 m. Located near km 58 of the N 20 national road.

Fig. 7. Telescopic-like alluvial fan in the Tambolar area (B in Fig. 5). Some terrace levels (T_0–T_3) are displayed. Terrace T_1 (arrow) has a thickness of 5.5 m.

thickness constitute the TAF. In this region there are other examples of telescopic-like alluvial fans in the Ingeniero Maury (IM), Puerta de Tastil (PT) and Corte Blanco (B) areas.

Alluvial segments: genesis

An alluvial fan could be formed when the tributaries deposit their bedload as they join the main river, whose base level at that time was low. The water level increased because the development of the alluvial fan acted as a dam (Malde 1968; Jarrett & Costa 1986; Clague & Evans 2000) resulting in a temporary lake upstream. When the alluvial fan ceased to grow, the water level of the main river increased until the dam was overtopped. Subsequently, as a result of upstream erosion, the complete incision of the dam was produced, facilitating the total drainage of the temporary lake. Finally, a base level similar to that of the main river was once again reached (Fig. 16).

After additional overflows an incision of the new alluvial segment was generated in front of the earlier one. An almost complete section of the new terrace was produced, attaining once again the lowest local base level, which was controlled by the main river. The repetition of this process lead to the development of an alluvial fan where the younger terraces were located in front of the earlier ones at progressively lower levels. The alluvial fans thus generated are termed telescopic-like, given that they occupy the space located in front of the remnants of the previous

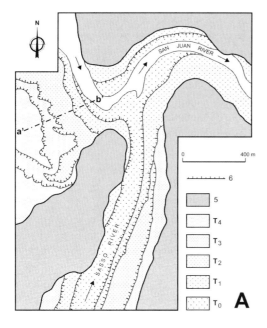

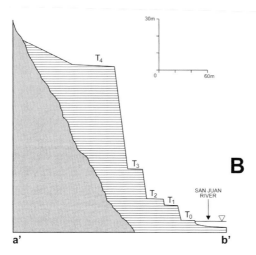

Fig. 8. Sasso (S in Fig. 5) telescopic-like alluvial fan. The cartography (**A**) shows the distribution of the terrace levels (T_0–T_4); (5) basement; and (6) terrace escarpment. The (a'–b') longitudinal section (**B**) displays the different thicknesses of each terrace.

segments. The surface of alluvial segments corresponds to the alluvial terraces that in many cases are erosional (Colombo *et al.* 2000) and not accumulative. The fact that the thick and coarse materials become thin and fine suggests that the sedimentary currents from the TAF reach the temporary lake directly (Fig. 17).

A number of episodes of incised valleys and diverse events of sedimentary infill are necessary to produce the generation of TAF and the associated alluvial terraces. The scenario could be as follows: (1) development of the fluvial and regional drainage net with the arrangement of the main rivers and their tributaries; (2) an incision episode of the rivers with valleys very similar to the current ones; (3) alluvial-fan generation due to sedimentary accumulation produced by the tributaries that fill the main river valley laterally; (4) a number of successive erosional episodes reflecting local variations in the base level – these episodes were due to TAF incision produced by the main river base-level variations; (5) coevally, a temporary lake formed upstream due to the obstruction of the main valley by the alluvial fan; and (6) the final incision resulted in the destruction of the dam and in the recovery of the valley by the main river.

The erosion of the dams took place whenever the retained waters of the temporary lakes overtopped the dam. Diverse levels of erosional terraces controlled by the different periods of erosion of the dams were produced. These terraces could also have developed as a result of the local reactivation of the main channel.

Longitudinal profiles: variability

Although the studied tributary rivers (Colombo *et al.* 2000) were incised in previously deposited sediments they did not reach the substratum. This could give rise to the following situations.

- The longitudinal profile of the alluvial terraces is divergent with respect to the main riverbed base level (Fig. 18A). Thus, the surfaces of the most modern erosional alluvial terraces have increasingly steep slopes with the result that the slopes continuously adapt themselves to the main river base level. If the main river were in an excavating episode, the local base level would be increasingly incised, resulting in the episodic development of new erosional terraces. This is the case for the other rivers in the region (Fig. 11), for example the Albarracín River.
- According to the longitudinal profile of the alluvial terraces, the new profile for a given terrace is less steep than the previous one. Thus, the longitudinal profile of the older erosive terrace surface has a steeper slope than that of the younger one. This suggests local tectonic activity. The longitudinal profile of the most recent erosional terrace is the current one (Fig. 18B).
- The longitudinal profile of the alluvial terrace is roughly parallel to that of the other terraces.

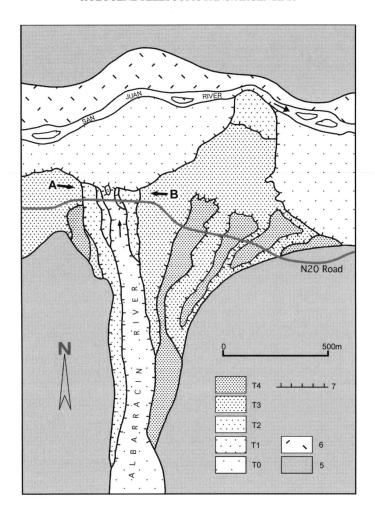

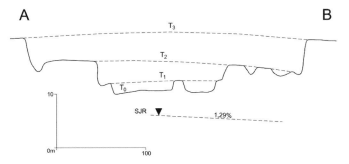

Fig. 9. Cartography of the Albarracín (A in Fig. 5) telescopic-like alluvial fan with the terrace levels (T_0–T_4) distinguished. Cross-section of terrace levels (T_0–T_4) displaying a characteristic convex-up topography. Note the slope of the (SJR) San Juan River. (5) basement; (6) San Juan River deposits; and (7) terrace escarpment (Colombo *et al.* 2000, modified).

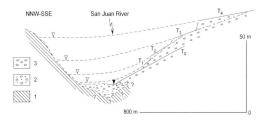

Fig. 10. Albarracín (A in Fig. 5) telescopic-like alluvial fan. Longitudinal section of terrace levels (T_0–T_4). The increase in slope of younger terraces is remarkable: (1) basement; (2) San Juan River sediments; and (3) Albarracín alluvial fan. The white triangle corresponds to different positions of the base level. The black triangle depicts the current base level.

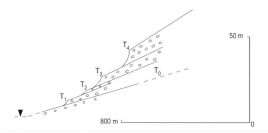

Fig. 11. Longitudinal section of terrace levels (T_0–T_4) displayed in the La Laja area (L in Fig. 5). The greater slope of old terraces with respect to the recent ones is remarkable. The triangle denotes current base level.

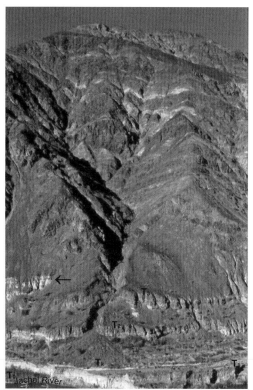

Fig. 12. The remnants of a telescopic-like alluvial fan in the Jáchal River valley related to the T_0–T_3 terraces. Note the interfingering of the T_3 terrace with white lacustrine sediments (arrow). The T_k is a new alluvial fan developed in recent times when the Jáchal River reached its current base level. The small-scale source area of telescopic-like alluvial fan is noticeable. Terrace T_1 has a thickness of 4.5 m.

This would display a certain parallelism between the terrace profiles (Fig. 18C) with the result that the geometric distribution could have a general origin rather than a local one. When a tectonic rising block exists in an area, a parallelism of the longitudinal profile of the terraces is produced (Amorosi *et al.* 1996).

In some cases the longitudinal profiles corresponding to the older terraces have steep slopes increased by tectonic tilting. This is the case of the terraces of the La Laja area (Colombo *et al.* 2000), which underwent (Fig. 11) considerable seismic activity in recent historical times (Uliarte *et al.* 1990).

In all the studied cases the variation in the local base level led to the development of the surfaces of the erosional alluvial terraces and to the generation of telescopic-like alluvial fans in the valleys of the main rivers that cross the Andean Ranges.

Discussion

These alluvial terraces were originally assumed to have been generated in a way similar to that of the Quaternary terraces of European rivers because of a significant variation in the general base level (Dawson & Gardiner 1987; Starkel 1990, 1993; Schumm 1993; Nash 1994). This may be a general marine base level, which could have undergone significant variations during the Quaternary due to successive glaciations and warm interglacial periods (Wood *et al.* 1993). Accumulation terraces are formed by conglomerates made up of poorly sorted angular and subangular clasts embedded in a poorly sorted sandy matrix usually with a not very well defined cross-stratification. These features suggest flows with sufficient energy (Costa 1988) to transport large clasts over a short time span, preventing the sorting of the clasts. This implies hydraulic episodes, which result in the development of upper flow regime sedimentary structures produced by sporadic high-energy flows of the flash-flood type (Baker *et al.* 1988; Komar 1988).

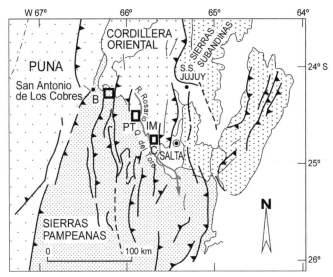

Fig. 13. NW Argentine area. The location of telescopic-like alluvial fans in the Ingeniero Maury (IM) and Punta del Tastil (PT) areas in the Quebrada del Toro region. The Cerro Blanco (B) telescopic-like alluvial fan is located near the Puna border. The Cordillera Oriental, Sierras Pampeanas and Sierras Subandinas are represented by different patterns where the main faults and overthrusts are displayed.

Given the variety of river terraces, it is unreasonable to ascribe their genesis to significant variations in the same general base level. Moreover, it should be borne in mind that these rivers are located along the eastern margin of the Argentine Andes more than 1000 km from the Atlantic coast. Thus, the main factor controlling the generation of terraces is not a marine base-level oscillation. Another type of generation should therefore be proposed. This would be of a general type with local conditions that would control the characteristics and number of terraces generated in each specific location.

The widespread incision of the fluvial valleys that cross the recent Andean structures implies antecedent phenomena. The riverbeds existed before the general elevation of the mountain ranges during the Andean orogeny. Their last orogenic movements were responsible for the generation of the deeply incised valleys. This could have occurred at the end of the Tertiary or at the beginning of the Quaternary, during which the last main episodes of elevation of the Andean Ranges took place (Ramos 1999).

The local variety of these TAFs and their large areal extent suggest that they were controlled mainly by climatic factors. Thus, the irregular pattern of rainfall probably gave rise to large alluvial fans where the tributaries join the main river valley (Clague & Evans 2000). This kind of alluvial fan is controlled the generation of temporary lakes in the main valleys (Colombo et al. 2000). The location of these lakes was conditioned by the lateral drainage basin related to the main tectonic structures responsible for the high elevated topography.

In the case of the Jáchal River, a number of episodes of fan generation formed a dam. This increased in height coevally with the lake development, resulting in the sedimentary infill of the generated accommodation space (Fig. 12). In the Jáchal River valley the maximum thickness of the lacustrine deposits is approximately 45–50 m. These deposits are made up of very-fine-grained materials with diverse sedimentary structures, for example at the base by stratified silts with thin sandy intercalations with cross-lamination of ripples, and in the upper parts by thick silts (Busquets et al. 2002). This suggests processes of dynamic dilution of fluvial water in a stable water body where the flow speed is diminished by friction and where a sedimentary decantation usually exists. Interfingered levels of medium-grained sand occur sporadically and granule levels appear rarely. These coarse levels could indicate limits of lacustrine sequences associated with diverse episodes of dam development. These sequences have thicknesses of 5–10 m (Busquets et al. 2002), which could be the thickness of each growth episode of the dam.

A TAF probably developed as a result of several events rather than one rapid event. The episodic growth is evidenced by the alluvial–lacustrine interfingering, whereas other sedimentary characteristics have been obscured by subsequent episodes of alluvial degradation. The intercalations of coarse materials (Fig. 17) suggest variations in the deposition of lacustrine materials, indicating significant changes in the depth of the lake. A number of lacustrine levels show diverse organic remains that have

Fig. 14. Terrace levels (T_0–T_3) displayed in the Chorrillos area (C in Fig. 13). Terrace T_2 has a thickness of 5 m.

Fig. 15. Terrace levels (T_0–T_3) in the Cerro Blanco locality (B in Fig. 13). Terrace T_2 has a maximum thickness of 2.5 m.

yielded some absolute dates (Busquets *et al.* 2002). In the Jáchal River valley (Totoralitos, T in Fig. 5) there are vegetation and snail remains that have given some ^{14}C dates ranging between 10 030 and 7090 years BP (Table 1). This suggests that the lakes developed during the Holocene, and were more or less continuously active as traps of mainly fine-grained sediments over approximately the last 10 000 years.

In the Quebrada del Toro two different outcrops of lacustrine deposits provide fossil remains (snails), allowing AMS (accelerator mass spectrometry) ^{14}C dating (P.M. Grootes, Kiel University). The first outcrop (Gólgota) is located approximately 2 km north of Ingeniero Maury and provides a date of 25 210 years BP. The second outcrop (Arroyo Colorado), which is situated approximately 2.5 km south of Ingeniero Maury, provides a date of 26 370 years BP. This implies that the dams and temporary

lakes developed during the late Pleistocene. Large rock avalanches produced by tectonic events have been proposed to explain the valley obstruction and the development of a temporary lake upstream (Hermanns & Strecker 1996; Trauth *et al.* 2000). Criteria used for differentiating the deposits of catastrophic landslide deposits are: homogeneous lithology of clasts and matrix; mass-flow facies; disorganized and convolute bedding; large-scale unity of emplacement; and morphological relations with the main valley. Although the majority of these characteristics are absent in the Quebrada del Toro, one large avalanche does not necessarily account for the dam development. The fact that the coarse-grained materials interfingered with fine-grained lacustrine ones suggests that these coarse-grained materials were accumulated episodically and coevally with temporary lake development. Thus, the coarse-grained

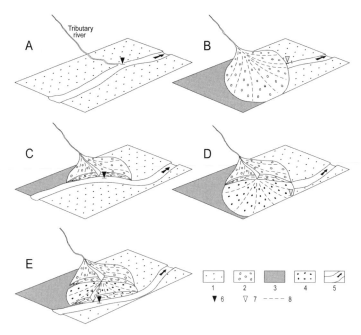

Fig. 16. Schematic model of the telescopic-like alluvial fan generation where a tributary (**A**) reaches the main riverbed (eg San Juan River). A fan (**B**) was developed at the junction of the tributary and the main river, producing a dam and generating a temporary lake upstream. When the water overtopped the dam, the subsequent erosion produced a large scar (**C**). Thereafter, when a new alluvial fan was generated (**D**), another dam coeval with the temporary lake was developed in the same fluvial valley. The repeating processes generated a number of alluvial terraces (**E**) that characterize the telescopic-like alluvial fans. (1) San Juan riverbed; (2) first alluvial fan deposits. (3) lacustrine; (4) second alluvial fan deposits; (5) San Juan River flow path; (6) low water level; (7) high water level; and (8) alluvial flow paths (Colombo *et al.* 2000, modified).

Fig. 17. Fine-grained lacustrine sediments (pale) interfingered with alluvial coarse-grained (dark) materials. The lacustrine sedimentation was coeval with the alluvial fan development. Quebrada del Toro near the Golgota area. This outcrop is located 2 km northwards of Ingeniero Maury (IM in Fig. 13). The displayed thickness is 22 m.

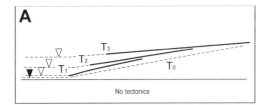

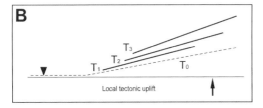

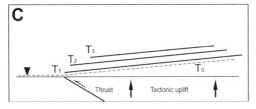

Fig. 18. Geometric distribution of longitudinal profiles of terrace levels T_0–T_3. (**A**) The increase in the slope of the younger terraces could be due to episodic falls in the local base level. (**B**) The increase in the slope of the older terrace levels could be caused by local tectonic activity. (**C**) The maintenance of the terrace slope could be produced by block tectonic uplift (Amorosi *et al.* 1996). Black triangle, current base level; white triangle, terrace base level (Colombo *et al.* 2000, modified).

Fig. 19. Terrace T_1 of the Albarracín (A in Fig. 5) telescopic-like alluvial fan contains (arrow) a man-made (ashlar) block, the remnant of an old bridge of the N 20 national road, which collapsed due to natural causes between 1968 and 1972.

Table 1. *Lacustrin levels dating*

Sample	Laboratory Symbol	Corrected pMC *	Yr BP
FGT – 201	KIA – 24304	4.34±0.09	25 210 ± 170
FAC – 3	KIA – 24303	3.75±0.09	26 370 +200/−190

The sample FGT corresponds to the Golgota outcrop. The sample FAC corresponds to the Arroyo Colorado section placed southwards of Ingeniero Maury locality. Both samples were collected in the Quebrada del Toro.

The KIA denomination corresponds to the laboratory symbology for the samples studied by the Prof. P.M. Grootes, Leibnitz Labor für Altersbestimmung und Isotopenforschung, Christian-Albrechts-Universität, Kiel, Germany.

* Indicates the per cent of modern (1950) carbon corrected for fractionation using the ^{13}C measurement.

materials were transported to the lake by highly turbulent waters rather than by disorganized and massive flows following large rock avalanches or landslides (Hovius *et al.* 1997; Mather *et al.* 2003) associated with tectonic activity (Silva *et al.* 1993).

The geometrical distribution of fine-grained (lacustrine) materials interfingered with coarse (alluvial) materials suggests their coeval accumulation (Fig. 17). The facies of coarse sediments indicates that their accumulation was due to high-energy turbulent flows reaching the lake level. The lacustrine facies was characterized by mud cracks and root casts that suggest a shallow lacustrine environment. This implies that the top of the dam was never very high. Therefore, an unstable equilibrium existed between the growth of TAF and the lacustrine development in the main valley.

Concluding remarks

Tectonics and seismicity are ruled out given their unsuitability as a general explanation for an area that exceeds 1000 km in length. Variations in the general base level due to the oscillation of the Atlantic marine surface are also excluded. However, climate could account for the wide geographical distribution of the TAF in the main valleys. Variations in heavy thunderstorms could give rise to some of these TAF and to the dams. Temporary lakes existed at 10 000 years BP in the San Juan River area (Colombo *et al.* 2000) and in the area of the Jáchal River (Busquets *et al.* 2002). Archaeological data yield evidence of the presence of human settlements on the coasts of some large lakes at 6500 years BP (Uliarte *et al.* 1990). The available palaeoclimatological data (Markgraf *et al.* 1986; Espizua 1993, 2000; Espizua & Bigazzi 1998) corroborate the considerable variation in rainfall and in its areal distribution. Dating reveals that temporary lakes were developed at approximately 25 000 years BP in the Quebrada del Toro area.

Regional sedimentary activities with an irregular distribution and a variable intensity can clearly be recognized in late Quaternary materials in South America (Stager & Mayewski 1997). These Holocene sedimentary accumulations, which are present over a large region, could have been controlled by one specific climatic factor, the activity of the El Niño Southern Oscillation (ENSO). The dynamics of this oscillation suggest (Markgraf *et al.* 1986; Villagrán & Varela 1990; Markgraf 2001; Markgraf & Seltzer 2001; Villa Martínez *et al.* 2003) that very intense and randomly distributed rainfall could cause floods that are locally very important. The ENSO is clearly evidenced in very recent sediments, becoming less evident in older ones.

The typical manifestations of the ENSO are related to interannual periods. By contrast, these situations are repeated in decadal periods ranging between 10 and 50 years. They could also occur over longer cycles (Dettinger *et al.* 2001), although the evidence is scant. It is possible to reconstruct the performance and rhythm of these cycles in recent sedimentary accumulations younger than 5000 years BP (Sandweiss *et al.* 1996, 1999; Markgraf & Seltzer 2001), where the evidence is very clear. However, one period between 5000 and 8000 years BP does not yield strong evidence (Enfield & Mestas-Núñez 2001). Records of this activity also exist in alluvial materials dated approximately at 15 000 years BP (Rodbell *et al.* 1999).

The ENSO produces an irregular distribution of the main regional rainfall, which leads to significant episodes of clastic sediment transport. This could give rise to dams generated by lateral alluvial fans, due to the significant and irregular increases in rainfall, resulting in the formation of temporary lakes (Malde 1968; Jarrett & Costa 1986; Costa & Schuster 1988; Clague & Evans 2000). This is consistent with the data provided by the analysis of the Holocene sediments located *inter alia* in the areas of the San Juan (Colombo *et al.* 1996, 2000) and Jáchal rivers.

In the Quebrada del Toro the facies assemblages and the geometrical relationships between the alluvial fans and lacustrine sediments resemble the ones described above. Thus, the generation of a dam could account for the episodic development of telescopic-like coarse-grained alluvial fan materials that interfinger with fine-grained lacustrine sediments. Intense and randomly distributed thunderstorms producing heavy discharges are crucial for developing the TAF in this area. The lakes could have been developed during the last event of the 'Minchin' wet period (Baker *et al.* 2001; Fritz *et al.* 2004) between 40 000 and 25 000 years BP, which also affected other areas in tropical and subtropical South America (Ledru *et al.* 1996; Turcq *et al.* 1997). Meteorological variations of the ENSO (Dettinger *et al.* 2000, 2001) rather than climatic changes (Bull 1991; Bourke & Pickup 1999; Hattingh & Rust 1999) account for the wide regional distribution of TAF.

A good illustration of the high frequency of meteorological variability is provided by the destruction of the old N 20 national road bridge (1968–1972), which crossed the main channel of the Albarracín alluvial fan (A in Fig. 5). Terrace T_1 contains a fragment of the bridge in the form of a man-made (ashlar) block (Fig. 19). This block was probably transported by the effects of the ENSO, which produced an intense regional activity between 1968 and 1975 (Ebbesmeyer *et al.* 1991, in Dettinger *et al.* 2001).

I am indebted to A.M. Harvey, P. Silva and other unknown reviewers whose comments have improved the manuscript. I wish to thank Dr R. Omarini of the Universidad Nacional de Salta for his valuable geological data, N. Sole de Porta for the pollen data, P. Busquets and E. Ramos of the Universitat de Barcelona, N. Heredia and L.R. Rodríguez of the IGME, J. Vergés and J. Álvarez-Marrón of the CSIC for their valuable collaboration in the field campaigns, and R. Cardó, the director of the Delegación de San Juan del Servicio Geológico Minero Argentino (SEGEMAR), for the indispensable logistical help in the field campaigns. This work was supported by the Proyecto BTE2002-04316-C03-01 DGI, del Ministerio de Educación y Ciencia Spain. Grup de Qualitat del Comissionat de Universitats i Recerca. Generalitat de Catalunya, 2001SGR -00074.

References

AMOROSI, A., FARINA, M., SEVERI, P., PRETI, D., CAPORALE, L. & DI DIO, G. 1996. Genetically related alluvial deposits across active fault zones: an example of alluvial fan; terrace correlation from the upper Quaternary of the southern Po basin, Italy. *Sedimentary Geology*, **102**, 275–295.

BAKER, V.R., KOCHEL, R.C. & PATTON, P.C. (eds). 1988. *Flood Geomorphology*. Wiley, Chichestes.

BAKER, P.A., RIGSBY, C.A., SELTZER, G.O., FRITZ, S.C., LOWENSTEIN, T.K., BACHER, N.P. & VELIZ, C. 2001. Tropical climate changes at millennial and orbital timescales on the Bolivian Altiplano. *Nature*, **409**, 698–701.

BLISSENBACH, E. 1954. Geology of alluvial fans in semiarid regions. *Bulletin of the Geological Society of America*, **65**, 75–190.

BOURKE, M.C. & PICKUP, G. 1999. Fluvial form variability in Arid central Australia. *In:* MILLER, A.J. & GUPTA, A. (eds) *Varieties of Fluvial Forms*. Wiley, Chichester, 249–271.

BOWMAN, D. 1978. Determination of intersection points within a telescopic alluvial fan complex. *Earth Surface Processes*, **3**, 265–276.

BULL, W.B. 1964. *Geomorphology of Segmented Alluvial Fans in Western Fresno County, California*. US Geological Survey, *Professional Paper*, **352-E**, 89–129.

BULL, W.B. 1968. Alluvial fans. *Journal of Geology*, **16**, 101–106.

BULL, W.B. 1977. The alluvial fan environment. *Progress in Physical Geography*, **1**, 222–270.

BULL, W.B. 1979. Threshold of critical power in streams. *Bulletin of the Geological Society of America*, **90**, 453–464.

BULL, W.B. 1991. *Geomorphic Responses to Climatic Change*. Oxford University Press, New York.

BULL, W.B. & MCFADDEN, L.D. 1977. Tectonic geomorphology north and south of Garlock fault, California. *In*: DOEHRING, D.O. (ed.). *Geomorphology in Arid Regions*. Binghamton, State University New York, 115–138.

BUSQUETS, P., COLOMBO, F., HEREDIA, N., RODRIGUEZ FERNANDEZ., L.R. SOLE DE PORTA, N. & ALVAREZ MARRON, J. 2002. El Holoceno del valle del río Jáchal, Precordillera andina (San Juan, Argentina): caracterización sedimentológica, estratigráfica y palinológica. *XV Congreso Geológico Argentino, Actas*, **1**, 346–351.

CLAGUE, J.J. & EVANS, S.G. 2000. A review of catastrophic drainage of moraine-dammed lakes in British Columbia. *Quaternary Science Reviews*, **19**, 1763–1783.

COSTA. J.E. 1988. Floods from dam failures. *In*: BAKER, R.C., KOCHEL, R.C. & PATTON, P.C. (eds) *Flood Geomorphology*. Wiley, Chichester, 439–464.

COSTA, J.E. & SCHUSTER, R.L. 1988. The formation and failure of natural dams. *Bulletin of the Geological Society of America*, **100**, 1954–1968.

COLOMBO, F., ANSELMI, G., BUSQUETS, P., CARDO, R., RAGONA, D. & RAMOS-GUERRERO, E. 1996. Aterrazamientos en el río Albarracín, cuenca del río San Juan: génesis y significado geodinámico (Provincia de San Juan, Argentina). *Geogaceta*, **20**, 1112–1115.

COLOMBO, F., BUSQUETS, P., RAMOS, E., VERGÉS, J. & RAGONA, D. 2000. Quaternary alluvial terraces in an active tectonic region: the San Juan River Valley, Andean Ranges, San Juan Province, Argentina. *Journal of South American Earth Sciences*, **13**, 611–626.

DAWSON, M.R. & GARDINER, V. 1987. River terraces: the general model and a palaeohydrological and sedimentological interpretation of the terraces of lower Severn. *In*: GREGORY, K.J., LEWIN, J., & THORNES, J.B. (eds) *Palaeohydrology in Practice*. Wiley, Chichester, 269–305.

DETTINGER, M.D., BATTISTI, D.S., GARREAUD, R.D., MCCABE, G.J. & BITZ, C.M. 2001. Interhemispheric effects of interannual and decadal ENSO–as climate variations on the Americas. *In:* MARKGRAF, V. (ed.) *Interhemispheric Climate Linkages*. Academic Press, San Diego, CA, 1–16.

DETTINGER, M.D., CAYAN, D.R., MCCABE, G.J. MARENGO, J.A. 2000. Multiscale hydrologic variability associated with El Niño/Southern Oscillation. *In:* DIAZ, H.F. & MARKGRAF, V. (eds) *El Niño and the Southern Oscillation, Multiscale Variability, Global and Regional Impacts*. Cambridge University Press, Cambridge, 113–146.

EBBESMEYER, C.C., CAYAN, D.R., MCKLEIN, D.R., NICHOLS, F.H., PETERSON, D.H. & REDMOND, K.T. 1991. 1976 step in the Pacific climate: Forty environmental changes between 1968–75 and 1977–84. *In*: BETANCOURT, J.L. & THARP. V.L. (eds) *Proceedings of the 7th Annual Pacific Climate (PACLIM) Workshop*. California Department of Water Resources, Interagency Ecological Studies Program Technical Report, **26**, 115–126.

ENFIELD, D.B. & MESTAS-NÚÑEZ, A.M. 2001. Interannual multidecadal climate variability and its relationship to global sea surface temperatures. *In:* MARKGRAF, V. (ed.). *Interhemispheric Climate Linkages*, Academic Press, San Diego, CA, 17–29.

ESPIZUA, L.E. 1993. Quaternary glaciations in the Rio Mendoza valley, Argentine Andes. *Quaternary Research*, **40**, 50–162.

ESPIZUA, L.E. 2000. Quaternary glacial sequence in the Rio Mendoza valley, Argentina. *In*: SMOLKA, P.P. & VOLKHEIMER, W. (eds) *Southern hemisphere paleo- and Neoclimates: Key Sites, Methods, Data and Models*. Springer, Berlin, 287–293.

ESPIZUA, L.E. & BIGAZZI, G. 1998. Fission-track dating of Punta de Vacas glaciation in the Rio Mendoza valley,

Argentina. *Quaternary Science Reviews*, **17**, 755–760.

FERRILL, D.A., STAMATAKOS, J.A., JONES, S.M., RAHE, B., MCKAGUE, H.L., MARTIN, R.H. & MORRIS, A.P. 1996. Quaternary slip history of the Bare Mountain Fault (Nevada) from the morphology and distribution of alluvial fan deposits. *Geology*, **24**, 559–562.

FRITZ, S.C., BAKER, P.A., *ET AL*. 2004. Hydrologic variation during the last 170,000 years in the southern hemisphere tropics of South America. *Quaternary Research*, **61**, 95–104.

HARVEY, A.M. 1984. Aggradation and dissection sequences on Spanish alluvial fans: influence on geomorphological development. *Catena*, **11**, 289–304.

HARVEY, A.M. 1987*a*. Alluvial fan dissection: relationships between morphology and sedimentation. *In*: FROSTICK, L. & REID, I. (eds) *Desert Sediments, Ancient and Modern*. Geological Society, London, Special Publications, **35**, 87–103.

HARVEY, A.M. 1987*b*. Patterns of Quaternary aggradational landform development in the Almeria Region, southeast Spain: a dry-region tectonically active landscape. *Die Erde*, **118**, 193–215.

HARVEY, A.M. 1990. Factors influencing Quaternary alluvial fan development in southeast Spain. *In*: RACHOCKI, A.H. & Church, M. (eds) *Alluvial Fans: A Field Approach*. Wiley, Chichester, 247–269.

HARVEY, A.M. 1996. The role of alluvial fans in mountain fluvial systems of southeast Spain: implications of climatic change. *Earth Surface Processes and Landforms*, **21**, 543–553.

HARVEY, A.M., FOSTER, G., HANNAM, J., & MATHER, A. 2003. The Tabernas alluvial fan and lake system, southeast Spain: applications of mineral magnetic and pedogenic iron oxide analyses towards clarifying the Quaternary sediment sequences. *Geomorphology*, **50**, 151–171.

HARVEY, A.M., SILVA, P.G., MATHER, A.E., GOY, J.L., STOKES, M. & ZAZO, C. 1999. The impact of sea-level and climatic changes on coastal alluvial fans in the Cabo de Gata ranges, southeast Spain. *Geomorphology*, **28**, 1–22.

HATTINGH, J. & RUST, I.C. 1999. Drainage evolution and morphological development of the Late Cenozoic Sundays River, South Africa. *In*: MILLER, A.J. & GUPTA, A. (eds) *Varieties of Fluvial Forms*. Wiley, Chichester, 145–166.

HERMANNS, R.L. & STRECKER, M.R. 1996. Structural and lithological controls on large Quaternary rock avalanches (sturzstroms) in arid nothwestern Argentina. *Bulletin of the Geological Society of America*, **111**, 938–948.

HOVIUS, N., STARK, C.P. & ALLEN, P.A. 1997. Sediment flux from a mountain belt derived by landslide mapping. *Geology*, **25**, 231–234.

IGARZABAL, A.P. 1991. Morfología de las provincias de Salta y Jujuy. *Revista Instituto de Geología y Minería de la Universidad Nacional de Jujuy*, **8**, 7–121.

IGARZABAL, A.P. & MEDINA, A.J. 1991. La cuenca torrencial del río Mojotoro; su evolución y riesgos derivados. Departamento La Caldera, Provincia de Salta. *Revista Instituto de Geología y Minería de la Universidad Nacional de Jujuy*, **8**, 123–144.

JANOCKO, J. 2001. Fluvial and alluvial fan deposits in the Hornád and Torysa river valleys: relationship and evolution. *Slovak Geological Magazine*, **7**, 221–230.

JARRETT, & COSTA, J.E. 1986. *Hydrology, Geomorphology and Dam-break modelling of the July 15, 1982 Lawn Lake Dam and Cascade Lake Dam failures, Lasimer County, Colorado*. US Geological Survey, Professional Paper, **1369**, 1–78.

KOMAR, P.D. 1988. Sediment transport by floods. *In*: BAKER, V.R., KOCHEL, R.C. & PATTON, P.C., (eds) *Flood Geomorphology*. Wiley, New York, 97–111.

LEDRU, M.P., BRAGA, P.I.S., SOUBIES, F., FOURNIER, M., MARTIN, L., SUGUIO, K. & TURCQ, B. 1996. The last 50 000 years in the Neotropics (Southern Brazil) evolution of vegetation and climate. *Palaeogeography, Palaeoclimatology, Palaeoecology*, **123**, 239–257.

MALDE, H.E. 1968. *The catastrophic late Pleistocene–Bonneville Flood in the Snake River plain, Idaho*. US Geological Survey, Professional Paper, **596**, 1–52.

MARKGRAF, V. (ed.). 2001. *Interhemispheric Climate Linkages*. Academic Press, San Diego, CA.

MARKGRAF, V. & SELTZER, G. O. 2001. Pole–Equator–Pole paleoclimates of the Americas integration: toward the big picture. *In*: MARKGRAF, V., (ed.) *Interhemispheric Climate Linkages*. Academic Press, San Diego, CA, 433–442.

MARKGRAF, V., BRADBURY, J.P. & FERNANDEZ, J. 1986. Bajada de Rahue, province of Neuquen, Argentina: An interstadial deposit in northern Patagonia. *Palaeogeography, Palaeoclimatology, Palaeoecology*, **56**, 51–258.

MATHER, A.E., GRIFFITHS, J.S. & STOKES, M. 2003. Anatomy of a 'fossil' landslide from the Pleistocene of SE Spain. *Geomorphology*, **50**, 135–139.

NASH, D.B. 1994. Effective sediment-transporting discharge from magnitude–frequency analysis. *Journal of Geology*, **102**, 79–95.

RAMOS. V.A. 1999. Rasgos estructurales del territorio argentino. *In*: CAMINOS, R. (ed.) *Geología Argentina. Instituto de Geología y Recursos Naturales Servicio Geológico Minero Argentino, Anales* (Buenos Aires), **29**, 715–759.

RODBELL, D.T., SELTZER, G.O., ANDERSON, D.M., ABBOTT, M.B., ENFIELD. D.B. & NEWMAN, J.H. 1999. An 15,000 year record of El Niño-driven alluviation in southwestern Ecuador. *Science*, **283**, 516–520.

SANDWEISS, D.H., RICHARDSON J.B., III, REITZ, E.J., ROLLINS, H.B. & MAASCH, K.A. 1996. Geoarcheological evidence from Peru for a 5,000 years BP onset of El Niño. *Science*, **273**, 1531–1533.

SANDWEISS, D.H., MAASCH, K.A., BURGER, R.L. & ANDERSON, D. 1999. Transition in the Mid-Holocene. *Science*, **283**, 499–500.

SCHUMM, S.A. 1993. River response to base level change: implications for sequence stratigraphy. *Journal of Geology*, **101**, 279–294.

SHAOPING, C. & GUIZHI, Y. 1999. Segmented variations in tectonic geomorphology of Datong–Yangyuan fault zone, NW Beijing, China. *Journal of Balkan Geophysical Society*, **2**(2), 46–62.

SILVA, P.G., GOY, J.L., SOMOZA, L., ZAZO, C. & BARDAJÍ, T. 1993. Landscape response to strike-slip faulting linked to collisional settings: quaternary tectonics and basin formation in the Eastern Betics, Southeastern Spain. *Tectonophysics*, **224**, 289–303.

STAGER, J.C. & MAYEWSKI, P.A. 1997. Abrupt early to mid Holocene climate transition registered at the Equator and the poles. *Science*, **276**, 1834–1836.

STARKEL, L. 1990. Fluvial environments as an expression of geoecological changes. *Zeitschrift für Geomorphologie Neue Folge, Supplement Band*, **88**, 17–28.

STARKEL, J.K. 1993. Hydrological changes in the last 20Ky. *Palaeogeography, Palaeoclimatology, Palaeoecology*, **103**, 488–502.

STOKES, M. & MATHER, A.E. 2000. Response of Plio-Pleistocene alluvial systems to tectonically induced base-level changes, Vera Basin, SE Spain. *Journal of the Geological Society, London*, **157**, 303–316.

TRAUTH, M.H., ALONSO, R.A., HASELTON, K.R., HERMANNS, R.L. & STRECKER, M.R. 2000. Climate change and mass movements in the NW Argentine Andes. *Earth and Planetary Science Letters*, **179**, 243–256.

TURCQ, B., PRESSINOTTI, M.M.M. & MARTIN, L. 1997. Paleohydrology and paleoclimate of past 33 000 years at the Tamaduá River, Central Brasil. *Quaternary Research*, **47**, 284–294.

ULIARTE, E.R., RUZYCKI, L. & PAREDES, J.D. 1990. Relatorio de geomorfología. *In*: BORDONARO, O. (ed.) *Relatorio de Geología y Recursos Naturales de la provincia de San Juan. XI Congreso Geológico Argentino*, **1**, 212–227.

VILLAGRÁN, C. & VARELA, J. 1990. Palynological evidence from increased aridity on the central Chilean coast during the Holocene. *Quaternary Research*, **34**, 198–207.

VILLA MARTÍNEZ, R., VILLAGRÁN, C. & JENNY, B. 2003. The last 7500 cal yr B.P. of westerly rainfall in Central Chile inferred from high-resolution pollen record from Laguna Aculeo (34°S). *Quaternary Research*, **60**, 284–293.

WALLACE, R.E. 1978. Geometry and rates of change of fault-generated range fronts, north-central Nevada. *US Geological Survey, Journal of Research*, **6**, 637–650.

WOOD, L.J., ETHRIDGE, F.G. & SCHUMM. S.A. 1993. The effects of base-level fluctuation on coastal-plain, shelf and slope depositional system: an experimental approach. *In*: POSAMENTIER, W., SUMMERHAYES, C.P., HAQ, B.U., ALLEN, G.P. (eds) *Sequence Stratigraphy and Facies Associations*. International Association of Sedimentologists, *Special Publications*, **18**, 43–53.

Morphometry and depositional style of Late Pleistocene alluvial fans: Wadi Al-Bih, northern UAE and Oman

ASMA AL-FARRAJ[1] & ADRIAN M. HARVEY[2]

[1]Department of Geography, University of the United Arab Emirates, PO Box 15551, Al-Ain, United Arab Emirates

[2]Department of Geography, University of Liverpool, PO Box 169, Liverpool L69 3BX, UK

(e-mail:amharvey@liv.ac.uk)

Abstract: Three types of alluvial-fan settings are recognized in the Wadi Al-Bih area of the Musandam Mountains, northern UAE and Oman; mountain-front fans, tributary-junction fans and steep hillslope debris cones. Three styles of fan geometry, only partly dependent on fan setting, can be recognized: telescopic fans, stacked fans and truncated fans. Each style, together with degree of confinement, reflects the topographic and geological context of the fan and its source area. The mountain-front fans are mostly unconfined fans with telescopic styles. The tributary-junction fans are confined fans, some with stacked or telescopic styles, others that have been truncated by base-level-induced toe trimming. Most of the debris cones are simple hillfoot debris cones. Standard morphometric analyses of fan areas and fan gradients in relation to drainage basin area yield results that compare with other studies, but the relationships differ between the three groups of fans, in part reflecting fan style, especially between mountain-front and tributary-junction fans. The morphometry of the debris cones is only poorly characterized by the morphometric relationships. Cone morphology most strongly reflects source-area lithology. Analysis of the residuals from the regression analyses suggests that the morphometric differences between the three groups reflect fan sedimentary processes, fan setting and fan style, particularly relating to confinement.

Alluvial-fan morphometry conventionally expresses the relationships between alluvial-fan properties, notably fan area and gradient, and feeder catchment properties especially drainage area (Bull 1977; Harvey 1997). The most commonly used relationships are normally expressed in the form of regression equations:

$$F = pA^q \qquad (1)$$

$$G = aA^b \qquad (2)$$

where A is drainage area, F is fan area (both in km^2) and G is fan gradient (Harvey 1997). For the fan-area relationship, characteristic values of q for dry-region fan groups are fairly consistent, and range between c. 0.7 and c 1.1.Values for p show a wide range between c. 0.1 and c. 2.1 (Harvey 1997), reflecting different catchment geologies (Hooke & Rohrer 1977; Lecce 1991) or fan histories. The fan-gradient relationship is inverse, but generally less clear-cut than the fan-area relationship, and shows characteristic values of b of between c. -0.15 and c. -0.35 (Harvey 1997). Values for a show a range of between c. 0.03 and c. 0.17, possibly reflecting lithology, depositional processes and base-level influences (Harvey et al. 1999).

When drainage basin relief or slope properties are also included in the regressions, they tend to have little influence in the fan-area case, but may modify the fan-gradient regressions presumably because sediment properties are influenced by basin steepness and, in turn, have more effect on fan gradients than on fan areas (Harvey 1987).

The primary factors influencing fan morphometry appear to be the supply of water and sediment from the feeder catchment. However, analysis of the differences between the regression relationships for different fan groups, or of the residuals from the regressions within fan groups, may indicate the effects of other factors (Silva et al. 1992). Of major importance is fan setting, controlling accommodation space and controlled by the long-term geomorphic history, involving such factors as tectonics and base-level conditions. Previous studies have demonstrated the influence on fan morphometric properties of tectonic setting (Silva et al. 1992; Calvache et al. 1997), accommodation space (Viseras et al. 2003) and base level (Harvey et al. 1999; Harvey 2002a). Fans commonly occur in two main topographic situatuions, at mountain fronts and at tributary junctions, creating distinct fan geometries in relation to confinement. However, there have been few studies that specifically deal with the influence of confinement on fan morphology and morphometry (Sorriso-Valvo et al. 1998; Harvey et al. 1999).

This paper deals with Quaternary alluvial fans in the Musandam Mountains, UAE and Oman,

From: HARVEY, A.M., MATHER, A.E. & STOKES, M. (eds) 2005. *Alluvial Fans: Geomorphology, Sedimentology, Dynamics.* Geological Society, London, Special Publications, **251**, 85–94. 0305-8719/05/$15 © The Geological Society of London 2005.

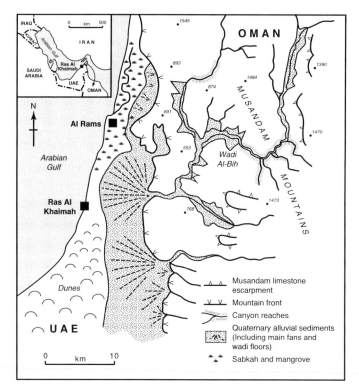

Fig. 1. Location map of the Musandam Mountains, UAE and Oman.

contrasting the morphometric properties of unconfined fans along the western mountain front of the Musandam range with those of confined tributary-junction fans along Wadi Al-Bih, a major drainage within the Musandam range (Fig. 1).

Study area

The Musandam Mountains (Fig. 1) dominantly comprise uplifted and deformed Mesozoic platform carbonate rocks (Glennie *et al.* 1974). They were deformed during the Tertiary in relation to the closure of Tethys by the collision between the Arabic and Asian plates. To the west they are strongly folded and thrust over Mesozoic basinal sediments to form the mountain-front zone, whereas to the east they are gently flexured. Uplift and deformation continued into the early Pleistocene, but there is no unequivocal evidence of ongoing late Quaternary deformation within the fan sediments of this part of the range (Al-Farraj 1996). The area is deeply dissected by Wadi Al-Bih and its tributaries, with major relief provided by two main massive limestones: the Jurassic–Cretaceous Musandam limestone forming a huge escarpment along the eastern watershed; and the Permo-Triassic Ghail limestone forming the

elevated plateau dissected by the canyon in the centre of the study area (Fig. 1). The intervening more subdued terrain is formed on rocks of the Upper Triassic Elphinstone Group, thinly bedded sandstones, shales and limestones.

The present climate is arid, with mean annual temperatures in excess of 25 °C, and annual rainfall less than 135 mm, occurring mostly in winter. Flash floods in Wadi Al-Bih occur once or twice a year, producing an annual runoff of less than 3 mm (Al-Farraj, 1996).

Alluvial fans are widespread in the Wadi Al-Bih area and occur in three distinct geomorphic settings: mountain-front fans, tributary-junction fans and valley-side debris cones (Figs 1, 2 & 3a–c). The latter issue from small, steep, bedrock channels (Figs 1 & 2), as opposed to the alluvial valley floors of the feeder channels of the other two groups. The mountain-front fans and the tributary- junction fans (Fig. 2) are fluvially dominant fans; composed mainly of fluvial sheet and/or channel gravels, although on some of the smaller fans debris flows may be interstratified with fluvial deposits in the proximal zones. The debris cones, by definition, are composed of debris flows (Al-Farraj 1996). The fans can be further differentiated according to their geometry and evolution into stacked, telescopic and

Fig. 2. Alluvial fans in the Wadi Al-Bih area. (**a**) Mountain-front fan. (**b**) Steep debris cone (centre) and lower angle tributary-junction fan (left), Musandam escarpment area.

trenched types. The mountain-front fans are unconfined fans. They toe out at stable base levels and are mostly telescopic in style (Bowman 1978) (Fig. 3d). The proximal fan surfaces are dissected by fanhead trenches from which the younger fan segments prograde distally. The tributary-junction fans are fans confined by the valley walls, and are strongly influenced by the local base level provided by the channel of Wadi Al-Bih. These fans include both telescopic and stacked fans (Fig. 3d & e), with some fans trenched throughout, the style being influenced by the local relationships to the channel of the wadi. Many of these fans have been truncated by 'toe cutting' (Leeder & Mack 2001) either by incision or by lateral migration of the main wadi (Fig. 3f). The debris cones are simple forms, mostly with stacked styles and little trenching.

Aims and methods

The aim of this paper is to examine the morphometric properties of the Musandam fans, and to assess the extent to which fan setting, particularly confinement, modifies the morphometric relationships.

The alluvial fans were initially mapped from air photographs onto topographic map enlargements at a scale of approximately 1:25 000. The mapping was checked and corrected in the field. Sixty-three fans were selected as representative of the range of fan environments: 16 mountain-front fans, 21 tributary-junction fans and 26 debris cones (Figs 3 & 4). For the morphometric data, fan areas were derived from the fan limits as mapped, and fan gradients were measured by hand levelling in the field. On the larger fans sample profiles were measured and on the smaller

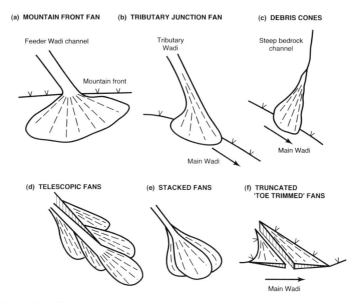

Fig. 3. Schematic illustration of fan styles.

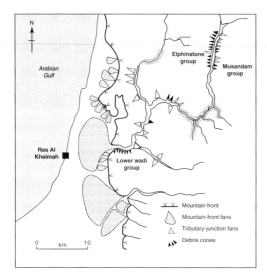

Fig. 4. Location of sample fans.

fans complete axial profiles were surveyed. The gradients quoted here are the axial gradients of the proximal half of each fan. The feeder drainage basins were deliminated on 1:50 000 topographic maps and drainage areas measured from the maps. Drainage basin slopes were calculated by dividing basin relief by basin length.

In addition to the collection of the morphometric data, other observations were made in the field on the geomorphology of the alluvial fans and their relationships with the Quaternary evolution of the

area (Al-Farraj 1996). The constituent fan surfaces have been correlated and related to wadi terrace surfaces by studies of surface modifications through pedogenesis and desert pavement development (McFadden *et al.* 1989) reported elsewhere (Al-Farraj & Harvey 2000). Three main late Quaternary phases of fan aggradation have been differentiated and appear to relate to climates wetter and/or colder than today's, interspersed with arid periods when, as happens today, dissection was the dominant geomorphic regime affecting the fans. How these sequences of fan evolution reflect late Quaternary climatic change and local base-level changes have also been described elsewhere (Al-Farraj 1996; Harvey 2002*b*, 2003).

Fan morphometry

Conventional fan morphometric analyses have been carried out, relating fan area and gradient to drainage basin area by regression analyses as in equations (1) and (2). This was done first to derive the overall regression relationships for all the sample, then separately for each fan group (mountain-front fans, tributary-junction fans and debris cones).

Drainage basin area–fan-area relationship

Using the same nomenclature as in equation (1), $F = pA^q$, the fan-area regression equations are as follows (see Fig. 5).

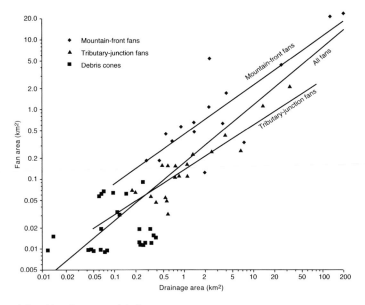

Fig. 5. Regression relationships: fan area and drainage area.

For the all-fans case:

$$F = 0.18\,A^{0.85} \qquad (3)$$

(correlation coefficient, $R = 0.86$; standard error of the estimate, SE $= 0.49$ log units).

For the tributary-junction fans:

$$F = 0.14\,A^{0.66} \qquad (4)$$

($R = 0.88$; SE 0.22).

For the mountain-front fans:

$$F = 0.45\,A^{0.72} \qquad (5)$$

($R = 0.85$; SE 0.39).

The relationship for the debris cones alone is not statistically significant.

For the three significant relationships involving fluvially dominant fans, the correlation coefficients are high and the values of q are similar to a number of those quoted in the literature (Harvey 1997), but lower than some of those quoted by Viseras *et al.* (2003). Note that in the 'all-fans' case the higher value of q might simply reflect the effects of aggregating the three groups; however, the higher value of q for the mountain-front group than for the tributary-junction group might reflect the effects of fan confinement at tributary junctions preventing the development of very large fans even from large drainage areas. The values of p are much lower for the tributary-junction case than for the mountain-front fans, clearly reflecting the affects of fan confinement on accommodation space. Compared with those in the literature, values of p for the Wadi

Al-Bih fans are in the lower part of the range quoted by Harvey (1997); this may be related to the age of Wadi Al-Bih fans that appear to be younger than most of those in the literature. In addition, many of Wadi Al-Bih fans are stacked fans, while many of those in the literature are telescopic fans, hence fan area for Wadi Al-Bih fans would tend to be lower.

Drainage basin area–fan-gradient relationship

Using the same nomenclature as in equation (2), $G = aA^b$, the fan-area regression equations are as follows (see Fig. 6).

For the all-fans case:

$$G = 0.092\,A^{-0.23} \qquad (6)$$

($R = -0.67$; SE $= 0.22$ log units).

For the tributary-junction fans:

$$G = 0.091\,A^{-0.12} \qquad (7)$$

($R = -0.50$; SE $= 0.13$).

For the mountain-front fans:

$$G = 0.065\,A^{-0.13} \qquad (8)$$

($R = 0.53$; SE $= 0.18$).

The relationship for the debris cones again is not stastically significant.

As would be expected, fan gradients show a negative relationship with drainage areas, but, as is often the case (Harvey 1997), the correlation

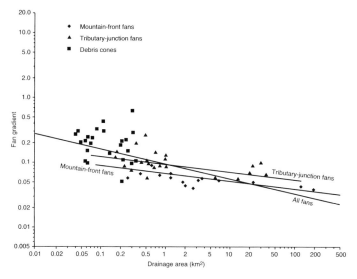

Fig. 6. Regression relationships: fan gradient and drainage area.

coefficients are poorer than for the fan-area regressions. For the all-fans case, the regression is similar to those in the literature (Harvey 1997). When the all-fans case is separated into the different groups, both mountain-front fans and tributary-junction fans have values of b lower than those in the literature, and the debris cones' relationship is not significant. The lower value of a for the mountain-front group than for the tributary-junction group may be related to the lithology of the drainage basins. The mountain-front fans tend to drain zones of folded, faulted and fractured rock that may affect sediment transport and deposition processes and produce relatively low gradient fans.

The role of drainage basin slope

On the basis of previous work we might expect drainage basin relief characteristics to influence fan area, and particularly fan gradient (Harvey 1987, 1997). However, drainage basin slope shows only weak relationships with either fan area and/or fan gradient. For the all-fans case (where S is drainage basin slope; drainage basin relief/drainage basin length):

For fan area: $F = 0.045 \, S^{-1.07}$ (9)
($R = -0.27$, SE $= 0.81$).

For fan gradient: $G = 0.16 \, S^{0.51}$ (10)
($R = 0.36$, SE $= 0.27$).

Even when drainage basin slope is considered as an extra variable within multiple regressions its effects are only limited.

For fan area: $F = 0.24 \, A^{0.88} \, S^{0.33}$ (11)
($R = 0.87$, SE $= 0.42$)

which shows little or no improvement on the simple drainage area to fan-area relationship.

For fan gradient: $G = 0.107 A^{-0.22} S^{0.16}$ (12)
($R = 0.68$, SE $= 0.22$)

again showing little or no improvement on the simple drainage area to fan-gradient relationship.

Fan area – fan gradient relationships

So far, fan area and fan gradient have been considered in relation to drainage basin properties, but they also can be used to summarize fan properties. Mean fan areas and fan gradients (Table 1) differ markedly between the three groups, as would be expected, with mountain-front fans on average much larger and less steep than the others, and debris cones much smaller and steeper (Student's t-tests show these differences to be highly significant). However, more important in characterizing the fan

Table 1. *Mean fan areas and gradients*

	n	Fan area (km^2)	Fan gradient
All fans	63	1.70	0.15
Mountain-front fans	16	6.39	0.064
Tributary-junction fans	21	0.28	0.012
Debris cones	26	0.044	0.22

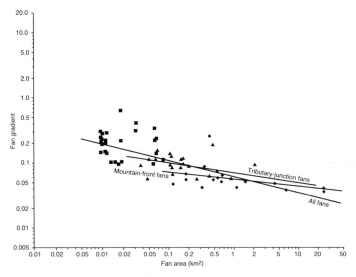

Fig.7. Regression relationships: fan area and fan gradient.

geometry could be the relationships between fan areas and fan gradients, expressed by regression relationships as follows (see Fig. 7).

For the all-fans case: $G = 0.062\ F^{-0.25}$ (13)
($R = -0.69$; SE = 0.21 log units).

For the tributary-junction fans:
$G = 0.069\ F^{-0.14}$ (14)
($R = -0.41$; SE = 0.13, significant only at the 5% level).

For the mountain-front fans:
$G = 0.055\ A^{-0.12}$ (15)
($R = -0.39$; SE = 0.18, significant only at the 10% level).

Again the relationship for the debris cones is not statistically significant.

These relationships, however, differentiate only between fan settings and do not pinpoint the role of confinement on fan morphometry.

Residuals from regressions

Another way to examine overall fan morphometric proprieties that can help to identify the role of specific factors is to consider the residuals from the two main regression relationships discussed earlier. This method has previously revealed morphometric differences that can be related to fan setting, such as tectonic context (Silva *et al.* 1992) and base-level control (Harvey *et al.* 1999), with some influence also from source-area geology (Harvey 2002a). On this basis we might expect the influence of confinement to be apparent.

Regression residuals from the two main morphometric regression relationships between drainage area and fan area (equation 3), and drainage area and fan gradient (equation 6), were plotted to establish whether differences can be identified between fan groups (Fig. 8).

Residuals from the fan-area regression give the clearest differentiation between groups, modified slightly by those from the fan-gradient regression. The mountain-front fans form a cluster with few outliers, indicating not only relatively large fans, but fans relatively large in relation to drainage area and with gradients generally relatively low in relation to drainage area. The three main outliers are not easy to explain. Two are very small fans and their positions may relate to assumptions of regression linearity disproportionately affecting values at the ends of the distribution. The stacked mountain-front fans are generally less steep than the trenched, telescopic fans.

In Figure 8, the tributary-junction fans tend to plot to the left and slightly above the mountain-front fans, indicating relatively steeper gradients and relatively smaller fan areas, clearly reflecting the confined nature of these fans. There are differences between the relatively small tributary-junction fans that join the main wadi downstream of the canyon reach and relatively large fans in the upper part of the wadi system. This may in part reflect the geology of the source areas, but also the relief of the tributary catchments. Downstream of the canyon the tributary catchments are dominantly on Bih and Ghail limestone, whereas the upper catchments drain the very steep relief of the Musandam escarpment (see Fig. 2).

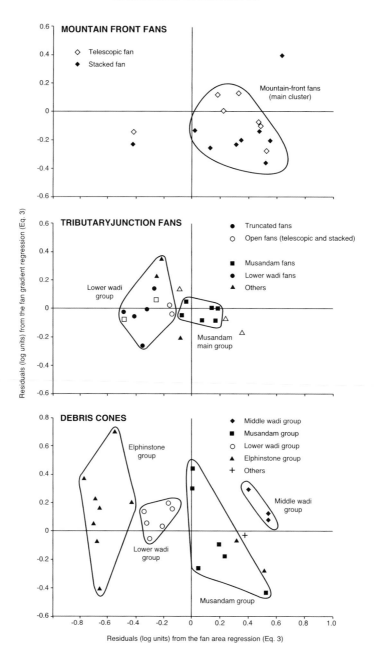

Fig. 8. Residuals from the two main regression relationships.

The plotting positions of the debris cones in Figure 8 show a wide scatter of generally steep and small fans in relation to drainage area. The scatter can be partly resolved by grouping the debris cones in relation to source area. The Elphinstone group, fed by low-relief terrain on rocks of the Elphinstone series, are all simple fans of late Quaternary age and most plot far to the left of Figure 8, with much smaller fan areas in relation to drainage area than the other fans. In contrast, the debris cones fed by the the steep high Musandam limestone escarpment all plot to the right of the debris cone cluster, indicating very large debris cones in relation to source area. The lower wadi and mountain-front group, fed by intermediate relief terrain mostly on the Bih and Ghail limestones, plot in intermediate positions. Most of the debris cones are simple stacked forms, but three larger telescopic cones in the middle reach plot with relatively large areas and steep slopes.

Discussion and Conclusions

Previous work on the morphometry of dry-region alluvial fans has demonstrated a general relationship between drainage area and fan area (Bull 1962; Hooke 1968; Hooke & Rohrer 1977; Harvey 1997, 2002c), and between drainage area and fan gradient (Bull 1962; Harvey 1997, 2002c).These studies identified regional differences in the regression relationships attributable, perhaps, to climatic differences, but especially to tectonic context, controlling accommodation space (Viseras et al. 2003) and to lithology of source-area bedrock. Few studies have specifically identified fan confinement as an important variable influencing fan morphometry (Sorriso-Valvo et al. 1998) or fan morphometric regressions (Harvey et al. 1999). It is clear from this study that fan confinement, as influenced by fan context, does influence the standard fan morphometric regression equations when the regressions for the mountain-front and tributary-junction fans are compared. This is especially true for the fan-area regressions (Fig. 5; equations 4 and 5); with the much higher values of p for the mountain-front fans reflecting the relatively large areas of these unconfined fans. There are also differences in fan gradient (Fig. 6; equations 7 and 8), expressed by the higher value of a for the tributary-junction fans, suggesting that confinement may lead to steeper fans.

Similarly, when the residuals from the all-fans regressions (equations 3 and 6) are examined, the plot of the residuals from the two main regressions (Fig. 8) does differentiate between the three groups of fans, confirming the trends identified above, but, further, offers some explanation of the wide range of plotting positions for the debris cones. Confinement is clearly important in differentiating mountain-front from tributary-junction settings, but other factors are also involved. For the mountain-front fans fan style appears to be reflected in the residual plotting positions in Figure 8, with stacked fans generally of relatively lower gradient than the large telescopic fans. However, for the tributary-junction fans truncation appears to have no influence, indicating that the effects of the base-level influence cannot be detected from the regression relationships. Lithology and, to some extent, relief characteristics appear to be the main differentiators within the debris cones. The Musandam group, fed by very steep catchments on Jurassic–Cretaceous limestone, are relatively large. Interestingly, these debris cones

plot in the same field in Figure 8 as the Musandam tributary-junction fans, suggesting the importance of relief and lithology. Similarly, both the debris cones and the tributary-junction fans from the lower wadi plot together in Figure 8, indicating relatively small fan areas for both fans and debris cones fed by catchments on the Permo-Triassic limestones. These two trends suggest that lithology is an important control of the volume of sediment supplied, hence influencing fan area. The plotting position of the Elphinstone debris cones, well to the right in Figure 8, suggests that fan history may be an important control. The Elphinstone beds are fairly weak, and could be expected to yield high volumes of sediment. Catchment relief is relatively low, but the most important control appears to be the youth of these debris cones. They have developed only after the wadi migrated away from the valley margin here, relatively recently in the history of the wadi (Al Farraj 1996).

Of the controls over fan style and morphometry, confinement appears to be of major importance, influencing both fan style and fan morphometry. Source-area geology and relief also appear to be important, especially in influencing fan morphometry. Base level appears to have only a minor influence, affecting fan style but with no obvious influence on morphometry.

The field work for this study was undertaken as part of the University of Liverpool PhD work of A. Al-Farraj, funded by the Education ministry of the United Arab Emirates which also funded the the field work of A.M. Harvey in the UAE. We are also grateful to His Highness, Al Shake Nahyan Bin Mubarak, to the University of the UAE, Al-Ain and to the staff of the London Embassy of the UAE for their support. We thank the staff of the Cartographics section of the Department of Geography, University of Liverpool for help in producing the illustrations. We thank C. Viseras and S. Jones for their constructive reviews.

References

AL-FARRAJ, A. 1996. Late Pleistocene geomorphology in Wadi Al-Bih, northern U.A.E. and Oman: with special emphasis on wadi terrace and alluvial fans. PhD Thesis, University of Liverpool.

AL-FARRAJ, A. & HARVEY, A.M. 2000. Desert pavement characteristics on wadi terrace and alluvial fan surfaces: Wadi Al-Bih, U.A.E. and Oman. Geomorphology, 35, 279–297.

BOWMAN, D. 1978. Determination of intersection points within a telescopic alluvial fan complex. Earth Surface Processes, 3, 265–276.

BULL, W.B. 1962. Relations of Alluvial Fan Size and Slope to Drainage Basin Size and Lithology, in Western Fresno County, California. US Geological Survey, Professional Paper, 430B, 51–53.

BULL, W.B. 1977. The alluvial fan environment. Progress in Physical Geography, 1, 222–270.

CALVACHE, M., VISERAS, C. & FERNÁNDEZ, J. 1997. Controls on alluvial fan development – evidence from fan morphometry and sedimentology; Sierra Nevada, SE Spain. *Geomorphology*, **21**, 69–84.

GLENNNIE, K.W., BOEUFF, M.G.A., HUGHES-CLARKE, M.W., MOODY-STUART, M., PILAAR, W.H.F. & REINHART, B.M. 1974. *Geology of the Oman Mountains*. Nederlands Geologisch Mijnbouwkundig Genootschap, Verhandelingen 31 (Delft).

HARVEY, A.M. 1987. Alluvial fan dissection: relationships between morphology and sedimentology. *In*: FROSTICK, L. & REID, I. (eds) *Desert Sediments Ancient and Modern*. Geological Society, London, Special Publications, **35**, 87–103.

HARVEY, A.M. 1997. The role of alluvial fans in arid zone fluvial systems. *In*: THOMAS, D.S.G. (ed.) *Arid Zone Geomorphology: Process, Form and Change in Drylands*, 2nd edn. Wiley, Chichester, 231–259.

HARVEY, A.M. 2002a. The role of base-level change in the dissection of alluvial fans: case studies from southeast Spain and Nevada. *Geomorphology*, **45**, 67–87.

HARVEY, A.M. 2002b. Factors influencing the geomorphology of dry-region alluvial fans: a review. *In*: PÉREZ-GONZÁLEZ, A., VEGAS, J. & MACHADO, M.J. (eds) *Aportaciofies a la geomorfologiá de Espana en el inicio del tercer Milenio*. Publicaciones del Instituto Geologico y Minero de Espana, Madrid, Serie Geología, 1, 59–72.

HARVEY, A.M. 2002c. Effective timescales of coupling within fluvial systems. *Geomorphology* **44**, 175–201.

HARVEY, A.M. 2003. The response of dry-region alluvial fans to late Quaternary climatic change. *In*: ALSHARHAN, A.S., WOOD, W.W., GOUDIE, A.S., FOWLER, A. & ABDELLATIF, E.M. (eds) *Desertification in the Third Millenium*. Balkema, Rotterdam, 83–98.

HARVEY A.M., SILVA, P.G., MATHER, A.E., GOY, J.L., STOKES, M. & ZAZO, C. 1999. The impact of Quaternary sea-level and climatic change on coastal alluvial fans in the Cabo de Gata ranges, southeast Spain. *Geomorphology*, **28**, 1–22.

HOOKE, R. LE B. 1968. Steady state relationships of arid region alluvial fans in closed basins. *American Journal of Science*, **266**, 609–629.

HOOKE, R. LE B. & ROHRER, W.L. 1977. Relative erodibility of source area rock types from second order variations in alluvial fan size. Bulletin of the *Geological Society of America*, **88**, 1177–1182.

LEEDER, M.R. & MACK, G.H. 2001. Lateral erosion ('toe-cutting') of alluvial fans by axial rivers: implications for basin analysis and architecture. *Journal of the Geological Society, London*, **158**, 885–893.

LECCE, S.A. 1991. Influence of lithologic erodibility on alluvial fan area, western White Mountains, California and Nevada. *Earth Surface Processes and Landforms*, **16**, 11–18.

MCFADDEN, L.D., RITTER, J.B. & WELLS, S.G. 1989. Use of multiparameter relative-age methods for age estimation and correlation of alluvial fan surfaces on a desert piedmont, Eastern Mojave Desert, California. *Quaternary Research*, **32**, 276–290.

SILVA, P.G., HARVEY, A.M., ZAZO, C. & GOY, J.L. 1992. Geomorphology, depositional style and morphometric relationships of Quaternary alluvial fans in the Guadalentín Depression (Murcia, southeast Spain). *Zeitschrift für Geomorphologie Neue Folge*, **36**, 325–341.

SORRISO-VALVO, M., ANTRONICO, L. & LE PERA, E. 1998. Controls on modern fan morphology in Calabria, Southern Italy. *Geomorphology*, **24**, 169–187.

VISERAS, C., CALVACHE, M.L., SORIA, J.M. & FERNÁNDEZ, J. 2003. Differential features of alluvial fans controlled by tectonic or eustatic accomodation space. Examples from the Betic Cordillera, Spain. *Geomorphology*, **50**, 181–202.

Climatic controls on alluvial-fan activity, Coastal Cordillera, northern Chile

ADRIAN J. HARTLEY[1], ANNE E. MATHER[2], ELIZABETH JOLLEY[3] & PETER TURNER[4]

[1] *Department of Geology & Petroleum Geology, Meston Building, King's College, University of Aberdeen, Aberdeen AB9 2UE, UK*
[2] *School of Geography, University of Plymouth, Plymouth PL4 8AA, UK*
[3] *BP Exploration Co. of Trinidad & Tobago, 5–5a Queens Park West, Port of Spain, Trinidad, West Indies*
[4] *School of Earth Sciences, University of Birmingham, Birmingham B15 2TT, UK*

Abstract A description of the distribution, drainage basin characteristics, surface morphology, depositional process and age of 64 alluvial fan systems from both flanks of the hyper-arid Coastal Cordillera of northern Chile between 22°15′S and 23°40′S is presented. The coastal fans on the western flank of the Coastal Cordillera are dominated by debris-flow deposits fed from steep catchments. Two drainage basin types are recognized: type A drainage basins are small (10–30 km^2) and do not cut back beyond the main coastal watershed; and type B drainage basins are large (up to 400 km^2) and cut inland beyond the coastal watershed. The western Central Depression fans on the eastern flank of the Coastal Cordillera are characterized by sheetflood deposition fed from relatively shallow catchments in small drainage basins (10–50 km^2). The surface morphology, sedimentation rates, a luminescence date and regional cosmogenic radionucleide data suggest that these fans have been inactive for at least the last 230 000 years and probably for much of the Neogene.

The principal control on fan activity in the study area is climate. The Coastal Cordillera forms an orographic barrier to recent El Niño-related precipitation events that are restricted to the western flank of the Coastal Cordillera. These events did not penetrate into the Central Depression as indicated by the inactive nature of the western Central Depression fans located 25 km east of the active coastal-fan catchments. This scenario is considered to have prevailed for much of the Neogene. Climate also controls rates of weathering on alluvial-fan surfaces. The coastal fog results in rapid salt weathering of clasts on coastal fans resulting in the production of fines, but does not penetrate into the Central Depression. Fault activity is important in controlling drainage basin size. The larger (type B) drainage basins are commonly focused on active faults that cut the coastal watershed, facilitating drainage basin expansion. Source-area lithology is not important in controlling depositional processes. Fans on both sides of the cordillera have the same basaltic andesite and granodiorite source lithologies, yet coastal fans are dominated by debris-flow and western Central Depression fans by sheetflood deposition. A combination of chemical weathering and stream power related to gradient are considered to account for the differences in process.

Understanding what the primary controls are on the development of alluvial-fan systems has long been an important goal in the geomorphological and sedimentological literature (e.g. Blair & McPherson 1994*a*, *b*; Whipple & Traylor 1996; Harvey *et al.* 1999*a*, *b*; Ritter *et al.* 2000; Harvey 2002; Viseras *et al.* 2003). However, it is often difficult to isolate the influence of specific variables such as climate, tectonics, source-area lithology, etc., on fan systems because these variables are commonly interlinked. Here we present a comparative study of alluvial-fan systems from the flanks of the Coastal Cordillera of northern Chile. The late Cenozoic climatic and tectonic history of the Coastal Cordillera is already well known (e.g. Okada 1971; Paskoff 1978, 1980; Hartley & Jolley 1995, 1999; Scheuber & Gonzalez 1999; Hartley & Chong 2002; González & Carrizo 2003; González *et al.* 2003). Alluvial systems on

both flanks of the cordillera are derived from the same lithologies, but the climatic and tectonic controls vary between the two areas. As such, the Coastal Cordillera of northern Chile represents a unique area in which to elucidate the relative influence of regional climate and tectonics in controlling Quaternary alluvial-fan development.

Aims

The aim of this paper is to describe the distribution, drainage basin characteristics, surface morphology, depositional process and age of alluvial-fan systems from both flanks of the hyper-arid Coastal Cordillera of northern Chile, between 22°15′S and 23°40′S (Fig. 1). Interpretations based on these descriptions are used to assess the roles of climate,

From: HARVEY, A.M., MATHER, A.E. & STOKES, M. (eds) 2005. *Alluvial Fans: Geomorphology, Sedimentology, Dynamics.* Geological Society, London, Special Publications, **251**, 95–115. 0305–8719/05/$15 © The Geological Society of London 2005.

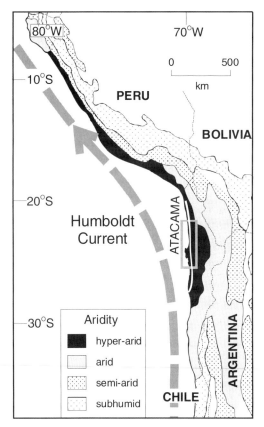

Fig. 1. Distribution of climate zones in South America and the location of the Humboldt Current. The box shows the outline of the area illustrated in Figures 2–4. The white line illustrates the extent of the Coastal Cordillera within the Atacama Desert.

source-area lithology and tectonic activity on fan development, with particular focus on the impact of the Coastal Cordillera on moisture distribution and fan activity. This study provides the first description of alluvial-fan systems from the Coastal Cordillera of northern Chile and is probably the largest regional study of alluvial-fan systems developed under a hyper-arid climatic regime. Previous work on fans in hyper-arid climate regimes is restricted to small-scale studies on hydrology (e.g. Ben-David Novak *et al.* 2004), soils and weathering (Amit *et al.* 1996; Berger & Cooke 1997), and depositional processes (e.g. Blair 1999). We describe 42 fans, from the western flank of the Coastal Cordillera that we refer to as coastal fans, and 22 from the eastern flank referred to as western Central Depression fans. Alluvial fans present within endorheic basins of the Coastal Cordillera are not discussed.

Climate, geology and geomorphology of the Coastal Cordillera

The Coastal Cordillera of northern Chile form a prominent topographic feature 700 km long and 20–50 km wide (Fig. 1). In the study area (between 22°15′S and 23°40′S) they attain heights of over 2000 m, with an average altitude of 1000 m (Figs 2 & 3). The cordillera form a topographic barrier between the Pacific Ocean and the Central Depression. The strong temperature inversion along the Pacific Coast between 800 and 1000 m prevents moist Pacific air masses from directly entering the Central Depression. The temperature inversion is caused by cold upwelling waters of the Humboldt Current and the stable location of the SE Pacific anticyclone (Rutllant & Ulriksen 1979; Trewartha 1981) (Fig. 1). The area is classified as hyper-arid (UNEP 1992) and precipitation in the Coastal Cordillera between 18° and 24°S averages less than 5 mm year (Houston & Hartley 2003). The Atacama Desert, which extends inland from the coast and encompasses much of northern Chile and southern Peru, receives less than 10 mm of precipitation per year (Fig. 1). Precipitation does occur below the temperature inversion; however, the Coastal Cordillera effectively prevent moist air masses penetrating eastwards into the Central Depression. For example, Quillagua located in the Central Depression 60 km NE of Tocopilla receives annual precipitation of 0.15 mm year (15 year mean, Houston & Hartley 2003). Precipitation events that affect the Coastal Cordillera are infrequent, but do have a significant impact on population centres located at the foot of the coastal scarp. In Antofagasta, for example, seven such events occurred between 1916 and 1999, and resulted in significant mud- and debris-flow deposits (Vargas *et al.* 2000).

The Coastal Cordillera largely comprise Jurassic andesites of the La Negra Formation (Garcia 1967) and associated granitic intrusions, in addition Lower Cretaceous conglomerates and limestones, and Palaeozoic metamorphic rocks are also present (Ferraris & Di Biase 1978) (Fig. 2). These lithologies are cut by a series of active N–S to NE–SW-trending faults (Fig. 2). Uplift and delineation of the cordillera as a separate morphotectonic unit (i.e. distinct from the adjacent Central Depression, Figs 2 & 3) is considered to have taken place in the Oligocene and continues to the present day (Hartley *et al.* 2000).

The western margin of the cordillera comprise the coastal scarp – a prominent break in slope 700–1000 m high that extends for 900 km along the coastline of northern Chile (Figs 2 & 3). The origin of the scarp has been the subject of some debate. Brüggen (1950) and Armijo & Thiele (1990) suggested that it represents an active fault, whereas Paskoff (1978, 1980) preferred an origin through marine erosion of a Miocene fault scarp(s).

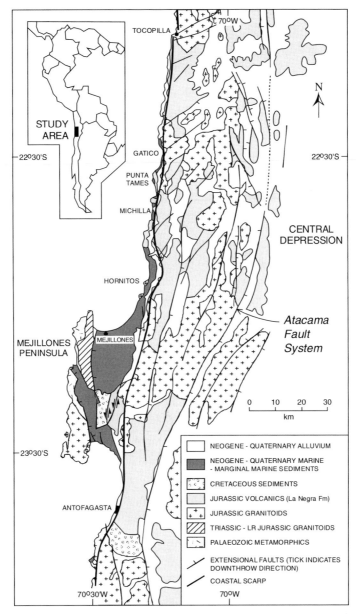

Fig. 2. Geological map of the studied part of the Coastal Cordillera of northern Chile. Coastal alluvial fans are present to the west of the Coastal scarp (based on Ferraris & Di Biase 1978; Boric *et al.* 1990). West of the Central Depression fans are developed to the east of the Atacama Fault Zone.

In contrast, Mortimer & Saric (1972) and Hartley & Jolley (1995) used the absence of active fault scarps and the irregular morphology of the scarp to suggest that it represents a weathered palaeo-cliffline, developed through marine erosion during post mid-Miocene uplift. A narrow coastal plain is present at the foot of the scarp and is defined as the area beneath the 200 m contour line.

To the east, the Coastal Cordillera is separated from the Central Depression by the 1000 km-long Atacama Fault Zone (AFZ) (Figs 2, 3b & 4). Where the fault is active, prominent scarps up to 3 m high can be observed, elsewhere the eastern margin of the Coastal Cordillera has a gentle topography with an irregular mountain front embayed by the alluvial fill of the Central Depression. The AFZ has a long

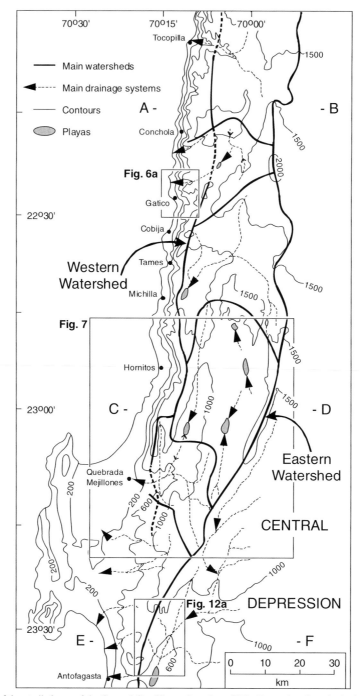

Fig. 3. (a) Map of the studied part of the Coastal Cordillera of northern Chile illustrating the principal watersheds, topography and drainage patterns. Contours at elevations of 200, 600, 1000, 1500 and 2000 m. The coastal scarp corresponds approximately to the 1000 m contour line. Location of Figures 7a, 8 and 12a are shown. Letters correspond to the locations of cross-sections shown in (b).

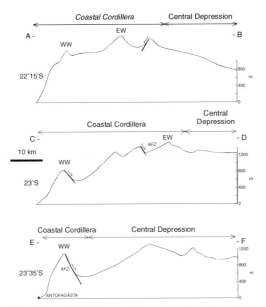

Fig. 3. *(contd.)* (**b**) Topographic cross-sections through the Coastal Cordillera: AFZ, Atacama Fault Zone; EW, eastern watershed; WW, Western watershed.

and complex history stretching from the Jurassic to the present day (e.g. Okada 1971; Hervé 1987; Scheuber & Andriessen 1990; Dewey & Lamb 1992; González & Carrizo 2003; González *et al.* 2003). Since the mid-Miocene, the AFZ has accommodated the differential uplift of the Coastal Cordillera relative to the Central Depression. Late Pliocene movement along the fault zone in the study area is considered to have been predominantly dip-slip with a small component of dextral strike-slip (Dewey & Lamb 1992; González & Carrizo 2003; González *et al.* 2003). East of the AFZ, the Central Depression has a general elevation of 1000 m in the west and rises to more than 2000 m in the east. It is 25–100 km wide. Up to 1100 m of Oligocene–Holocene strata are preserved within this basin (Hartley *et al.* 2000). Sedimentary rocks in the Central Depression include alluvial-fan, fluvial, lacustrine, playa and nitrate deposits with interbedded volcanic ashes.

Coastal alluvial fans

Distribution

The studied alluvial fans occur at the slope break between the coastal scarp and coastal plain along the western margin of the Coastal Cordillera (Fig. 5). The coastal plain extends from 23°30′S to 22°15′S, and is most extensively developed in the southern part of the study area between the Mejillones Peninsula and the Coastal Cordillera (Fig. 4). Traced northwards it narrows from 15 km wide at Quebrada Mejillones to 0–1.5 km north of Conchola (Fig. 3a). The radial length of alluvial fans ranges from 6.5 km (Quebrada Mejillones) in the south where fans terminate on the coastal plain to 0.5 km in the north where fan toes are truncated at the coastline (Table 1). Exposed thicknesses of alluvial-fan sediments reach a maximum of 35 m at the coastline. Total fan thickness estimates assuming a horizontal datum at the top Pleistocene marine terrace or the coastal plain and the gradient of the fan surfaces (measured in the field) extrapolated back to the fan apex gives values of up to 440 m (Table 1). Interfan areas are either covered by aeolian sands or comprise marine planation surfaces developed on bedrock or on late Pleistocene marine deposits exposed on the coastal plain.

Drainage basin characteristics

The source-area lithologies for the coastal fans comprise either Jurassic granodiorites and/or basaltic andesites (Fig. 2; Table 1). In some instances alluvium stored in the catchment area may also form source material. Two main drainage basin types can be recognized: type A, which breach the coastal scarp but not the western watersheds; and type B, which breach both the coastal scarp and the western watershed. Type A drainage basins dominate between Gatico and Quebrada Mejillones (Fig. 3a) where they typically have steep feeder canyons, and cover an area of between 10 and 30 km^2 (Figs 3a & 6–8). Type B drainage basins are up to 400 km^2 in area and frequently have feeder canyons that are located on or adjacent to active faults (Figs 3a & 6–8).

Surface morphology

The surface slope angles of the fans can be divided into two segments: a relatively steep fan-cone segment with an average slope of 11° (ranges from 19° to 4°) that comprises 20–40% of the fan surface, and a fan-apron segment with an average slope angle of 5° (ranges from 10° to 1.5°) comprising 60–80% of the fan surface (Table 1). The transition between the two segments is gradational. Both incised and non-incised alluvial fan-surface morphologies can be recognized (Fig. 9). Incised types include: fans displaying proximal incision, fans displaying distal incision, and those displaying both proximal and distal incision which when linked are referred to as fully trenched. Fans showing proximal incision have fanhead trenches 4–20 m deep and 10–50 m wide that extend up to 400 m downslope from the fan apex where they pass into channels

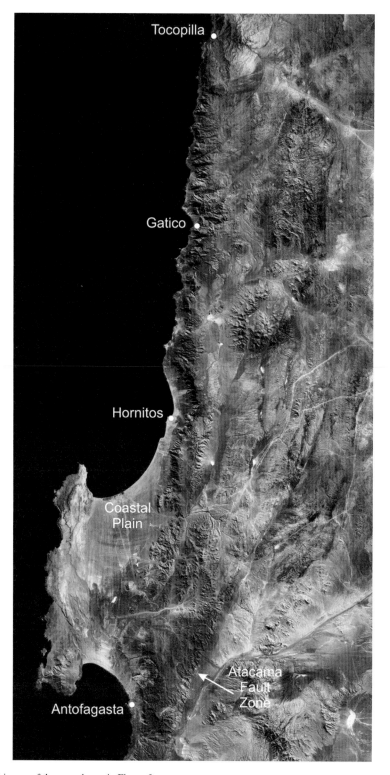

Fig. 4. Landsat image of the area shown in Figure 3a.

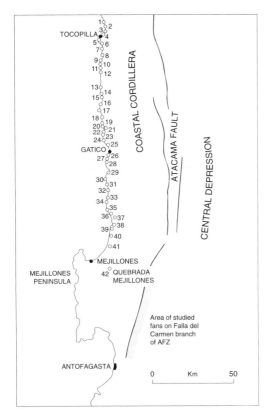

Fig. 5. Location of the studied coastal fans between Quebrada Mejillones in the south and just north of Tocopilla in the north. Location of the studied western Central Depression fans highlighted by the shaded area east of the Atacama Fault Zone. Numbers are the same as those in Table 1.

2–3 m deep that in turn pass into a broad array of shallow channels (0.3–2 m deep) which dominate the remaining downslope portion of the fan surface. Fans displaying distal incision have 1–5 m-deep gullies cut at the fan toes. Gullies are cut through alluvium into underlying Plio-Pleistocene marine sediments and incise landwards via head-cutting from Pleistocene marine terraces (e.g. Figs 6b & 8b). Fully entrenched fans ($n = 5$; Table 1) have a single major channel cut directly from the fan apex to the coast (Figs 7–9). The channel ranges from 10 to 25 m in depth and from 20 to 50 m in width with vertical cut banks. In plan view they have a low sinuosity. All drainage (with the exception of small tributaries) is directed into the channel resulting in bypassing of the fan surface. There are no active fan deltas as reworking by shallow-marine processes dominates at the coastline (cf. Hartley & Jolley 1999). Alluvial fans that display no evidence for incision ($n = 18$) are dominated by a radial distributary system of shallow channels. These have a

maximum depth of 2 m in the fan apex that pass downwards into wide, shallow washouts 0.3–1 m deep over the fan apron.

On all fans, the oldest abandoned fan surfaces show evidence for strong salt weathering of clasts and the development of an ochre-coloured, salt-cemented, fine-grained veneer of sediment. In some cases boulders and cobbles have been almost completely destroyed by salt weathering. Younger surfaces have an ochre colouring and comprise pebble–cobble-sized detritus.

Depositional processes

Three different processes can be identified from observations of recent deposits, preserved deposits of sections exposed within channels cutting the fan surface and in the marine terraces that truncate the fan toes.

Debris-flow deposits form the majority of all the studied sections (approximately 85%), and have an original depositional dip of 4°–19° to the west. They can be divided into two types: levée and lobe deposits (Fig. 10). Levée deposits comprise very poorly sorted angular cobble–boulder-grade material (Fig. 5). Bed boundaries where distinguishable are marked by a thin 5–10 cm-thick fine-grained (silt) horizon. Beds display rapid slope-parallel changes in thickness (0.2–2 m over 3 m), although downslope sections show less dramatic thickness changes (generally a few tens of centimetres to over 5 m). Beds are clast-rich and locally may be clast-supported particularly on the steeper slopes of the fan-cone area. They have a coarse-grained sand–pebbly sandstone matrix that can form up to 60% of the bed. Clasts may be randomly oriented, show vertical long-axis alignment, poorly defined imbrication or a crude bedding-parallel fabric (Fig. 10). Grading is generally absent. Levée deposits are particularly well developed on the high slope-angle fan-cone area where they form a series of ridges up to 0.5–2 m high separated by small channels 0.5–4 m wide.

Lobe deposits comprise very poorly sorted angular gravels of pebble–boulder-grade material (Fig. 10). Beds range from 0.1–2 m in thickness. Contacts may be sharp (often irregular) or diffuse. Beds are predominantly clast-rich, although in some examples the matrix may form up to 60% of the bed. Matrix-supported deposits are restricted to the fan apron. The matrix comprises coarse-sand or granule-grade material, mud and silt form only minor proportions (usually <10%). Grading is generally absent, although occasional normal grading may be observed. Clasts are generally randomly orientated, although occasionally a crude imbrication or bedding-parallel clast fabric may be developed in clast-supported beds. Lobe deposits are well

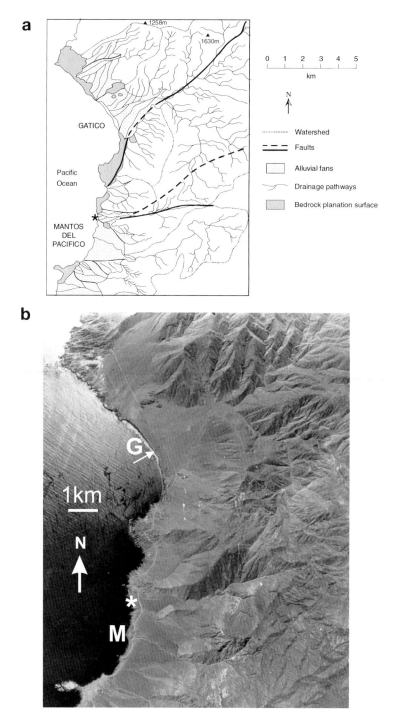

Fig. 6. (a) Map and **(b)** photographs of drainage basin development and fan morphology in the Gatico area (see Fig. 3 for location.). Aerial photograph shows different fan-surface morphologies: a single major channel (M, Mantos del Pacifico), several large channels (G, Gatico) and areas with no marked channelization. Note the gullies headcutting from the marine terrace just to the right of the G (arrowed). The impact of faulting is illustrated by the fan marked with an asterisk, which has a narrow, steep internal catchment associated with a single fault.

developed on the fan-apron area where they have a relatively smooth surface topography with only small gullies up to 0.5 m deep and 2 m wide dissecting the surface.

The unsorted, disorganized, muddy, sandy, bouldery, cobble–gravel texture of these deposits is diagnostic of debris-flow deposition (e.g. Pierson 1980; Johnson 1984). Clast-supported deposits record deposition from non-cohesive debris flow (e.g. Blair & McPherson 1994*b*), and dominate the fan cone and large parts of the fan apron. Matrix-supported deposits reflect deposition from cohesive flows (e.g. Wells & Harvey 1987) and only occur on the fan apron. Debris-flow deposits dominate steeper fan segments, particularly those derived from type A drainage basins, and are also important in fans sourced from type B drainage basins.

Hyperconcentrated flow deposits represent approximately 10% of the studied fan sections. They include angular, clast-supported small pebble–cobble gravels with a very poorly sorted coarse–very-coarse sand–granule matrix (Figs 10 & 11). Beds range from 10 to 50 cm in thickness and boundaries are often diffuse. Thin (up to 5 cm thick) granule and small pebble units occur between the more clast-rich intervals and define bed boundaries. Beds with larger clasts typically display imbrication occasionally with clast long axes steeply inclined–vertical (Fig. 10). Smaller clasts tend to be imbricated or show a bedding-parallel fabric, although commonly no obvious fabric is present. In plan view the deposits have a lobate form with steep, marginal relief of the order of 10–50 cm. Observations from recent deposits indicate that the matrix is infiltrated following deposition during low stage flow. These beds display similarities with both non-cohesive debris-flow (clast-supported, structureless) and sheetflood (imbrication, grain-size segregation) deposits, and are thought to have been deposited by flows transitional between true Newtonian flows (normal river flow) and non-Newtonian or plastic flow (debris flow). They are interpreted as hyperconcentrated flow deposits similar to those described by Wells & Harvey (1987).

Sheetflood deposits comprise angular, clast-supported conglomerates and pebbly sandstones with a coarse sand–granule-grade matrix, and form approximately 5% of the studied sections. Beds range from 5 to 45 cm in thickness and boundaries are sharp–irregular. Imbrication and bedding-parallel clast fabrics are common and beds are frequently normally graded. In the more distal (coastal) sections, gravel–sandstone couplets can be observed (Fig. 10). These beds are considered to result from poorly confined sheetflow on the fan surface (e.g. Blair & McPherson 1994*a*, *b*). Sheetfloods often rework the top of debris flows in the fan cone and proximal apron areas, but become increasingly important on

the lower reaches of fans, particularly on fans in the southern part of the study area that terminate on the coastal plain and are derived from a type B catchment.

Age

The age of the alluvial fan deposits can be constrained by their relationship to Pleistocene marine terrace deposits. The age of terrace deposits has been determined from marine macrofauna, U–Th disequilibrium, amino-acid racemization and electron spin resonance techniques on marine molluscs (for details see Radtke 1985, 1987; Leonard & Wehmiller 1991; Ortlieb *et al.* 1996). Alluvial-fan deposits can be seen to overlie terraces ranging from 400 000 to 80 000 years in age (Ortlieb 1995), and in many cases modern-day fan activity results in deposition at the coastline or significant catchment activation and fan deposition. At Hornitos, Ortlieb *et al.* (1996) noted that the majority of fans have prograded across terraces produced during isotopic stage 9 (300–330 ka) and that some have prograded over younger terraces which correspond to isotope stages 5 (100–125 ka) and 7, although the latter are not widespread (Fig. 8).

Western Central Depression alluvial fans

Distribution

The eastern margin of the Coastal Cordillera is well defined in the southern part of the study area where it is associated with the Falla del Carmen segment of the AFZ (Fig. 7). North of 23°S, however, the AFZ becomes less well defined and splits into a series of faults resulting in a gradational transition into the Central Depression (Fig. 4). The fans that have been studied in detail have catchments that drain eastwards across the Falla del Carmen strand of the AFZ and form a bajada that terminates against an ephemeral, SW-draining, axial fluvial system (Figs 3a, 5 & 7). The radial length of the alluvial-fans ranges from 1000–3000 m, with larger radii fans corresponding to those with the largest drainage basins (Fig. 12).

Drainage basin characteristics

The source-area lithologies for the western Central Depression fans comprise either Jurassic granodiorites and/or basaltic andesites (Fig. 2). Frequently, alluvium stored in the catchment area may also form source material, particularly as many of the catchments are backfilled (Fig. 13). In the studied fans adjacent to the Falla del Carmen, basaltic andesites

Table 1. Summary of the characteristics of the studied coastal fans including: the nature of the drainage basin (type A or B and the number of feeder channels); fault control on drainage basin development; fan radii (measured from the head of the fan apex to the termination point which is at the coast for fans 1–29 and on the coastal plain for fans 30–42); incision and channelization of the fan surface; fan slope angle (in °) and percentage of the fan surface at that slope angle in parentheses, downslope changes in gradient are shown from right to left; estimated thicknesses of alluvial fan deposits calculated from slope angles, distance from apex and thickness of deposits (shown in parentheses) as measured above the late Pleistocene marine terrace on which the fan either terminates on the coastal plain or where it is exposed in a marine terrace at the coastline

Locality	Type B	Type A	Fault	Radius (m)	Surface slope angle	Channels	Thickness (m)	Source
1 (Ana)	2	3	x (SE–NW)	800	7 (40%); 4 (60%)	Minor	93 (20)*	gran
2	1		X	1000	12 (25%); 5 (75%)	Minor	135 (16)*	gran
3 (Brava)	2		X	900	12 (25%); 6 (75%)	Minor	131 (12)*	gran
4		2	x (N–S)	750	12 (20%); 6 (80%)	None	114 (19)	gran
5		4	x (N–S)	750	12 (25%); 6 (75%)	None	129 (30)	gran/and
6 (Algondonales)		2		900	18 (23%); 11 (40%); 8 (37%)	None	204 (25)	gran/and
7		4	x (ENE–WSW)	800	19 (20%); 14 (20%); 8 (60%)	Major	183 (10.5)	andesite
8	1			750	8 (40%); 4 (60%)	None	98 (25)	andesite
9	1			500	20 (100%)	None	190 (7.5)	andesite
10	1		x (NE–SW)	600	11 (33%); 7 (67%)	Minor	96 (9)	andesite
11 (Blanca)	1		x (NE–SW)	1100	7 (40%); 5 (60%)	Minor	128 (16)	andesite
12	1		x (NE–SW)	1200	13 (25%); 9 (75%)	Major	230 (18)*	andesite
13		5		750	16 (25%); 10 (75%)	None	210 (20)*	andesite
14		1		800	18 (25%); 7 (75%)	None	197 (17)*	andesite
15		1		1000	5 (100%)	Minor	88 (14)	andesite
16 (Alala)		1		500	18 (100%)	None	266 (20)	andesite
17	1		x (NE–SW)	2000m	12 (33%); 6 (66%)	Minor	300 (18)*	andesite
18 (Conchola)	1			2000	18 (30%); 7 (70%)	None	387 (20)	andesite
19	1		x (NE–SW)	2500	11 (35%); 4 (65%)	None	276 (10)*	andesite
20 (Punta Ampa)	1		x (NE–SW)	3000	9 (33%;) 6 (66%)	Minor	373 (5)	andesite
21	1		x (NE–SW)	2200	9 (30%); 5 (70%)	Major	316 (22)*	andesite
22	1	2	x (NE–SW)	2500	9 (30%); 5 (70%)	Major	282 (10)*	gran/and

Site			Orientation	Elevation	Composition	Faulting	Azimuth	Lithology
23 (Chinos)	1	3		2500	7 (100%)	None	307 (8)*	gran/and/Q
24		1		3200	8 (30%); 5 (70%)	Minor	339 (8)	gran/and/Q
25 (Gatico)	2	1	x (NE–SW)	2500	6 (34%); 3 (66%)	Minor	184 (6.2)	gran/and/Q
26 (M. del Pacifico)		1		2000	8 (35%); 4 (65%)	Major	221 (19.6)	gran/and
27 (Cobija)		2		3700	8 (25%); 5 (75%)	None	373	gran/and
28 (Tamira)		2	x (NE–SW)	3300	8 (25%); 4 (75%)	Minor	290 (1)	gran/and/Q
29 (Tames)	1	1		3200	8 (34%); 5 (66%)	Minor	337	gran/and/Q
30		1		2600	12 (40%); 4 (60%)	None	330	andesite
31 (N.Michilla)		5		3500	8 (23%); 6 (77%)	None	394	andesite
32 (Michilla)	1	3		3000	6 (100%)	Minor	315	andesite/Q
33 (S.Michilla)		2	x (ENE–WSW)	3600	6 (63%); 3 (37%)	None	315	gran/and/Q
34		3		3700	7 (33%); 5 (66%)	None	367	gran/and/Q
35 (Yayes)		1		2500	10 (100%)	Minor	441	andesite
36		4	x (NNE–SSW)	3400	7 (33%); 4 (66%)	Minor	297	andesite
37		8		3800	10 (38%); 1.5 (62%)	Minor	289	andesite
38 (Horno)	1		x (NNE–SSW)	3900	6 (40%); 3 (60%)	Minor	287	andesite/Q
39 (Hornitos)	1		x (NNE–SSW)	4100	6 (100%)	Minor	431	andesite/Q
40 (Chacaya)	1		x (NNE–SSW)	3100	12 (20%); 2 (80%)	Minor	219	gran/and/Q
41 (Chacaya Sur)		1		3300	12 (25%); 2 (75%)	None	261	gran
42 (Q.Mejillones)	1	4	x (NNE–SSW)	6500	4 (20%); 2 (80%)	None	272	gran/and/Q

* Indicates that the base of the alluvial fan section could not be observed and that thicknesses could therefore be greater.

Q. Quaternary alluvium; And, andesite; gran, granodiorite.

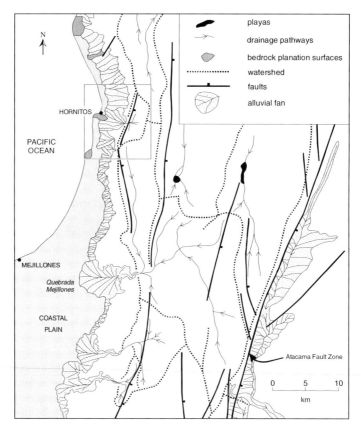

Fig. 7. (a) Map of drainage development in the Coastal Cordillera between 22°45′ and 23°15′S (see Fig. 3 for location). The fan developed at the mouth of Quebrada Mejillones is the largest of the coastal fans and has the largest drainage basin. Note the common occurrence of fault-parallel drainage. Capture of separate fault-controlled sub-basins by a combination of headward erosion–overspill is considered to have resulted in drainage expansion, accessing large amounts of alluvium for supply to the alluvial fan. Note the small drainage basin size of fans developed on the western side of the Central Depression and the likelihood that, should the Quebrada Mejillones system expand further eastwards, it will capture drainage from the Central Depression. The box shows the outline of Figure 8a.

form the principle source lithology. Drainage basin size ranges from approximately 10 to 50 km² (Fig. 12). Drainage basins extend between 3 and 6 km west of the Falla del Carmen strand of the AFZ to the regional watershed that trends subparallel to the fault. The elevation difference from the top of the watershed to the apex of the alluvial fan within an individual drainage basin is up to 500 m.

Surface Morphology

The surface slope of the alluvial fans ranges from 1.5° to 4° and does not change downslope except locally where a fault scarp cuts the fan surface. The surface morphology shows little variation and is represented by a gently undulating fan surface with a maximum relief of 1 m associated with broad, shallow channels (Fig. 13). Headward eroding

gullies are present where the active fault scarp cuts the fan surface (Fig. 13); however, these rarely link back to channels in the catchment. As such, gully development is restricted to reworking of existing fan deposits without activation of the catchment. Fan surfaces are covered with a well-developed desert pavement composed predominantly of pebble-grade material with occasional cobbles, although channel areas contain only small pebbles and coarse-grained sand (Fig. 13).

Where exposed in small gullies and channels an up to 1 m thick gypsic soil may be present. The soil shows many of the characteristics of gypcretes described by Watson (1988, 1989) from the Quaternary of Tunisia and Namibia. The desert pavement that represents the Ao horizon is underlain in some areas by a vesicular horizon (Av) up to 1 cm thick. Beneath either the Av horizon or directly below the pavement a gypsic silty–sandy loam horizon

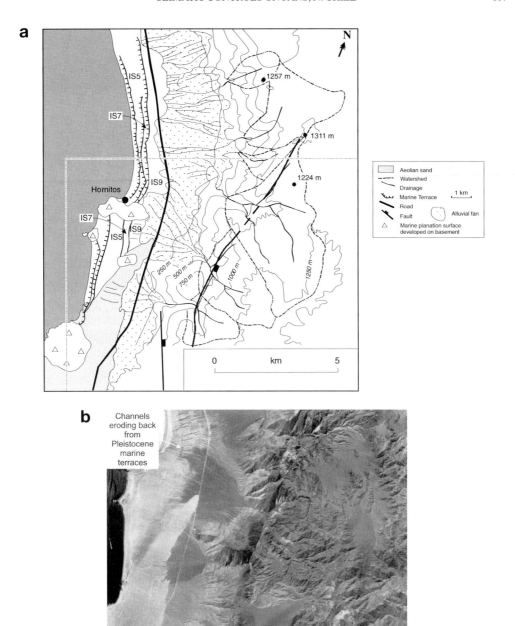

Fig. 8. (**a**) Map and (**b**) aerial photograph of the Hornitos area illustrating the influence of a NNE–SSW-trending fault on catchment development. Note the absence of major channels on fan surfaces. IS5, IS7 and IS9 correspond to marine planation surfaces formed during isotope stages 5, 7 and 9 (as identified by Ortlieb *et al.* 1996). Note the gullying developed post-marine terrace formation.

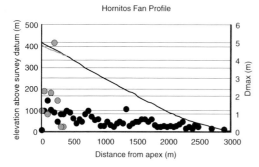

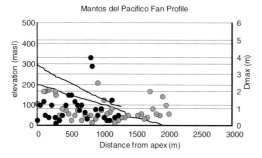

Fig. 9. Long profiles of coastal fans from Hornitos (non-trenched) and Mantos del Pacifico (trenched). The maximum clast sizes are also shown. Light shaded circles correspond to clasts measured in the trenched sections of the fans. See Figs 6 and 8 for locations.

between 10 and 50 cm thick, containing occasional clasts up to 5 cm in diameter, is present, and is considered to represent the B horizon of a desert soil. Beneath this a hard, well-cemented gypsic horizon is present between 5 and 25 cm thick with scattered clasts throughout, and is considered to represent the C horizon. Western Central Depression alluvial fans north of the study area (NE of Tocopilla) have extensive nitrate-, sodium sulphate- and halite-rich soil horizons (Chong 1988). No obvious recent activity was observed on the fan surfaces.

Depositional processes

Depositional processes are inferred from a number of sections present in small gravel quarries and channels, although they are frequently masked by gypsum precipitation and weathering. Where observed the deposits comprise pebbly sandstones and occasional angular, clast-supported gravel intervals with a coarse-sand–granule-grade matrix (Fig. 13). Beds range from 5 to 50 cm in thickness. Boundaries may be sharp or diffuse. Diffuse boundaries occur where gypsic soil horizons similar to that described above from the present-day surface have disturbed the underlying stratification.

Imbrication and bedding-parallel clast fabrics are common where not disturbed by gypsic cements, and a crude, bedding-parallel stratification can be observed in most beds. The characteristics of the western Central Depression alluvial fans indicate that the principal depositional process was poorly confined sheetflow. Poorly sorted, largely structureless, clast-supported conglomerates form less than 10% of the observed sections. They are considered to represent either hyperconcentrated or clast-rich debris-flow deposits (cf. Wells & Harvey 1987; Blair & McPherson 1994*a*, *b*). Depositional events were separated by significant time intervals as indicated by the development of gypsic soil horizons (e.g. Fig. 13b).

Age

The western Central Depression fans are largely inactive at the present day and have been for some time, as indicated by pedified fan surfaces including gypcrete development, the relatively fine grain size, low topographic relief of fan surfaces, pavements developed within gullies and the fact that they are cemented by nitrates in the NE part of the study area (see Chong 1988 for more details). Berger (1993) used the optically stimulated thermoluminescence technique to date a sample taken from an alluvial fan in the Central Depression adjacent to the study area. The sample was taken from a channel system cut into a dissected surface. She concluded that the fan surface was older than 230 000 years (the upper age limit of the technique). Another feature indicating limited Quaternary sedimentation is the accumulation of only 2 m of alluvial-fan deposits above a 3 Ma ash exposed in the hanging wall of the Falla del Carmen strand of the AFZ north of the Salar del Carmen (Naranjo 1987). Other dated ashes in the footwall of the AFZ range from 2.9–5 Ma (González & Carrizo 2003) and are overlain by 3 m or less of alluvial-fan sediments. Another factor that suggests limited fan activity is the lack of erosion of fault scarps associated with the AFZ. Although the age of the scarps can only be constrained as post-2.9 Ma, the scarps are well defined and are blanketed by a pervasive gypcrete (Fig. 7).

Further information on the age of alluvial-fan surfaces in the Central Depression is provided by exposure dating of pebbles and boulders from surface pavements. Recent ^{21}Ne cosmogenic isotope data from surface pebbles in the central part of the Coastal Cordillera at $19°30'$ S reveal exposure ages that range from 36.5–8 Ma, with a peak at 23 Ma (Dunai *et al.* 2005). Cosmogenic ^{21}Ne exposure ages of 9 Ma have also been reported from the Atacama Desert (Owen *et al.* 2003), and recent cosmogenic ^{3}He data from surface boulders in the Central Depression, between

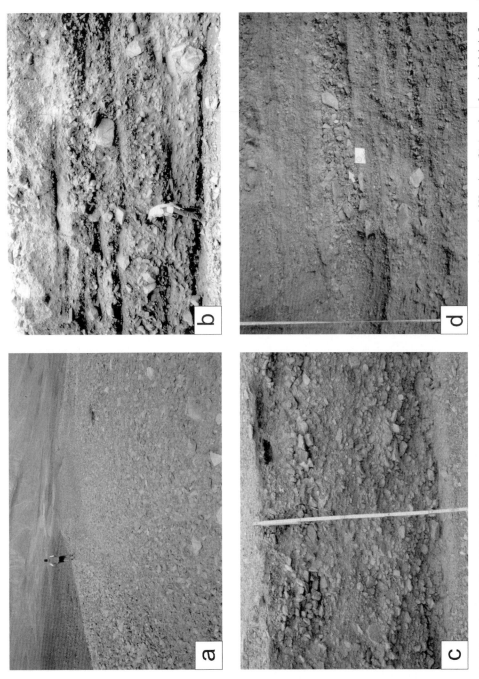

Fig. 10. Field photographs of the Coastal fans. (**a**) View of a recent (1991) debris-flow lobe at Chacaya Sur, 2 km south of Hornitos. (**b**) A series of stacked debris flows with thin reworked tops at Gatico. (**c**) Stacked hyperconcentrated flow deposits from the distal end of the Gatico fan. (**d**) Sheetflood deposits from a small gravel pit on the distal end of the Hornitos alluvial fan.

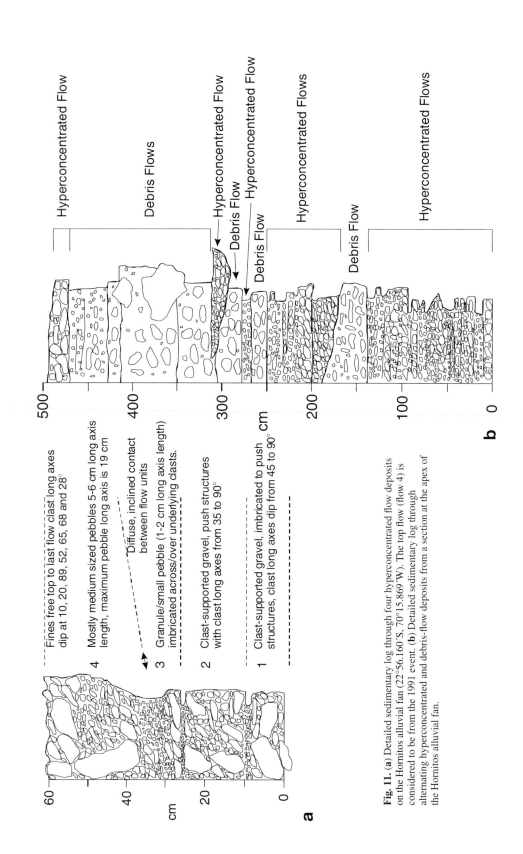

Fig. 11. (a) Detailed sedimentary log through four hyperconcentrated flow deposits on the Hornitos alluvial fan (22°56.160´S, 70°15.869´W). The top flow (flow 4) is considered to be from the 1991 event. (b) Detailed sedimentary log through alternating hyperconcentrated and debris-flow deposits from a section at the apex of the Hornitos alluvial fan.

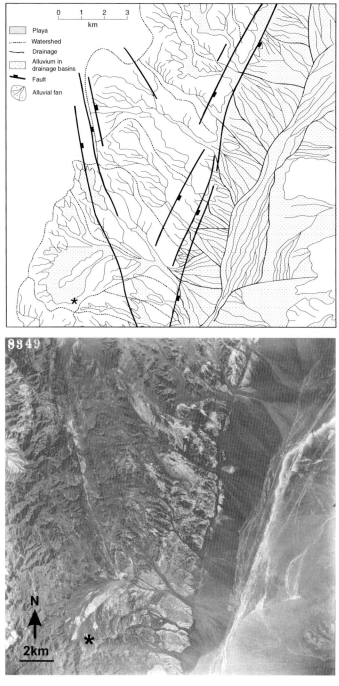

Fig. 12. (**a**) Detailed map of drainage basins and alluvial fans on the western side of the Central Depression associated with the Falla del Carmen segment of the Atacama Fault Zone. For the location see Figure. 3. (**b**) Aerial photograph covering the boxed area outlined in Figure 6a. Note that the asterisk is located in the same place on both diagrams.

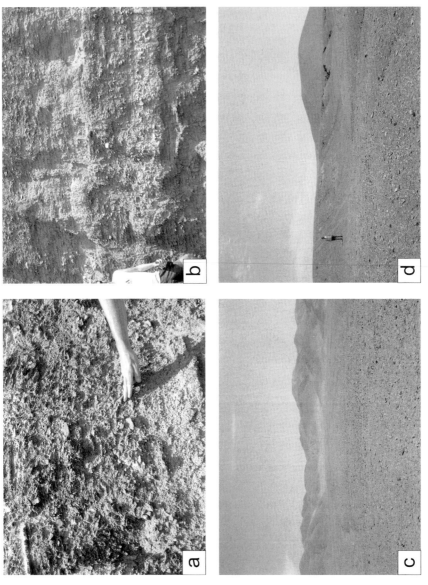

Fig. 13. Field photographs of the western Central Depression fans. (**a**) Detail from a gravel quarry illustrating poorly sorted, angular pebble-sized conglomerates with very-coarse-sand–granular matrix. Note a crude bedding-parallel fabric. Original depositional structures are difficult to identify due to the development of an extensive gypsum cement. (**b**) A series of stacked coarse-grained pebbly sandstone beds 20–80 cm thick. Bedding is defined by variable amounts of gypsum cement. Finer grained intervals (e.g. at the level of the head of the figure for scale) represent gypcrete horizons. (**c**) View of a typical fan catchment, note the backfilled nature, the lack of channels and gullies, and the well-developed pavement. (**d**) View of the Salar del Carmen segment of the Atacama Fault Zone, note the limited gullying on the fault scarp and the development of a pavement in the foreground and on the scarp itself.

19° and 22°S, have yielded ages of between 24 and 14 Ma (L. Evenstar pers. comm. 2005). These exposure ages are reported from sites to the north and east of the fans studied here; however, they do appear to indicate that there has been little erosion in at least parts of the Coastal Cordillera and Central Depression for much of the Neogene.

Comparison between coastal and western Central Depression fans

A number of surface morphological features suggests that the western Central Depression fans are much less active than the coastal fans. These include the limited topographic relief of the western Central Depression fans, the development of an extensive pavement across the catchment and entire fan surfaces including shallow gullies, and the presence of a well-developed gypcrete. The lack of activity is also indicated by the luminescence date of more than 230 000 years. from an adjacent fan in the Central Depression. In contrast, the vast majority of the coastal fans studied show evidence of recent debris-flow or sheetflood activity (particularly following the 1991 storm event); indeed, seven runoff events were recorded from coastal fans at Antofagasta between 1916 and 1999 (Vargas *et al.* 2000).

Drainage basin area varies significantly for the coastal fans depending on whether they have type A or B drainage basins. Type A drainage basins typically have steep gradients and are generally small (10–30 km^2) as they are developed only west of the main watershed. This drainage basin size is similar to that of the western Central Depression fans (10–50 km^2) whose drainage basins have not expanded beyond their adjacent watershed either. In contrast, type B drainage basins for the coastal fans breach the adjacent watersheds, frequently utilizing faults, resulting in a much greater aerial extent. Indeed, expansion of these westerly-draining alluvial systems frequently encroaches on the drainage basins of the western Central Depression fans (e.g. Quebrada Mejillones in Fig. 7).

Differences in fan-surface gradient and dominant depositional processes between the two fan types are marked. The coastal fans are dominated by debris flows that are deposited on relatively steep fan-surface gradients, generally more than 5°. In contrast, the western Central Depression fans are dominated by more sheetflood-style deposition on relatively shallow fan-surface gradients, generally less than 4°. An interesting feature is that there is no major difference in source-area lithology between the coastal and western Central Depression fans, they both erode the same lithologies (Fig. 1). A comparison of fan radii for both the coastal and western Central Depression fans would reveal little

due to toe cutting by wave erosion on the coastal fans or by an axial fluvial system, as occurs in the Central Depression.

Discussion: controls on alluvial-fan systems

The first-order control on alluvial-fan systems adjacent to the Coastal Cordillera of northern Chile is uplift of the cordillera itself. The Coastal Cordillera is the only subaerial part of the South American continental crust in direct contact with the subducted Nazca Plate (Allmendinger *et al.* 2005), uplift can therefore be directly related to plate coupling at the subduction zone (e.g. Hartley *et al.* 2000; Lamb & Davis 2003). As the present-day tectonic regime became established in the Oligocene, alluvial-fan activity may extend back to this period, although the oldest known fan deposits associated with the Coastal Cordillera and adjacent areas are lower–middle Miocene in age.

A second-order control on fan development, again related to tectonics, is drainage basin size. The majority of fans on both sides of the Coastal Cordillera are supplied by relatively small drainage basins (<50 km^2) that have not expanded beyond their immediately adjacent watershed. However, coastal fans with external drainage basins are considerably larger than both those with an internal drainage basin and those of the western Central Depression fans. This is considered to be due to: (1) the regional topographic gradient across the Coastal Cordillera; and (2) fault activity within the cordillera.

A further tectonic control that may help to explain the difference in dominant depositional process between the two sides of the Coastal Cordillera is related to the elevation differences between catchment watersheds and the fan apex. The coastal fans supplied by internal drainage basins have a similar drainage basin size to the western Central Depression fans and are sourced from the same lithology, yet have an elevation difference of 1100 m compared to that of 500 m for the western Central Depression fans. This elevation difference will result in steeper gradients in the catchments and greater transport capacity, such that debris-flow transport is likely to be sustained beyond the catchment with deposition onto the fan surface. In contrast, the lower gradients of the western Central Depression catchments only appear to have occasionally been capable of sustaining debris-flow transport to the alluvial fan.

Climatic controls on alluvial-fan system development in the Coastal Cordillera are manifested in a number of ways. It is clear that the western Central Depression main bajada surface is effectively fossilized, and has been so for at least the last 230 000 years and probably for substantially longer as suggested by the cosmogenic exposure ages reported from the Central Depression (e.g. Dunai *et al.* 2005) to the north and east of the study area. In contrast, the coastal fans 10–25 km to the west have been active throughout the Pleistocene to the present day. This extreme variation in fan activity must be related to precipitation within the fan catchments and may possibly be explained by studying the causes of historical fan catchment activity. Vargas *et al.* (2000) in a study of historical mud- and debris-flow events in the Antofagasta area established a relationship between fan activity and El Niño events caused by movement of the SE Pacific anticyclone. Recent precipitation events took place during El Niño years in the austral winter and occurred when southerly airflows brought frontal precipitation from extra-tropical cyclones (Vuille 1999). Notably, these events only activated catchments on the western flank of the Coastal Cordillera and did not penetrate into the Central Depression. It is likely that this climatic scenario has prevailed at least through the late Pleistocene (or possibly back into the Miocene) to the present day, depending on the age of the western Central Depression fans. In addition, there has obviously been a significant climate change in the Pleistocene or earlier from a 'wetter' to drier climate that has resulted in the effective fossilization of the western Central Depression fan systems.

Another important impact of climate on alluvial-fan development is weathering. The rapid breakdown and extensive weathering of clasts on the coastal fans is due to salt weathering that is related to the coastal fog or *Camanchaca*. In contrast, penetration of the *Camanchaca* into the Central Depression is much less frequent, such that salt-related weathering, although important, is not as intense as occurs on the coastal margin. Weathering will also have a significant impact on the production of fines and may help to account for the dominance of debris-flow activity on the coastal fans and limited debris-flow deposition on the western Central Depression fans despite both being sourced from the same bedrock lithology.

A significant observation concerned with the catchment characteristics on both sides of the Coastal Cordillera is that the coastal-fan catchments display little or no evidence of backfilling whereas the western Central Depression catchments are backfilled. This indicates that sediment-transport mechanisms in the western Central Depression fan catchments were insufficient to transport sediment out of the system. In addition, runoff in these catchments was more susceptible to transmission losses such that the extent of any floods that were generated was limited. In contrast the greater elevation difference between the catchment watershed and the fan apex and the lack of backfilled alluvium in the coastal-fan catchments means that once flows are generated they are likely to be sustained and deliver sediment to the fan apex.

Conclusions

A study of 64 alluvial-fan systems on the flanks of the Coastal Cordillera of northern Chile has allowed the general impact of tectonics, climate and source-area lithology on controlling alluvial-fan system development to be constrained.

The coastal fans west of the Coastal Cordillera are dominated by debris-flow deposits, and are fed by steep catchments. Two drainage basin types are recognized: type A drainage basins are small (10–30 km²) and do not cut back eastwards beyond the main western watershed; type B drainage basins are large (up to 400 km²) and cut back beyond the coastal watershed accessing sub-basins within the Coastal Cordillera. The western Central Depression fans developed on the eastern flank of the Coastal Cordillera are characterized by relatively small drainage basins (10–50 km²). The surface morphology, gypcrete development, sedimentation rates, a luminescence date and cosmogenic exposure age dating from elsewhere in the Coastal Cordillera and Central Depression suggest that these fans have not been active for over 230 000 years and possibly back to the Miocene, and are effectively fossilized.

Regional tectonic activity is responsible for the uplift of the Coastal Cordillera and generation of topography required for alluvial fan development, and is driven by direct coupling of the Nazca and South American plates at the subduction zone. Regional uplift coupled with the trenchward collapse of the leading edge of the South American Plate results in a decrease in elevation of 1500–1000 m from west to east across the Coastal Cordillera, promoting westerly-flowing drainage systems.

Local tectonic activity associated with specific faults has a significant impact on drainage basin development. Drainage is frequently focused on active faults that facilitates breaching of the western watershed and drainage basin expansion. Continued expansion of drainage basins results in significant westerly drainage development, ultimately linking directly to the Central Depression.

Climate controls alluvial-fan activity. Coastal alluvial fans have been active throughout much of the Pleistocene and are active at the present day, whereas the western Central Depression fans are inactive despite being located 25 km west of the coastal-fan catchments. Recent observations of precipitation events on the Coastal Cordillera during the austral winter of El Niño years have shown that precipitation was restricted to the western flank of the Coastal Cordillera and did not penetrate into the Central Depression. It is suggested that this scenario has prevailed for at least the last 230 000 years and possibly into the Miocene. Climate also controls rates of weathering on alluvial-fan surfaces. The coastal fog results in rapid salt weathering of clasts on coastal fans but, as it has limited penetration into the Central Depression, it does not significantly impact the western Central Depression fans.

Source-area lithology is not an important control on depositional processes on the Coastal Cordillera fans. Fans on both sides of the cordillera have the same source lithology: basaltic andesite and granodiorite, yet coastal fans are dominated by debris flow and western Central Depression fans by sheetflood. A combination of weathering and stream power related to differences in gradient and the presence or absence of alluvium within the catchment are considered to account for the differences in process.

The authors would like to thank Prof. G. Chong, Universidad Católica del Norte, Antofagasta for logistical support and scientific advice. AJH would like to acknowledge a British Council travel grant. AEM would like to acknowledge Royal Society and British Council travel grants. In addition we would like to thank L. Evenstar (University of Aberdeen) and F. Stuart (Scottish Universities Environmental Research Centre, East Kilbride) for discussing the results of recent cosmogenic ^{3}He exposure age data from northern Chile.

References

ALLMENDINGER, R.W., GONZÁLEZ, G., YU, J., HOKE, G. & ISACKS B. 2005. Trench-parallel shortening in the Northern Chilean Forearc: Tectonic and climatic implications. *Bulletin of the Geological Society of America*, **117**, 89–104.

AMIT, R., HARRISON, J.B.J., ENZEL, Y. & PORAT, N. 1996. Soils as a tool for estimating ages of Quaternary fault scarps in a hyper-arid environment – The southern Arava Valley, the Dead Sea Rift, Israel. *Catena*, **28**, 21–45.

ARMIJO, R. & THIELE, R. 1990. Active faulting in northern Chile: ramp stacking and lateral decoupling along a subduction plate boundary? *Earth and Planetary Science Letters*, **98**, 40–61.

BEN-DAVID NOVAK, H., MORIN, E. & ENZEL, Y. 2004. Modern extreme storms and rainfall thresholds for initiating debris flows on the hyper-arid western escarpment of the Dead Sea, Israel. *Bulletin of the Geological Society of America*, **116**, 718–728.

BERGER, L.A. 1993. *Salts and surface weathering features on alluvial fans in northern Chile*. PhD Thesis, University College London.

BERGER, L.A. & COOKE, R.U. 1997. The origin and distribution of salts on alluvial fans in the Atacama Desert, northern Chile. *Earth Surface Processes and Landforms*, **22**, 581–600.

BLAIR, T.C. 1999. Sedimentary processes and facies of the waterlaid Anvil Spring Canyon alluvial fan, Death Valley, California. *Sedimentology*, **46**, 913–940.

BLAIR, T.C. & MCPHERSON, J.G. 1994a. Alluvial fans and their natural distinction from rivers based on morphology, hydraulic processes, sedimentary processes, and facies assemblages. *Journal of Sedimentary Research*, **A64**, 451–490.

BLAIR, T.C. & MCPHERSON, J.G. 1994b. Alluvial fan processes and forms. *In*: ABRAHAMS, A.D. & PARSONS, A.J. (eds) *Geomorphology of Desert Environments*. Chapman & Hall, London, 354–402.

BORIC, R., DIAZ, F. & MAKSAEV, V. 1990. *Geología y Yacimientos Metaliferous de la Region de Antofagasta*. Servicio Nacional de Geología y Minera Chile, Bolletin, **40**.

BRÜGGEN, J. 1950. *Fundamentos de la Geología de Chile*. Instituto Geografico Militar, Santiago.

CHONG, G. 1988. The Cenozoic saline deposits of the Chilean Andes between 18° and 27°S. *In*: BAHLBURG, H., BREITKREUZ, C. & GEISE, P. (eds) *The Southern Central Andes*, Springer, Berlin, 137–151.

DEWEY, J.F. & LAMB, S.H. 1992. Active tectonics of the Andes. *Tectonophysics*, **205**, 79–96.

DUNAI, T.J., LÓPEZ. G.A.G. & JUEZ-LARRÉ, J. 2005. Oligocene–Miocene age of aridity in the Atacama Desert revealed by exposure dating of erosion sensitive landforms. *Geology*, **33**, 321–324.

FERRARIS, F. & DI BIASE, F. 1978. *Carta Geologica de Chile, Hoja Antofagasta*. Instituto de Investigaciones Geologicas de Chile, Carta, **30**.

GARCIA, F. 1967. *Geología del Norte Grande de Chile. Simposium sobre el Geosinclinal Andino*. Sociedad Geológica de Chile, **3**.

GONZÁLEZ, G. & CARRIZO, D. 2003. Segmentación, cinemática y cronologa relativa de la deformacióntardia de la Falla Salar del Carmen, Sistema de Falas de Aatacama (23°40′S), norte de Chile. *Revista Geológica de Chile*, **30**, 223–244.

GONZÁLEZ, G., CEMBRANO, J., CARRIZO, D., MACCI, A. & SCHNEIDER, H. 2003. The link between forearc tectonics and Pliocene-Quaternary deformation of the Coastal Cordillera, northern Chile. *Journal of South American Earth Sciences*, **16**, 321–342.

HARTLEY, A.J. & CHONG, G. 2002. A late Pliocene age for the Atacama Desert: Implications for the desertification of western South America. *Geology*, 30, 43–46.

HARTLEY, A.J. & JOLLEY, E.J. 1995. Tectonic implications of Late Cenozoic sedimentation from the Coastal Cordillera of northern Chile (22–24°S). *Journal of the Geological Society, London*, **152**, 51–63.

HARTLEY, A.J. & JOLLEY, E.J. 1999. Unusual coarse, clastic, wave-dominated shoreface deposits, Pliocene to Middle Pleistocene, northern Chile: implications for coastal facies analysis. *Journal of Sedimentary Research*, **69**, 105–114.

HARTLEY, A.J., CHONG, G., TURNER, P., MAY, G., KAPE, S.J. & JOLLEY, E.J. 2000. Development of a continental forearc: a Neogene example from the Central Andes, northern Chile. *Geology*, **28**, 331–334.

HARVEY, A.M. 2002. The role of base-level change in the dissection of alluvial fans: case studies from southeast Spain and Nevada. *Geomorphology*, **45**, 67–87.

HARVEY, A.M., SILVA, P.G., MATHER, A.E., GOY, J.L., STOKES, M. & ZAZO, C. 1999a. The impact of Quaternary sea-level and climatic change on coastal alluvial fans in the Cabo de Gata ranges, southeast Spain. *Geomorphology*, **28**, 1–22.

HARVEY, A.M., WIGAND, P.E. & WELLS, S.G. 1999b. Response of alluvial fan systems to the late Pleistocene to Holocene climatic transition: contrasts between the margins of pluvial Lakes Lahontan and Mojave, Nevada and California, USA. *Catena*, **36**, 255–281.

HOUSTON, J. & HARTLEY, A.J. 2003. The central Andean west-slope rainshadow and its potential contribution to the origin of hyper-aridity in the Atacama Desert. *International Journal of Climatology*, **23**, 1453–1464.

HERVÉ, M. 1987. Movimiento sinistral en el Cretácico Inferior de la Zona de Falla Atacama, al norte de Paposo (24°S), Chile. *Revista Geológica de Chile*, **31**, 37–42.

JOHNSON, A.M. 1984. Debris flow. *In:* BRUNSDEN, D. & PRIOR, D.B. (eds) *Slope Instability*. Wiley, New York, 257–361.

LAMB, S. & DAVIS, P. 2003. Cenozoic climate change as a possible cause for the rise of the Andes. *Nature*, **425**, 792–797.

LEONARD, E. & WEHMILLER, J.F. 1991. Geochronology of marine terraces at Caleta Michilla, northern Chile: implications for uplift rates. *Revista Geológica de Chile*, **18**, 81–86.

MORTIMER, C. & SARIC, N. 1972. Landform evolution in the coastal region of Tarapaca Province. *Revue de Geomorphologie Dynamique*, **21**, 162–170.

NARANJO, J.A. 1987. Interpretación de la actividad Cenozoica superior a lo largo de la zona de Falla Atacama, norte de Chile. *Revista Geológica de Chile*, **31**, 43–55.

OKADA, A. 1971. On the neotectonics of the Atacama Fault Zone region – Preliminary notes on Late Cenozoic faulting and geomorphic development of the Coast Range of northern Chile. *Bulletin of the Department of Geography, University of Tokyo*, **3**, 47–65.

ORTLIEB, L. 1995. *Late Quaternary Coastal Changes in Northern Chile*. Field Guide for International Geological Correlation Program Project 367, Late Quaternary Records of Coastal Change, Annual Meeting, Antofagasta, Chile, 1995.

ORTLIEB, L., ZAZO, C., GOY, J.L., HILLAIRE-MARCEL, C., GHALEB, B. & COURNOYER, B. 1996. Coastal deformation and sea-level changes in the northern Chile subduction area (23°S) during the last 330 ky. *Quaternary Science Reviews*, **15**, 819–831.

OWEN, J., NISHIIZUMI, K., SHARP, W., EWING, S. & AMUNDSEN, R. 2003. Investigations into the numerical ages of post-Miocene fluvial landforms in the Atacama Desert, Chile. *EOS, Transactions of the American Geophysical Union*, **84**, Fall Meeting, Supplement, Abstract T31C-0857.

Paskoff, R. 1978. Sur l'évolution geomorphologique du grand escarpement côtier du désert Chilien. *Geographique Physical Quaterniere*, **32**, 351–360.

PASKOFF, R. 1980. Late Cenozoic crustal movements and sea-level variations in the coastal area of northern Chile. *In*: MORNER, N.A. (ed.) *Earth Rheology, Isostasy and Eustasy*. Wiley, New York, 487–495.

PIERSON, T.C. 1980. Erosion and deposition by debris flows at Mt. Thomas, North Canterbury, New Zealand. *Earth Surface Processes*, **5**, 227–247.

RADTKE, U. 1985. Chronostratigraphie und Neotektonik mariner Terrassen in Nord- und Mittel Chile – erste ergibnisse. *Quatro Congresso Geologico Chileno, Antofagasta*, **4**, 436–458.

RADTKE, U. 1987. Palaeosea-levels and discrimination of the last and the penultimate interglacial fossiliferous deposits by absolute dating methods and geomorphological investigations. *Berliner Geographische Studien*, **25**, 313–342.

RITTER, J.B., MILLER, J.R. & HUSEK-WULFORST, J. 2000. Environmental controls on the evolution of alluvial fans in Buena Vista Valley, North Central Nevada, during late Quaternary time. *Geomorphology*, **36**. 63–87.

RUTLLANT, J. & ULRIKSEN, P. 1979. Boundary layer dynamics of the extremely arid northern Chile: the Antofagasta field experiment. *Boundary Layer Meteorology*, **17**, 45–55.

SCHEUBER, E. & ANDRIESSEN, P.A.M. 1990. The kinematic and geodynamic significance of the Atacama Fault Zone, northern Chile. *Journal of Structural Geology*, **12**, 243–257.

SCHEUBER, E. & GONZÁLEZ, G. 1999. Tectonics of the Jurassic–Early Cretaceous magmatic arc of the north Chilean Coastal Cordillera (22–26°S): A story of crustal deformation along a convergent plate boundary. *Tectonics*, **18**, 895–910.

SCHEUBER, E., BOGDANIC, T., JENSEN, A. & REUTTER, K.J. 1994. Tectonic development of the north Chilean Andes in relation to plate convergence and magmatism since the Jurassic. *In:* REUTTER, K.J., SCHEUBER, E. & WIGGER, P.J. (eds) *Tectonics of the Southern Central Andes*. Springer, Berlin, 121–140.

TREWARTHA, G.T. 1981. *The Earth's Problem Climates*. University of Wisconsin Press, Madison, WI.

UNEP 1992. *World Atlas of Desertification*. Edward Arnold, Sevenoaks, UK.

VARGAS, G., ORTLIEB, L. & RUTLLANT, J. 2000. Aluviones históricos en Antofagasta y su relación con eventos El Niño/Oscilación del Sur. *Revista Geológica de Chile*, **27**, 157–176.

VISERAS, C., CALVACHE, M.L., SORIA, J.M., & FERNANDEZ, J. 2003. Differential features of alluvial fans controlled by tectonic or eustatic accommodation space. Examples from the Betic Cordillera, Spain. *Geomorphology*, **50**, 181–202.

VUILLE, M. 1999. Atmospheric circulation over the Bolivian Altiplano during dry and wet periods and extreme phases of the Southern Oscillation. *International Journal of Climatology*, **19**, 1579–1600.

WATSON, A. 1988. Desert gypsum crusts as palaeoenvironmental indicators: A micropetrographic study of crusts from southern Tunisia and the central Namib Desert. *Journal of Arid Environments*, **15**, 19–42.

WATSON, A. 1989. Desert crusts and rock varnish. *In*: Thomas, D.S.G. (ed.) *Arid Zone Geomorphology*. Belhaven Press, London, 25–55.

WELLS, S.G. & HARVEY, A.M. 1987. Sedimentologic and geomorphic variations in storm generated alluvial fans, Howgill Fells, northwest England. *Bulletin of the Geological Society of America*, **98**, 182–198.

WHIPPLE, K.X. & TRAYLER, C.R. 1996. Tectonic control of fan size: The importance of spatially variable subsidence rates. *Basin Research*, **8**. 351–366.

Differential effects of base-level, tectonic setting and climatic change on Quaternary alluvial fans in the northern Great Basin, Nevada, USA

ADRIAN M. HARVEY

Department of Geography, University of Liverpool, PO Box 147, Liverpool L69 3BX, UK
(e-mail: amharvey@liv.ac.uk)

Abstract: Mountain-front alluvial fans in the northern Great Basin were affected by interactions between the tectonic setting, late Quaternary climatic changes and climatically induced base-level changes through fluctuations in pluvial lake levels. Four fan groups were studied on the margins of and near pluvial Lake Lahontan with varying geology and tectonic settings. All fans were affected by climatically led variations in sediment supply, but only those on steep mountain fronts adjacent to deep lakes were affected by base-level changes during lake desiccation.

Relationships between fan segments and dated lake shorelines, augmented by soil and desert pavement characteristics, have enabled fan segments older and younger than the lake highstands to be identified. Major periods of fan aggradation occurred prior to the last glacial maximum and during the Holocene, with little or no fan deposition occurring during and after the last glacial maximum, at the time of high lake levels. The interactions between tectonics, climate and base-level change have produced distinctive fan geometric relationships between older and younger fan segments, also expressed in the morphometric properties of the fans. Tectonics primarily influence the fan setting, particularly the accommodation space, and interact with sediment supply rates partly related to source-area geology. The climatic signal is present in all four groups of fans, but is modified locally by a base-level signal only where deep pluvial lakes abutted to relatively high levels on fans on relatively steep mountain fronts.

Many studies have demonstrated how the geomorphology of alluvial fans reflects interactions between three sets of factors (Bull 1977; Blair & McPherson 1994; Harvey 1997, 2002*a*): (i) those influencing fan context (tectonics, gross topography, accommodation space); (ii) those influencing water and sediment delivery to the fan, and the processes operating on the fan (basin geology and relief, climate); and (iii) those influencing the relationship between the fan and adjacent environments, of which base level is the most important.

Different combinations of these factors have been used to explain both spatial and temporal variations in fan geomorphology. For example, spatial differences between fans or groups of fans have been attributed to tectonics and gross topography (Blair & McPherson 1994) and accommodation space (Viseras *et al.* 2003), itself often related to the tectonic context. In addition, spatial differences have been attributed to rock resistance (Bull 1962; Hooke 1968, Hooke & Rohrer 1977; Lecce 1991). Such factors have also been cited in explanations of the evolution of fans over time, particularly tectonics (Calvache *et al.* 1997), climatic change (Bull 1991; Harvey 2003) and base-level change (Harvey 2002*b*). Indeed, the interactions between these factors has become one of the main themes of research into the geomorphology of alluvial fans (Bowman 1988; Frostick & Reid 1989; Ritter *et al.* 1995).

Within the northern Great Basin, Nevada, USA, range-front fans occur in a variety of tectonic settings. They are supplied with sediment from catchments of a range of geology and relief characteristics. The present-day climate is continental arid, with precipitation at low elevations of less than 200 mm, and July and January monthly mean temperatures of 20 and 0 °C, respectively (Houghton *et al.* 1975). Throughout the area the fans were subject to geomorphic change in response to variations in water and sediment supply controlled by late Quaternary climatic changes (Harvey *et al.* 1999*b*; Ritter *et al.* 2000; Harvey 2002*a*). However, different groups of fans were affected by different base-level conditions provided by the fluctuations in the levels of late Pleistocene pluvial lakes within the basins (Harvey 2002*b*).

This paper examines the geomorphology of four sets of mountain-front alluvial fans, representative of conditions in the northern Great Basin. Within the context of differing catchment geology and relief properties, the interactions between the response to late Quaternary climatic change and differential base-level influences are identified, and the effects on fan morphology are assessed.

Study area

Four groups of fans on the margins of and near pluvial Lake Lahontan were selected for study (Fig 1). The

From: HARVEY, A.M., MATHER, A.E. & STOKES, M. (eds) 2005. *Alluvial Fans: Geomorphology, Sedimentology, Dynamics.* Geological Society, London, Special Publications, **251**, 117–131. 0305–8719/05/$15 © The Geological Society of London 2005.

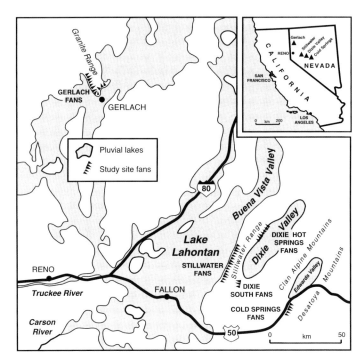

Fig. 1. Location map of the studied fans in the northern Great Basin, Nevada (locations of pluvial lakes adapted from Mifflin & Wheat 1979).

groups necessarily included a range of catchment geologies, but the primary reason for selection was the different relationship between the fan groups and the pluvial lake margins, resulting in different base-level conditions (Table 1). Cold Springs fans (for location see Fig. 1) toe out onto the low-gradient valley floor of the Edwards Creek Basin, and are detached from the effects of local base-level change provided by the late Quaternary fluctuations of pluvial Lake Edwards. Two subgroups of fans were studied in the Dixie Valley (for location see Fig. 1), one in the southern part of the valley, too far from the centre of the basin to have been affected by fluctuations of pluvial Lake Dixie, and another group in the central part of the basin (Dixie Hot Springs), directly affected by lake-level fluctuations. In some locations maximum lake levels reached the mountain front, elsewhere they reached only onto the piedmont. A large group of fans was studied on the Stillwater mountain front (for location see Fig. 1) adjacent to the SE shore of pluvial Lake Lahontan. Maximum lake levels reached the mountain front in some parts; elsewhere they reached the piedmont. The final group of fans are at Gerlach (for location see Fig. 1), adjacent to the narrower Black Rock–Smoke Creek Desert arm of Lake Lahontan. Lake levels drowned part of the mountain front here; elsewhere they reached the piedmont.

The basis for establishing the fan sequence lies with the dated sequence of pluvial lake shorelines (Table 2). Lake Lahontan reached its maximum level of 1340 m at about 13 ka BP (during the Sehoo stage of Morrison 1991) (Mifflin & Wheat 1979; Davis 1982; Grayson 1993; Adams & Wesnousky 1998, 1999), forming distinct upper shorelines. Desiccation took place during the early Holocene with a possible late lake stage to 1220 m (Morrison 1991). A similar late Quaternary chronology is assumed for the smaller lakes, Lake Dixie and Lake Edwards (Table 2) (Mifflin & Wheat 1979).

The lake shorelines provide timelines for establishing the broad chronology of alluvial-fan sedimentation. Major periods of fan aggradation within the northern Great Basin occurred prior to the last glacial maximum and during the Holocene, with little or no fan deposition at the time of high lake levels during the latest Pleistocene (Harvey et al. 1999b). This is in contrast with the Zzyzx fans in the Mojave Desert further south (Harvey & Wells 2003), where the Pleistocene–Holocene climatic transition was a period of major geomorphic activity. At that time the lower hillslopes and the fan surfaces in the northern Great Basin were covered by juniper woodland rather than desert scrub as in the Mojave, and it is unlikely that the effects of incursions of tropical storms into the Mojave would have

Table 1. *Study areas: geology and base-level conditions (for locations see Fig. 1). Geological information from Willden & Speed (1974)*

Stillwater fans (see Fig. 2):

Catchment geologies:	Late Tertiary basalts (Table Mountain fans) Early Tertiary rhyolites (central fans) Mesozoic low-grade metasedimentary rocks (northern fans)
Base-level control:	Lake Lahontan (main part of lake)

Cold Springs fans (see Fig. 2):

Catchment geologies:	Early Tertiary rhyolites and rhyodacites
Base-level:	Stable – above the elevation of Lake Edwards

Dixie Valley fans (see Fig. 4):

Catchment geologies:	Early-mid Tertiary rhyolites and tuffs (plus small areas of Tertiary granites – within the catchments of the southern fan group)
Base-level control:	Lake Dixie (only affecting the Dixie Hot Springs fan group: the southern group of fans are above the elevation of Lake Dixie, and base levels have been stable)

Gerlach fans (see Fig. 5):

Catchment geology:	Granite
Base-level control:	Lake Lahontan (northern arm)

Table 2. *Lake Shoreline data (after Mifflin & Wheat 1979; Davis 1982; Bell & Katzer 1987; Morrison 1991; Grayson 1993; Adams & Wesnousky 1998, 1999)*

Lake Lahontan (Stillwater and Gerlach fans)
22–13 ka: rise to maximum lake level (1340 m)
13–10.5 ka: rapid fall in lake level to 1235 m
Early Holocene desiccation, possibly late lake stage to 1220 m

Lake Dixie (Dixie Valley fans)
Chronology assumed to be similar to Lake Lahontan
(Late Pleistocene maximum – 1110 m – **Influences Dixie Hot Springs area, but no influence on southern Dixie Valley fans**)

Lake Edwards (Cold Springs fans)
Chronology assumed to be similar to Lake Lahontan
(Late Pleistocene maximum – 1625 m – **but no influence on base level for Cold Springs fans**)

---

penetrated as far north as northern Nevada (Harvey *et al.* 1999*b*).

The older fan segments are cut by the high shorelines and clearly predate the last pluvial lake highstand. They include several stages set into each other and, on the basis of the maturity of the soils on even the youngest of these segments, must predate the last glacial maximum by a considerable period. Various estimates put their probable age as at least 30 ka (Chadwick *et al.* 1984; Bell & Katzer 1987), and were probably formed over periods dating back to the last interglacial in excess of 100 ka, following the previous lake highstand (the Eetza stage of Morrison 1991).

The younger fan segments truncate the lowest shorelines and are clearly Holocene in age. In the field several younger fan segments can be identified and have been ascribed to the mid and late Holocene (Ritter *et al.* 2000).

Methods

Fan surfaces were mapped, initially from air photographs then in the field, and were assigned to age groups on the basis of morpho-stratigraphic relationships with the oldest and youngest dated lake shorelines (see below). Mapping was first undertaken on the Stillwater fans, where clear-cut relationships exist between fan surfaces and the lake shorelines. In that area fan-surface differentiation and local correlation was achieved through establishing soil and desert pavement properties of the fan surfaces older than and younger than the shorelines, using a multiparameter approach based on that of McFadden *et al.* (1989). The details are given elsewhere (Harvey *et al.* 1999*b*; Harvey 2002*b*).

On many fans it was possible in the field to identify several fan segments older than the highest shoreline, and others younger than the youngest shoreline, and in some cases small fan segments whose ages lie between those of the oldest and the youngest shorelines (Harvey *et al.* 1999*b*). For the purposes of this paper, for comparisons to be made between several groups of fans, it is sufficient to group the fan segments into two: those older and those younger than the shorelines.

On the Stillwater fans these two groups can be differentiated on the basis of post-depositional surface modification by soil and desert pavement development. Where the older surfaces are not buried by younger aeolian silts they show mature desert pavements (see Al-Farraj & Harvey 2000) of interlocked angular rock fragments, with a dark coating of desert varnish (see Harvey & Wells 2003). The soils show mature profiles characterized by a thick Av horizon (McFadden *et al.* 1998), over a well-developed Bt horizon (characteristic Munsell colour: 7.5YR 4–5/4–6), underlain by a stage II carbonate Bk horizon (terminology after Gile *et al.* 1966, modified by Machette 1985).

Two sets of surfaces younger than the shorelines can be identified, a set post-dating recession of the lake, characterized by soil profile development much less mature than on the older surfaces, and a youngest set on which little or no soil development is apparent. On the former, a weakly developed pavement may be present, with weak pale varnish. The soils show an Av horizon over a thin B horizon (characteristic Munsell colour 10YR 6–7/3–4) and only stage I carbonate accumulation.

Fan segments were then mapped on the other fans on the basis of relationships with lake shorelines, where present, and by comparison with the soils on the Stillwater fans. From the field maps, the fans were classified on the basis of the spatial patterns of erosion and deposition in relation to the sequence of base-level change. Several fan geometric styles were recognized. In order to assess relative influences of contextual factors and base-level controlled geometric style, a conventional morphometric analysis was carried out. For that analysis data were derived on fan and drainage basin properties, from a total of 106 fans (Stillwaters, 18 fans; Cold Springs, 48 fans; Dixie Valley, 20 fans; Gerlach, 20 fans). Fan gradients, representative of the proximal half of each fan, were measured in the field for the smaller fans, and derived from topographic maps for the largest fans (sample gradients were then checked in the field). Fan areas were derived directly from the field maps. Drainage basin properties (drainage basin area, relief and mean slope) were derived from the published topographic maps (scale 1:24 000). The analysis was a conventional regression analysis of the influence of drainage basin properties on fan properties, using residuals from the regressions to assess the relative influence of contextual and base-level controls.

Fan morphology

Stillwater fans (see Fig. 2)

These fans have been described previously (Harvey *et al.* 1999*b*; Harvey 2002*b*), so only a brief description is given here. The fans occupy the western

mountain front of the Stillwater Range, on a relatively steep shoreface of pluvial Lake Lahontan facing the widest and deepest section of the lake, where wave action would have been high. The shorelines, both erosional and depositional, are exceptionally well developed, in some places cutting the mountain front, elsewhere cutting older fans on the piedmont.

The mountain front is fault controlled, but apart from a minor fault trace across the piedmont and a small fault scarp near the toe of the central–northern fans there is little obvious sign of Holocene active tectonics. The fan segments have been mapped and, in addition to multistage older and younger fan segments, small intermediate segments dating from the time of the lake can be recognized (Harvey *et al.* 1999*b*; Harvey 2002*b*).

The fans occur in several groups (Fig. 2; Table 1). In the south, the Table Mountain fans are debris-flow-dominated fans fed by late Tertiary basalt source areas. Here the shoreline is at the mountain front and older fan segments are restricted to fan proximal areas within the mountain catchments. Following lake desiccation, incision of the fans took place by fanhead trenching and the younger segments prograde distally to form a generally telescopic fan structure (Bowman 1978).

The central fans are large fans with exposures in the constituent sediments suggesting sheetflood deposition. They occupy a less steep embayment within the mountain front, fed by early Tertiary rhyolites. The shoreline zone cuts across the piedmont, truncating the older fan segments in midfan (Fig. 3a). Substantial older fan remnants remain above lake level. Deep post-lake fanhead trenches have cut into them from which the younger fan segments prograde distally.

The northern fans occupy a steep mountain front on Mesozoic low-grade metasedimentary rocks. The deposits suggest sheetflood dominance, but also include debris flows in the smaller fans. The shoreline again approaches the mountain front with preservation of only small, truncated older fan remnants near some fan apices. There are deep fanhead trenches from which there is extensive distal progradation of the younger fan segments.

Cold Springs fans (see Fig. 2)

These fans have also been described previously (Harvey 2002*b*), so again only a short description is given here. They are sourced by early Tertiary rhyolite and rhyodacite terrain within the Desatoya Mountains (Table 2). They occur at a faulted mountain front, but there is no evidence of Holocene tectonic activity affecting late Quaternary fan evolution. Local base levels have been stable; the fans toe out onto the low-gradient floor of Edwards Creek Valley

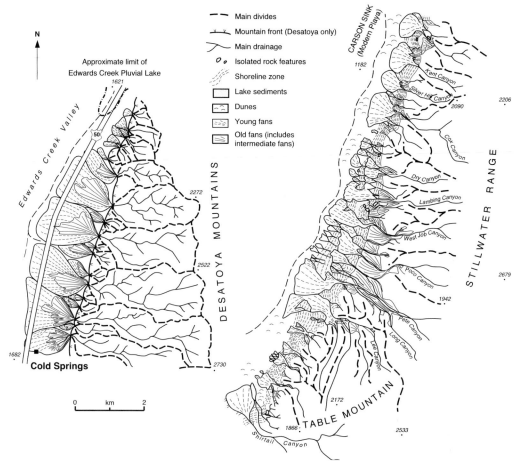

Fig. 2. Summary maps of (**a**) the Stillwater fans and (**b**) the Cold Springs fans (modified after Harvey 2002*a*). Spot heights in m.

well above any zone that might have been affected by fluctuations of the level of pluvial Lake Edwards. There is a range of fan types, including: (a) large, low-angle, unconfined, sheetflood-dominated master fans fed by the major drainages; (b) smaller debris-flow fans generally confined between the master fans, but including some small fans at the northern end of the study area that are fed by small lower relief catchments; and (c) a large number of small hillfoot debris cones. The larger fans show older fan surfaces in proximal locations, cut by shallow fanhead trenches from which there are extensive prograding distal younger fan surfaces (Fig. 3b).

Dixie Valley fans (see Fig. 4)

These fans are fed by terrain located dominantly in highly erodible early–mid Tertiary rhyolites and

tuffs of the eastern Stillwater Range (Table 2). This is a complex tectonically active mountain front with numerous Holocene fault scarps cutting both younger and older fan segments. In the southern part of the valley the faults occur in two groups, one along the mountain front, including the trace of the fault produced by the 1954 earthquake (Bell & Katzer 1987), and the other across the piedmont. In the central part of the valley these faults extend north to become the mountain-front faults.

Variations in late Quaternary base levels were provided by fluctuations in the levels of pluvial Lake Dixie, affecting fans in the northern part of the study area, but in the southern part of the area the fans toe out onto the low-gradient valley floor well above maximum lake level. Hence, in that area late Quaternary base-level changes have been non-effective.

Two sets of fans have been studied, one in the southern part of the valley and one at Dixie Hot

Fig. 3. Photographs of example fan types. (**a**) Central Stillwaters fans. Note the high shoreline of pluvial Lake Lahontan, as a beach ridge in the foreground and as a low cliff in the background; dissected older fans to the left, prograding younger fans to the right. (**b**) Cold Springs fans. Older fan segments at the fan apices, younger segments prograde towards the stable base level of the low-gradient valley floor. (**c**) Cottonwood Canyon fan, Dixie Hot Springs. Note the fault trace truncating the older fan segments; younger fan segments in the foreground and to the right. (**d**) Gerlach west-side fans. Note the shoreline zone on the mountainside. Young fans issue from canyons cut into the shoreline.

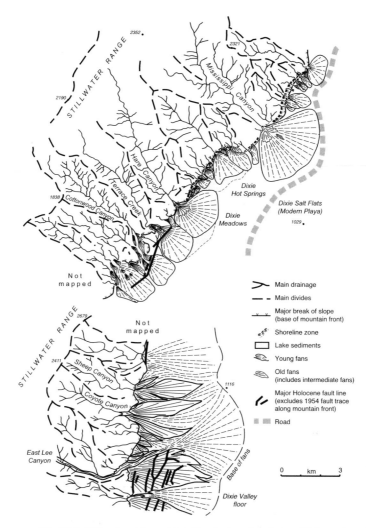

Fig. 4. Summary maps of the Dixie Valley fans (above, Dixie Hot Springs fans; below, Dixie southern fans). Spot heights in m.

Springs in the central part of the valley (Fig. 4). These fans have not previously been described, but Bell & Katzer (1987) described those in the intervening IXL Canyon area, and Chadwick *et al.* (1984) described Terrace Creek fan, one of those at Dixie Hot Springs (Fig. 4). The southern group are large sheetflood-dominated fans with extensive proximal older fan segments. These are cut by fanhead trenches from which younger fan segments prograde distally. The fans toe out to stable base levels on the valley floor. The fans are cut by Holocene faults, both at the mountain front, including the 1954 fault trace, and across the midfan area.

The Dixie Hot Springs fan group shows a range of relationships with base levels provided by pluvial Lake Dixie. At the south end are several large sheetflood-dominated fans that preserve older proximal fan segments truncated by faults near the mountain front (Fig. 3c). On these fans maximum shoreline levels were in the midfan but show no evidence of stimulating incision. Indeed, all the shorelines are buried by extensive younger fan segments both in the midfan and distally. North of the Hare Canyon fan the shoreline reached the mountain front, and there a series of small debris-flow-dominated fans occurs on which there is no preservation of older fan segments. The complete fan surfaces comprise young fan segments. To the north are several fans that preserve evidence of the maximum shoreline in the midfan, with some preservation of proximal older segments. There is evidence of only limited post-lake incision, but only in the proximal-fan locations, and extensive younger fan segments blanket large portions of these fans.

The Dixie Hot Springs fans, as a whole, show very limited post-lake dissection, but very extensive younger fan segments. Many of the fans are stacked aggrading fans rather than telescopic fans. This is also true of those in the IXL Canyon area (Bell & Katzer 1987), where younger fan sediments dominate the proximal-fan zones. This fan style suggests the interaction of three controlling factors: (a) tectonic control of depression in front of the mountain front creating proximal accommodation space; (b) a high rate of sediment input from the highly erodible volcanic rocks; and (c) no base-level effects.

Gerlach fans (see Fig. 5)

These fans occur on both sides of the peninsula where the Granite Range extends south into the Black Rock and Smoke Creek Desert segments of the northern arm on pluvial Lake Lahontan. Catchment geology is almost entirely early Tertiary granite, which weathers to easily erodible gruss, but late Tertiary basalts, faulted against the granite, form spurs on the western mountain front (Table 2).

There is no obvious evidence of Holocene faulting within the fans.

Late Quaternary base-level control was provided by the northern arm of pluvial Lake Lahontan, narrower than the main body of the lake, but, nevertheless, providing a considerable fetch to the west side of the peninsula. Maximum lake levels drowned the mountain fronts at the southern end of the peninsula and throughout the eastern side, but to the NW of the study area the shorelines cross the piedmont.

There has been no previously published work on these fans. Fans on both mountain fronts have been studied. On those on the eastern mountain front and on those on the southern part of the western mountain front, where lake levels rose above the mountain fronts, there is no preservation of the older fan segments, and the lower hillslopes preserve only shoreline features (Fig. 3d) and lake sediments. The fans comprise only younger fan segments, deposited after recession of the lake (Fig. 5).

To the NW the mountain front rises above lake levels, and shoreline features can be traced across the piedmont. Large fans preserve older fan segments above lake level, which are cut by deep proximal fanhead trenches from which younger fan segments prograde in the midfan and distally. The fanhead trenches do not appear to be related to base-

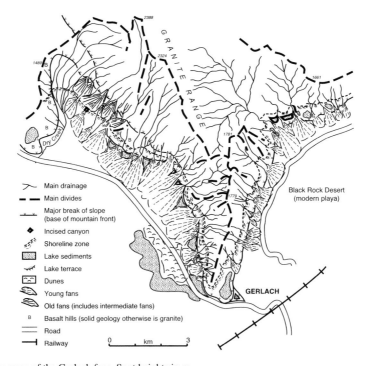

Fig. 5. Summary maps of the Gerlach fans. Spot heights in m.

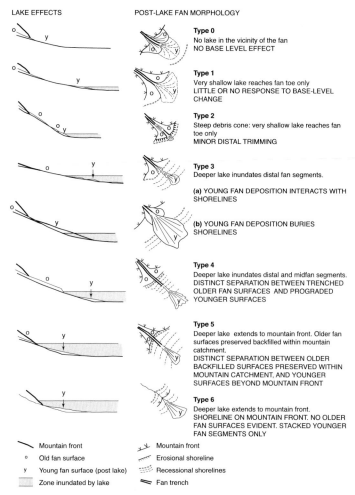

Fig. 6. Schematic classification of fan morphologies in relation to the influence of base level, following desiccation of pluvial lakes. For an explanation see the text.

level change, as they are detached from the shoreline zone and a secondary incision occurs around the shoreline zone itself (Fig. 5).

Fan morphologies – relation to base level

The degree of base-level control on the evolution of the fans appears to vary amongst the fan groups. Its influence appears to relate to: (a) the elevation and position of the highest shoreline in relation to the mountain front; (b) the gradient of the mountain front zone; (c) the depth of lake and the gradient of the zone inundated, i.e. the absolute relief generated by base-level fall; (d) the effectiveness of the coastal processes, dependent on the fetch and the wave

energy; and (e) the sediment supply through the fan environment, both during the lake highstand and afterwards.

On the basis of these factors, and taking into account observations on these fans, fans at other margins of Lake Lahontan (Ritter *et al.* 2000) and at the margins of other pluvial lakes in the American South-west notably the Zzyzx fans on the margins of Lake Mojave, California (Harvey *et al.* 1999*b*; Harvey & Wells 2003), a general model of seven broad types of fan geometry related to base-level control by falling pluvial lake levels can be recognized (Fig. 6). This general model would be applicable in situations typical of the Great Basin or Mojave Deserts, where fan sedimentation during the lake highstands is low.

- Type 0 (Fig. 6) represents fans with stable base levels. Fan geometry would reflect the context (tectonic control?) and the sequence of sediment supply (climatic control), generally leading to a mildly telescopic fan style, though a stacked aggrading style might result under the appropriate tectonic conditions of increasing accommodation space and under high sediment supply rates. Examples of type 0 are Cold Springs fans and Dixie Valley southern fans.
- Types 1 and 2 would occur in relation to a very shallow lake, whose shorelines just reach the fan toe. On larger fans the gradients would be insufficient to trigger incision as base level fell. Fan geometry would be dominated by context and climatic control. Only on smaller steeper fans might the gradients be sufficient to trigger incision, even then this would be restricted to the fan toes. Neither type have been seen in these study areas, but are typical of the Zzyzx fans adjacent to pluvial Lake Mojave, California (Harvey & Wells 2003), a shallow pluvial lake that just inundated the most distal fan zones.
- Type 3 would occur where a deeper lake inundates distal and midfan segments, but wave energy is too small to cause toe trimming (see Leeder and Mack 2001) by coastal erosion (see Harvey et al. 1999a; Harvey 2002b), or the gradients are insufficient to trigger incision as base levels fall. Two types of fan geometry may result. Under conditions of low sediment supply there may be interactions between shorelines and fan deposits, as described by Ritter et al. (2000) on some fans in the Buena Vista Valley, a shallow sheltered arm of Lake Lahontan (for the location see Fig. 1). Under conditions of higher sediment supply depositional shoreline features may be buried by aggrading fan sediments, as on some of the larger fans at Dixie Hot Springs.
- Type 4 would occur where a deeper lake inundates steeper distal and midfan segments, but where coastal processes are effective enough to cause erosion of the older fan segments and/or midfan, and shoreface gradients are sufficient to trigger incision, as base level falls. The resulting fan geometry would produce a distinct separation between trenched older proximal fan surfaces and prograded telescopic younger surfaces. Examples include the Stillwaters central fans: Gerlach NW fans, and the northernmost fans studied in Dixie Valley.
- Type 5 relates to where a deeper lake extends to the mountain front inundating all fan surfaces beyond the mountain front. Again, there would be a distinct separation between the older fan segments, preserved only as truncated backfilled remnants within the mountain catchments, and the younger surfaces extending beyond the

mountain front. Examples are the Stillwater, Table Mountain fans and some of the northern Stillwater fans.
- Type 6 would occur where a deep lake extends to the mountain front, inundating all older fan surfaces and the lower part of the mountain catchment. In these circumstances the highest shoreline would be higher than the mountain front, no older fan segments would be preserved and younger fan segments would tend to produce a stacked geometry burying any distal remnants of the older fan surfaces, as on small fans at Dixie Hot Springs and all the small fans at Gerlach (except the NW fans).

Fan morphometry

In order to examine further the relative influences of contextual and base-level factors on gross fan morphology, a conventional analysis of fan morphometry (Harvey 1997) was carried out relating fan area and gradient to drainage area, followed by an analysis of the residuals from the standard regression equations.

The basic relationships can be expressed as follows:

$$F = pA^q \tag{1}$$

$$G = aA^b \tag{2}$$

where A is the drainage area, F is the fan area (both in km^2) and G is fan gradient. Both regressions for all four study areas and for the all fans case are significant at the 99.9% level, and all give values for p, q, a and b within the ranges quoted in the literature (Harvey 1997). The results suggest that, overall, there is little gross difference between the four study areas (Table 3; Fig. 7). Although there is a suggestion that the Gerlach fans may tend to be larger per drainage area than the others, this and other apparent differences among the groups may simply be the result of different data spreads between the four groups. It appears that the all-fans case provides satisfactory regression relationships both for fan area and fan gradient to describe the general morphometric relationships for all fans within the northern Great Basin.

One approach that can be used to tease out the relative importance of different controlling factors is multiple rather than bivariate regression. Harvey (1987, 1992) demonstrated that incorporation of a measure of basin relief or slope, together with drainage area, may significantly improve the explanation level for fan gradient. For the all-fans case here such regressions were calculated taking into account mean basin slope (S, basin relief/basin length) for both fan area and fan gradient.

Table 3. Regression statistics*

| Fan group | For the fan-area regressions, $F = pA^q$ (equation 1) | | | | | For the fan-gradient regressions, $G = aA^b$ (equation 2) | | | | |
	n	p	q	R	SE (log units)	n	a	b	R	SE (log units)
Stillwater fans	18	0.406	0.684	0.920	0.206	18	0.091	−0.139	−0.780	0.079
Cold Springs fans	48	0.330	0.836	0.939	0.251	48	0.102	−0.120	−0.686	0.104
Dixie fans	20	0.356	0.968	0.944	0.305	20	0.103	−0.173	−0.801	0.117
Gerlach fans	20	0.535	0.963	0.946	0.287	20	0.096	−0.075	−0.490	0.115
All fans	106	0.376	0.861	0.956	0.264	106	0.095	−0.135	−0.782	0.107

*A is the drainage area, F is the fan area (km²), G is the fan gradient.

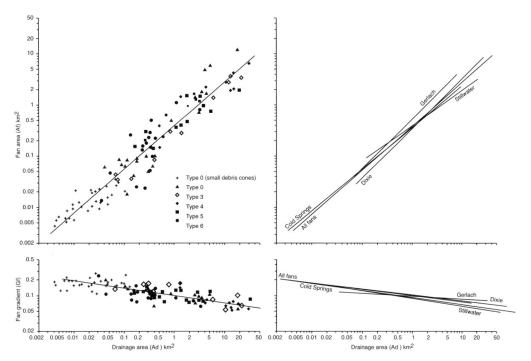

Fig. 7. Fan-regression relationships. Above, fan-area to drainage-area regressions; below, fan-gradient to drainage-area regressions, for statistical data see Table 3. Graphs to the right illustrate the regression relationships for each fan group.

For the fan-area regressions:

$$F = 0.376 A^{0.861}$$
$$(R = 0.956, \text{ standard error (SE)} = 0.262) \quad (3)$$

$$F = 0.389 A^{0.864} S^{0.026}$$
$$(R = 0.956, SE = 0.264). \quad (4)$$

In this case equation (4), the multiple regression, shows no improvement on equation (3), the simple bivariate regression, suggesting that either basin relief or gradient has little effect on fan area or what effect it has is hidden by any relationship between drainage area and fan area.

For the fan-gradient regressions:

$$G = 0.095 A^{-0.135}$$
$$(R = 0.782, SE = 0.107) \quad (5)$$

$$G = 0.158 A^{-0.091} S^{0.373}$$
$$(R = 0.820, SE = 0.099). \quad (6)$$

In this case equation (6), the multiple regression, does show a significant improvement in both a higher level of correlation and a reduced standard error over equation (5), the simple bivariate regression. This suggests that basin relief or slope characteristics do influence the delivery of sediment in terms of volume or calibre, which in turn influences the depositional gradient of the fan surface.

Another approach that has been used in the interpretation of alluvial-fan morphometric regression analyses has been the analysis of residuals from the two primary regressions (Silva *et al.* 1992; Harvey *et al.* 1999a; Harvey 2002b). These previous studies demonstrated that regression residuals tend to pick up the influence of contextual factors, including tectonics (Silva *et al.* 1992), fan confinement and base-level influences (Harvey *et al.* 1999a; Harvey 2002b). In this study, residuals from the two primary all-fans regressions have been plotted (Fig. 8), considered both by fan group and by fan type. It must be remembered that residuals are relative and not absolute, and that they are sensitive to the assumptions of regression linearity as well as to any inherent errors of measurement. Nevertheless, they present interesting patterns.

For the fan groups (Fig. 8a), as had been previously noted (Harvey 2002b), the Table Mountain fans are relatively small and steep compared with the other Stillwater fans. This might relate to their distinctive sedimentology rather than to any contextual or base-level characteristic. For the Cold Springs fans there is a wide scatter, but the master fans form a distinct and tight cluster, of relatively large fans of average relative gradient. The Dixie Valley fans tend to be relatively steep, and there is a clear separation between the relatively large southern fans and the relatively small Dixie Hot Springs fans. This may be a reflection of fan style with contrasts between the telescopic southern fans and the stacked style characteristic of most of the Dixie Hot Springs fans. For the Gerlach fans there is a clear distinction between the relatively large fans on the western mountain front and the relatively small fans on the east. This difference appears to be irrespective of fan type and therefore probably relates to accommodation space.

When the residuals are considered by fan type (Fig. 8b), both contextual and base-level influences are apparent. Type 0 master fans, including both Cold Springs master fans and Dixie southern fans, tend to be relatively large. Residuals for the secondary fans and the debris cones at Cold Springs show little pattern. Types 3 and 4 tend to show separate plotting positions, with type 3 fans (Dixie Hot Springs) relatively smaller and steeper than type 4 fans (Stillwaters, Dixie, Gerlach). This may be partly due to the stacked nature of the Dixie Hot Springs fans (context and sediment supply), but may also be related to the extreme progradation characteristic of type 4 fans. With the exception of the Stillwater Table Mountain fans (described above),

type 5 fans all tend to be relatively large, again possibly in response to the progradation associated with base-level induced dissection. Type 6 are all Holocene fans, which developed after lake recession, and therefore have not really been affected by base-level fall during their history. The plotting positions cover a wide range, but one which is clearly resolved into three distinct clusters, Dixie Hot Springs, Gerlach east and Gerlach west. These differences probably reflect contextual factors, particularly tectonic setting, influencing accommodation space and fan style as stacked or mildly prograding fans.

Discussion and Conclusions

The morphology of the late Quaternary alluvial fans of the northern Great Basin, represented by both the style and the morphometric properties of the fan groups studied here, reflects the interactions of several sets of factors: (a) contextual factors – gross topography, tectonics and accommodation space; (b) climatic, geological or relief-led variations in sediment supply and erosion, transport and depositional processes; and (c) base-level conditions.

These results throw some light on the particular roles of tectonics, climate and base level, a discussion central to much of the recent alluvial fan literature (e.g. Bowman 1988; Frostick & Reid 1989; Blair & McPherson 1994; Ritter *et al.* 1995; Harvey 2002a). Climate appears to be the primary control over the fan sequences, with periods of excess sediment supply leading to fan sedimentation through aggradation or progradation. There is a regionally consistent temporal sequence of fan sedimentation throughout the northern Great Basin (Chadwick *et al.* 1984; Bell & Katzer 1987; Harvey *et al.* 1999a; Ritter *et al.* 2000; Harvey 2002a), with pulses of sedimentation associated with periods of more arid climates during the Holocene and prior to the last glacial maximum, perhaps extending back to the last interglacial. Fan stability and/or fanhead entrenchment occurred during the intervening periods of lesser sediment supply. The effects of this climatic signal are modified spatially on particular groups of fans in response to catchment geology, fan context (tectonics, accommodation space) and base-level conditions.

On the southern Dixie Valley and Cold Springs fans Holocene base levels were stable and, given the available accommodation space, fan morphology is a simple product of the climatic signal (fan type 0).

At Dixie Hot Springs a combination of erodible geology and active tectonics produces a distinctive stacked Holocene fan style, and, even where Lake Dixie shorelines did cut the fans in midfan, falling

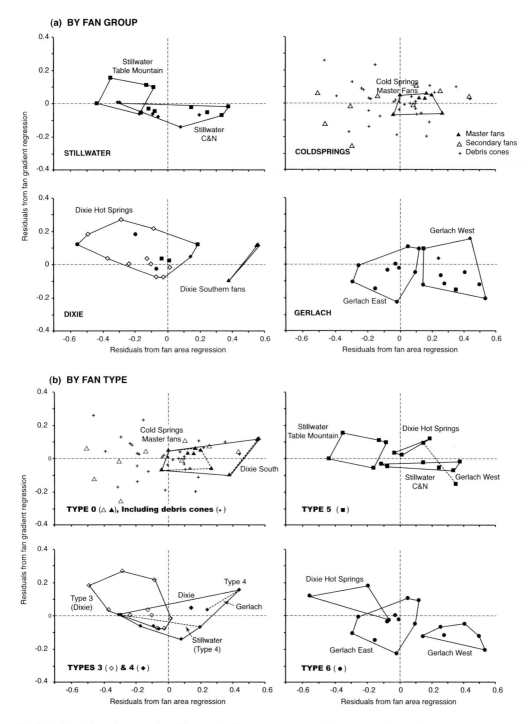

Fig. 8. Residuals from the regressions illustrated in Figure 7. (**a**) Grouped by fan group: fan type indicated by symbol (see below). (**b**) Grouped by fan type (see text and Fig. 6).

base levels had only a passive effect on Holocene fan evolution by creating more accommodation space. Highly erodible geology has led to major Holocene sedimentation and burial of any midfan shoreline features (fan type 3).

On the smallest fans at Dixie Hot Springs and on all fans at Gerlach except the most NW fans, the fans are of Holocene age only, deposited in accommodation space created by the lake-level fall. Hence, base-level change had only a passive role (fan type 6).

It is only on the Stillwater mountain front, and to a lesser extent the NW fans at Gerlach, that base-level change had a direct influence on fan evolution by causing erosional modification of the older fan segments and stimulation of dissection as lake levels fell (fan types 4 and 5).

These findings accord with those from other fans at the margins of pluvial lakes in the American West, where base-level change is seen in many cases as no more than a passive influence on fan evolution, for example in the Buena Vista Valley, a shallow arm of Lake Lahontan (Fig. 1) (see Ritter *et al.* 2000) (fan type 3a), or on the Zzyzx fans on the margins of Lake Mojave (see Harvey *et al.* 1999*b*; Harvey & Wells 2003) (fan types 1 and 2). Similarly, even on fans on the margins of larger and deeper lakes such as Lake Lisan, Israel, the effects of base-level change may be of secondary importance to those resulting from the direct influence of climatic change (Bowman 1988; Frostick & Reid 1989).

The expression of the relative influence of base-level and other factors on fan morphology may be expressed by fan style, especially by the spatial relationships between depositional zones of successive ages. Morphometric analysis, especially through consideration of the residuals from the two primary fan morphometric regression equations, does bring out the influence of base level, but that influence may be masked by other factors, notably by accommodation space.

It appears that within the context of tectonics and gross topography, the primary control on sequences of fan evolution is sediment supply. Variations in sediment supply may be directly climatically controlled. The morphological effects may be modified by base-level change, which itself may be a secondary response to climatic change. Base-level change has a direct influence on fan evolution at the margins of pluvial lakes, only, as on the Stillwater mountain front, where shoreline processes are strong and where the offshore gradient exposed by falling lake levels is high.

The field work for this study was undertaken while the author based at the Desert Research Institute, Reno, with the support of a Fulbright Scholarship. I acknowledge the help of the graphics section of the Department of Geography, University of Liverpool, particularly S. Mather in the preparation of the illustrations. I thank D. Bowman and A.M. Hartley for their constructive reviews.

References

ADAMS, K.D. & WESNOUSKY, S.G. 1998. Shoreline processes and the age of the Lake Lahontan highstand in the Jessup embayment, Nevada. *Bulletin of the Geological Society of America*, **110**, 1318–1332.

ADAMS, K.D. & WESNOUSKY, S.G. 1999. The Lake Lahontan highstand: age, surficial characteristics, soil development, and regional shoreline correlation. *Geomorphology*, **30**, 257–392.

AL-FARRAJ, A. & HARVEY, A.M. 2000. Desert pavement characteristics on wadi terrace and alluvial fan surfaces: Wadi Al-Bih, U.A.E. and Oman. *Geomorphology*, **35**, 279–297.

BELL, J.W. & KATZER, T. 1987. *Surficial geology, Hydrology, and the Quaternary Tectonics of the IXL Canyon Area, Nevada, As Related to the 1954 Dixie Valley Earthquake*. Nevada Bureau of Mines and Geology, Bulletin, **102**.

BLAIR, T.C. & MCPHERSON, J.G. 1994. Alluvial fan processes and forms. *In*: ABRAHAMS, A.D. & PARSONS, A.J. (eds) *Geomorphology of Desert Environments*. Chapman & Hall, London, 354–402.

BOWMAN, D. 1978. Determination of intersection points within a telescopic alluvial fan complex. *Earth Surface Processes*, **3**, 265–276.

BOWMAN, D. 1988. The declining but non-rejuvenating base-level – the Lisan Lake, the Dead sea, Israel. *Earth Surface Processes and Landforms*, **13**, 239–249.

BULL, W.B. 1962. *Relations of Alluvial Fan Size and Slope to Drainage Basin Size and Lithology, in Western Fresno County, California*. US Geological Survey, Professional Paper, **430B**, 51–53.

BULL, W.B. 1977. The alluvial fan environment. *Progress in Physical Geography*, **1**, 222–270.

BULL, W.B. 1991. *Geomorphic Responses to Climatic Change*. Oxford University Press, Oxford.

CALVACHE, M., VISERAS, C. & FERNÁNDEZ, J. 1997. Controls on alluvial fan development – evidence from fan morphometry and sedimentology; Sierra Nevada, SE Spain. *Geomorphology*, **21**, 69–84.

CHADWICK, O.A., HECKER, S. & FONSECA, J. 1984. *A soils Chronosequence at Terrace Creek: Studies of late Quaternary Tectonism in Dixie Valley, Nevada*. US Geological Survey, Open File Report, **84–90**.

DAVIS, J.O. 1982. Bits and pieces: the last 35,000 years in the Lahontan area. *In*: MADSEN, D.D. & O'CONNELLL, J.F. (eds) *Man and Environment in the Great Basin*. Society for American Archaeology, Paper, **2**.

FROSTICK, L.E. & REID, I. 1989. Climatic versus tectonic controls of fan sequences: lessons from the Dead Sea, Israel. *Journal of the Geological Society, London*, **146**, 527–538.

GILE, L.H., PETERSON, F.F. & GROSSMAN, R.B. 1966. Morphological and genetic sequence of carbonate accumulation in desert soils. *Soil Science*, **101**, 347–360.

GRAYSON, D.K. 1993. *The Desert's Past: A Natural History of the Great Basin*. Smithsonian, Washington, DC.

HARVEY, A.M. 1987. Alluvial fan dissection: relationships between morphology and sedimentology. *In*: FROSTICK,

L. & REID, I. (eds) *Desert Sediments Ancient and Modern*. Geological Society, London, Special Publications, **35**, 87–103.

HARVEY, A.M. 1992. The influence of sedimentary style on the morphology and development of alluvial fans. *Israel Journal of Earth Sciences*, **41**, 123–137.

HARVEY, A.M. 1997. The role of alluvial fans in arid zone fluvial systems. *In*: THOMAS, D.S.G. (ed.) *Arid Zone Geomorphology: Process, Form and Change in Drylands*, 2nd edn. Wiley, Chichester, 231–259.

HARVEY, A.M. 2002*a*. Factors influencing the geomorphology of dry-region alluvial fans: a review. *In*: PEREZ-GONZALEZ, A., VEGAS, J. & MACHADO, M.J. (eds) *Aportaciones a la geomorfologia de Espana en el Inicio del Tercer Milenio*. Publicaciones del Instituto Geologico y Minero de Espana, Madrid, Serie Geologia, **1**, 59–72.

HARVEY, A.M. 2002*b*. The role of base-level change in the dissection of alluvial fans: case studies from southeast Spain and Nevada. *Geomorphology*, **45**, 67–87.

HARVEY, A.M. 2003. The response of dry-region alluvial fans to late Quaternary climatic change. *In*: ALSHARAN, A.S., WOOD, W.W., GOUDIE, A.S., FOWLER, A. & ABDELLATIF, E.M. (eds) *Desertification in the Third Millenium*. Balkema, Rotterdam, 83–98.

HARVEY, A.M. & WELLS, S.G. 2003. Late Quaternary variations in alluvial fan sedimentologic and geomorphic processes, Soda Lake basin, eastern Mojave Desert, California. *In*: ENZEL, Y., WELLS, S.G. & LANCASTER, N. (eds) *Paleoenvironments and Paleohydrology of the Mojave and Southern Great Basin Deserts*. Geological Society of America, Special Paper, **368**, 207–230.

HARVEY, A.M., SILVA, P.G., MATHER, A.E., GOY, J.L., STOKES, M. & ZAZO, C. 1999*a*. The impact of Quaternary sea-level and climatic change on coastal alluvial fans in the Cabo de Gata ranges, southeast Spain. *Geomorphology*, **28**, 1–22.

HARVEY, A.M., WIGAND, P.E. & WELLS, S.G. 1999*b*. Response of alluvial fan systems to the late Pleistocene to Holocene climatic transition: Contrasts between the margins of pluvial Lakes Lahontan and Mojave, Nevada and California, USA. *Catena*, **36**, 255–281.

HOOKE, R. LE, B. 1968. Steady state relationships of arid region alluvial fans in closed basins. *American Journal of Science*, **266**, 609–629.

HOOKE, R. LE, B. & ROHRER, W.L.1977. Relative erodibility of source area rock types from second order variations in alluvial fan size. *Bulletin of the Geological Society of America*, **88**, 1177–1182.

HOUGHTON, J.G., SAKAMOTO, C.M. & GIFFORD, R.O. 1975. *Nevada's Weather and Climate*. Nevada Bureau of Mines and Geology, Special Publications, **2**.

LEEDER, M.R. & MACK, G.H. 2001. Lateral erosion ('toe-cutting') of alluvial fans by axial rivers: implications for basin analysis and architecture. *Journal of the Geological Society, London*, **158**, 885–893.

LECCE, S.A. 1991. Influence of lithologic erodibility on alluvial fan area, western White Mountains, California and Nevada. *Earth Surface Processes and Landforms*, **16**, 11–18.

MACHETTE, M.N. 1985. Calcic soils of the southwestern United States. *In*: WEIDE, D.L. (ed.) *Soils and Quaternary Geology of the Southwestern United States*. Geological Society of America, Special Paper, **203**, 1–21.

MCFADDEN, L.D., MCDONALD, E.V., WELLS, S.G., ANDERSON, K., QUADE, J. & FOREMAN, S.L. 1998. The vesicular layer and carbonate collars of desert soils and pavements: formation, age, and relation to climatic change. *Geomorphology*, **24**, 101–145.

MCFADDEN, L.D., RITTER, J.B. & WELLS, S.G. 1989. Use of multiparameter relative-age methods for age estimation and correlation of alluvial fan surfaces on a desert piedmont, eastern Mojave Desert, California. *Quaternary Research*, **32**, 276–290.

MIFFLIN, M.D. & WHEAT, M.M. 1979. *Pluvial Lakes and Estimated Pluvial Climates of Nevada*. Nevada Bureau of Mines and Geology, Bulletin, **94**.

MORRISON, R.B. 1991. Quaternary stratigraphic, hydrologic and climatic history of the Great Basin, with emphasis on Lakes Lahontan, Bonneville and Tecopa. *In*: MORRISON, R.B. (ed.) *Quaternary Nonglacial Geology; Coterminous US*. The Geology of North America. Geological Society of America, Boulder, CO, **K-2**, 283–320.

RITTER, J.B., MILLER, J.B., ENZEL, Y. & WELLS, S.G. 1995. Reconciling the roles of tectonism and climate in Quaternary alluvial fan evolution. *Geology*, **23**, 245–248.

RITTER, J.B., MILLER, J.R. & HUSEK-WULFORST, J. 2000. Environmental controls on the evolution of alluvial fans in the Buena Vista Valley, North Central Nevada, during late Quaternary time. *Geomorphology*, **36**, 63–87.

SILVA, P.G., HARVEY, A.M., ZAZO, C. & GOY, J.L. 1992. Geomorphology, depositional style and morphometric relationships of Quaternary alluvial fans in the Guadalentin Depression (Murcia, southeast Spain). *Zeitschrift für Geomorphologie*, NF, **36**, 325–341.

VISERAS, C., CALVACHE, M.L., SORIA, J.M. & FERNÁNDEZ, J. 2003. Differential features of alluvial fans controlled by tectonic or eustatic accommodation space. Examples from the Betic Cordillera, Spain. *Geomorphology*, **50**, 181–202.

WILLDEN, R. & SPEED, R.C. 1974. *Geology and Mineral Resources of Churchill County, Nevada*. Nevada Bureau of Mines and Geology, Bulletin, **83**.

Reconciling the roles of climate and tectonics in Late Quaternary fan development on the Spartan piedmont, Greece

RICHARD J.J. POPE[1] & KEITH N. WILKINSON[2]

[1] Division of Geographical Sciences, School of Education, Health and Sciences, University of Derby, Kedleston Road, Derby, DE22 1GB, UK (e-mail: R.J.Pope@Derby.ac.uk)

[2] Department of Archaeology, University College Winchester, Winchester SO22 4NR, UK

Abstract: The evolution of five alluvial fan systems is discussed in relation to chronology and possible tectonic and climatic triggering mechanisms. Two types of fan have evolved on the Spartan piedmont, Greece. First relatively large, low-angle fans, comprising four segments (Qf1–Qf4) composed of debris-flow and hyperconcentrated-flow deposits, with fluvial sediments restricted to the upper deposits of the distal segments. Second small, steep telescopically segmented fans, which consist of three segments (Qf1–Qf3), formed predominantly by debris-flow and hyperconcentrated-flow deposits. Morphological analysis of surface soils coupled with mineral magnetic and extractable iron (Fe_d) analyses of B-horizons suggest that individual segments can be correlated across the piedmont and have equivalent age. Luminescence dating of fine-grained deposits suggests that Qf1 segments formed during marine isotope stage (MIS) 6, Qf2 segments during MIS 5, Qf3 segments during MIS 4–2, and Qf4 segments during MIS 2 and 1. Tectonics has exerted a limited influence on fan systems. Regional uplift provides the gross relief conducive for fan development. The locations of fans were determined by transfer faults of Tertiary age, while Quaternary faulting initiated short phases of fan incision. Climate change as manifested by cycles of aridity and low vegetation cover during stadials, and humidity and deciduous woodland during interglacials and interstadials, played a key role in fan evolution during the later Middle and Upper Pleistocene. Aggradation occurred during stadials, with minor deposition and intermittent erosion during most interstadials, and entrenchment during the interglacials and longer interstadials. Deposition during the Holocene is limited in extent.

Range-front alluvial fan systems occupy a strategically important location between upland drainage systems and lowland basins, and have been seen as sensitive, long-term recorders of deposition within the piedmont zone (e.g. White *et al.* 1996; Harvey 2002; Pope *et al.* 2003). Major phases of aggradation, evidenced by the various depositional styles exhibited by sedimentary units, may form multiple fan segments (Hooke 1972; Blair and McPherson 1994; Pope 2000). Erosional phases, evidenced by reworking of fan sediment or fan surface incision, may coincide with periods of reduced sediment supply or increased runoff (e.g. Denny 1967; Harvey *et al.* 1999). Fan aggradation and erosion are controlled by numerous environmental factors including climate (e.g. Lustig 1965; Dorn 1994; Harvey & Wells 1994; White *et al.* 1996; Macklin *et al.* 2002), tectonic regime (e.g. Bull 1964; Rockwell *et al.* 1985; Leeder *et al.* 1988), contiguous environment (e.g. Wells *et al.* 1987) and, for the Holocene period, human land use (e.g. Demitrack 1986; Pope *et al.* 2003). In this paper we review the first two of these controls in relation to alluvial fans on the western side of the Sparta Basin, Peloponnese, Greece. We do this in the light of recent re-assessments of fan morphologies and surface soils, and new tectonic, palaeoclimatic and chronometric data.

Tectonic and climatic contexts of fan evolution

In areas where fan development has taken place against a backdrop of active tectonics and major climate change, it has proved difficult to clearly differentiate the respective roles of both controls on fan evolution (e.g. Leeder *et al.* 1988; Dorn 1994; Ritter *et al.* 1995; Harvey *et al.* 1999). Such difficulties reflect: (i) the absence of well-established local climatic and tectonic histories (e.g. White *et al.* 1996; Pope 2000); and (ii) uncertainty over the depositional and erosional histories of fan systems due to poorly constrained chronological frameworks (e.g. Harvey 1990; White & Walden 1994; Pope & Millington 2000; Harvey *et al.* 2003). Both factors have contributed to uncertainty over the dominant control of fan development within the Sparta Basin (e.g. Hempel 1984; Schneider 1986; Pope 1995).

Tectonic activity in the form of epeirogenic uplift creates and maintains relief (e.g. Calvache *et al.* 1997), and partially influences long-term sediment transfer rates to fan systems. Uplift also produces the effect of a gradual but continuous long-term fall in the base level of erosion which, depending on sediment availability and stream power, may potentially bring about major fan incision (Harvey 2002). Normal

From: HARVEY, A.M., MATHER, A.E. & STOKES, M. (eds) 2005. *Alluvial Fans: Geomorphology, Sedimentology, Dynamics.* Geological Society, London, Special Publications, **251**, 133–152. 0305–8719/05/$15 © The Geological Society of London 2005.

faulting may control the location of individual fan systems through influencing the location of principal upland drainage channels (e.g. Blair & McPherson 1994). In comparison, range-front faulting has been seen to trigger local incision, primarily at the fanhead or in proximal fan areas (e.g. Beaty 1961; Harvey 2002), or modify the overall gradient of proximal surfaces (e.g. Hooke 1967; Harvey 1990). In extreme cases, faulting not only vertically displaces fan surfaces, but also may cause distortion or deformation of thickly bedded depositional sequences (e.g. Leeder et al. 1988; Gerson et al. 1993). Climate controls the sediment and water supply within upland drainage systems, which has implications for the discharge (Q) to sediment load (Q_s) ratio (Langbein & Schumm 1958). One mechanism by which it does this is by determining the nature of the vegetation in the catchment. Highly vegetated catchments are characterized by a low sediment supply and usually by low discharge, with both variables potentially increasing as vegetation cover is reduced (Frenzel et al. 1992). Variation in the $Q:Q_s$ ratio potentially determines whether fan systems are dominated by aggradation or degradation (Harvey 2002), and the nature of deposition, i.e. by debris flow or stream flow (e.g. Wells & Harvey 1987). The latter influences key fan attributes including overall gradient (e.g. Blair & McPherson 1994; Harvey 1997) and area (e.g. Pope 1995).

The Sparta Basin

The southern portion of the Peloponnese is dominated by a series of NNW–SSE-trending extensional faults, which define a series of mid-Miocene asymmetric grabens (Poulimenous & Doutsos 1997). The Sparta Basin occupies a central position within the most easterly graben system (Pe-Piper & Piper 1985), forming a 25 km-long, partially blocked depression, which is drained by a major axial drainage system, the River Evrotas (Fig. 1). The position of the eastern margin of the basin is uncertain, but probably coincides with a broad limestone ridge marking the Parnon Mountains (Pope 1995). Five kilometres east of the modern town of Sparta, the basin sediments have been significantly incised by a forerunner of the present Evrotas river system, revealing a 25–40 m-thick Neogene sequence comprising gravels, gravelly-sands, sandy-silts and silty-clays (Wilkinson & Pope 2003). Shallow borehole data suggest that the Neogene fills extend to a depth of at least 280 m below ground surface (Yperisia Eggeion Beltioseon unpublished data), and progressively increase to depths of several hundred metres in the western half of the basin (Piper et al. 1982). The western margin of the basin is clearly defined by the eastern edge and lower portion of the Taygetos mountain range. The Taygetos range comprises

geological units of the Gavrovo–Tripolitsa zone of the west Hellenic nappe (Aubouin et al. 1976; Jacobshagen et al. 1978), and was initially uplifted and folded during the Hellenide orogeny. The eastern half of the Taygetos Mountains is composed of phyllitic basement rocks of (?)Permo-Triassic age, which are overlain by middle Triassic–upper Eocene dolomitic and crystalline limestones. The western half is dominated by the phyllitic-quartzite series, which consists of phyllites, schists and quartzites of Permian age (Jacobshagen et al. 1978).

Subduction and associated underplating along the Hellenic trench during the middle–late Miocene is thought to have initiated a phase of continuous uplift throughout the Peloponnese (Angelier et al. 1982), with the Taygetos range and Sparta Basin experiencing up to 0.4 mm year^{-1} of vertical displacement (Le Pichon & Angelier 1981). By way of comparison, Armijo et al. (1991) suggest that the 750 m-high triangular facets along the eastern edge of the Taygetos range reflect sustained uplift since the start of the Quaternary, equating to an average uplift rate of 0.29 mm year^{-1}. Dated marine terraces along the Lakonian coast suggest that regional rates of uplift since the late Pleistocene vary between approximately 0.22 and 0.25 mm year^{-1} (Kelletat et al. 1976; Keraudren & Sorel 1987; Doutsos & Piper 1990). Seismic activity that accompanied late Miocene uplift culminated in a series of dominantly WSW–ENE-trending transfer faults (Lyon-Caen et al. 1988), which effectively subdivided parts of the eastern flanks of the Taygetos range into a series of fault-bounded blocks, and controlled the spacing of several steep drainage systems. During the Holocene, shallow seismic activity became concentrated along the eastern edge of the Taygetos range (Dufaure 1977). It is thought that up to three shallow earthquakes during the early, middle (Armijo et al. 1991), and late Holocene (464 BC) (Ducat 1983) culminated in a segmented, 21 km-long range-bounding normal fault, and localized uplift of the immediate range-front area (Pope 1995).

The present climate of the Sparta Basin is of a sub-arid Mediterranean type. Average annual rainfall in the piedmont is 816 mm year^{-1}, and average annual temperatures range between 28°C in July and 9°C in January (Millington et al. 1990), although winter temperatures in Taygetos frequently fall below freezing and snow covers the mountain peaks between December and March. Until recently the lack of reliable climate proxies for the eastern Mediterranean meant that crude generalizations were made in the discussion of Late Quaternary palaeoclimate of southern Greece. It has now been established that the rapidly oscillating Upper Pleistocene climate events of the North Atlantic region seen, for example, in the Greenland Ice Core Project (GRIP) ice cores also characterized the eastern Mediterranean (Watts et al.

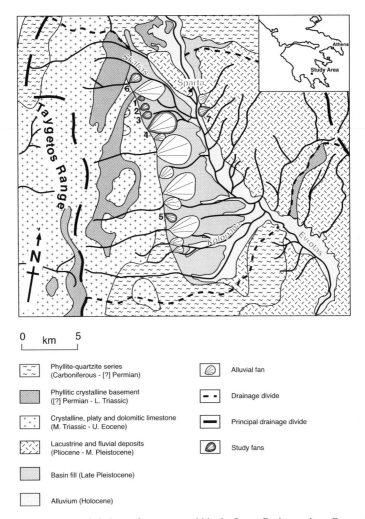

Fig. 1. Transverse drainage systems and piedmont fan systems within the Sparta Basin, southern Greece. The inset illustrates the location of the study area with respect to the rest of Greece. The study fans are: (1) South Parorion; (2); Kamares; (3) St Johns; (4) St Saviours; and (5) North Xilocambion. Other fan systems mentioned are: (6) Mystras; and (7) North Menelaion.

1996). Tzedakis' (1999) climate model for southern Greece suggests a humid climate and a mixed vegetation throughout most of marine isotope stage (MIS) 5, punctuated by short-lived arid spells characterized by the growth of grasses and herbs (*Artemisia* and *Amaranth*) during parts of MIS 5d, 5a and all of 5b. During MIS 4 drier conditions prevailed with woodland becoming progressively replaced by grass and herb communities. MIS 3 is marked by a rapidly oscillating climate resulting in frequent change from woodland to open conditions, whereas in MIS 2 deciduous woodland is reduced to a relict community as the

climate rapidly becomes arid. The Late Glacial stage of late MIS 2 is characterized by a single interstadial dating to approximately 14–13 ka cal BP (14–13 ka calibrated before present) during which deciduous oak woodland expanded (Allen 1990). Short, sharp stadials occurred both before and after this event. The Holocene pollen record for the Peloponnese suggests a warm, relatively damp climate dominated by undifferentiated oak woodland. The area occupied by the latter rapidly declined between 6.3 and 5.6 ka cal BP, an 'event' that is interpreted as resulting from human forest clearance (Greig & Turner 1974; Turner &

Greig 1975). However, Greek lake-level data indicate that arid climates in the eastern Mediterranean evolved from *c.* 5.7 ka cal BP, suggesting an alternate, climatic cause for the reduction of woodland (Harrison & Digerfeldt 1993).

The main alluvial fan zone of the Sparta Basin occupies a prominent position on the western margins of the Lakonian graben between the rivers Skatias and Kolopana (Fig.1). Sediment transfer from source areas located along the eastern flanks of the Taygetos range culminated in the formation of 15 fans. At the fanhead, the total thickness of deposits is estimated to vary between 95 and 150 m. This rapidly decreases-downfan, recording a maximum thickness of 4 m around the distal-fan margins (Yperisia Eggeion Beltioseon unpublished data). Previous research suggests that the Spartan fans consist of multiple discrete segments, which vary between 17 and 310 ha in area (Pope 1995). The surface of each segment is characterized by a uniform surface gradient (Pope 1995; Pope & Millington 2000), while surfaces are capped by distinctive post-incisive soil chronosequences (Pope *et al.* 2003). A narrow single thread trench extends across the whole surface of each fan resulting in coupling of mountainous transverse drainage systems with the main axial drainage system (Pope & Millington 2002).

Methods

To build on elements of previous work in the Spartan Basin (Pope & Millington 2000; Pope *et al.* 2003; Wilkinson & Pope 2003), data from two previously investigated fans (St Johns and North Xilocambion) were re-examined, while a detailed field and laboratory examination was made of three additional fans (St Saviours, Kamares and South Parorion) (Fig. 2).

For each selected fan complete proximal–distal profiles were surveyed to identify subtle breaks in slope and estimate average surface gradients. A number of sedimentary characteristics including clast size and lithology, the degree of sorting, the presence (or absence) of matrix-support, and internal structure of representative proximal, medial and distal sections were described and logged to determine vertical and horizontal textural trends.

For each study fan up to 40 test pits were excavated on individual fan segments in order to describe soil profiles and key soil properties (colour, texture, structure, parent material and $CaCO_3$ accumulation). Then between 10 and 40 samples were collected from the B- or the A/B-horizon of fan surface soils, at a depth of approximately 7–10 cm (Table 2). The laboratory methods subsequently used to obtain redness ratings, mineral magnetic and secondary iron parameters have been fully described by Pope & Millington (2000) and Pope *et al.* (2003).

A chronological framework for fan evolution is provided by a number of luminescence dates on samples collected

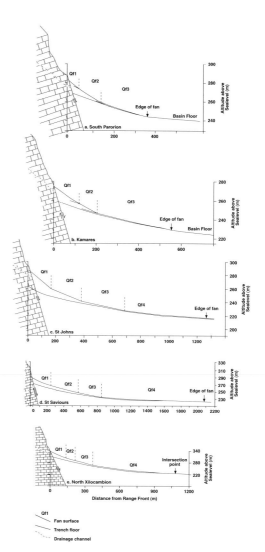

Fig. 2. Morphological sections illustrating overall fan shape, spacing of fan surfaces (Qf1, Qf2, Qf3 and Qf4) and trench characteristics.

Table 1. *Average fan gradients for discrete fan surfaces*

	Average surface gradient (°)			
	Qf1	Qf2	Qf3	Qf4
SP	13.6	10.8	4.75	–
K	12.6	9.9	5.5	–
SJ	9.8	6.6	4.4	2.6
SS	6.4	3.8	3.5	1.7
NX	9.0	6.4	5.3	2.0

SP, South Parorion; K, Kamares; SJ, St Johns; SS, St Saviours; NX, North Xilocambion.

using the methodology of Wintle (1991) from coarse-silt and fine-sand facies exposed in the fan trenches. Thermoluminescence (TL) dating was carried out at the University of Gloucestershire Geochronology Laboratory, while optically stimulated luminescence (OSL) measurements were made at the Nordic Laboratory for Luminescence Dating at the University of Aarhus. The fine-grained facies that were sampled for luminescence dating are thought to have formed during low-energy deposition of material derived both from upland drainage systems and reworked fan sediment, and/or as ebb flow following higher energy depositional events. Moreover, given that soils have often developed in the sampled fine-grained units, the luminescence dates not only provide ages for periods of silt and sand deposition, but also provide maximum ages for phases of landscape stability. In addition to the luminescence measurements, tentative maximum ages for several sedimentary units were derived on the basis of *in situ* archaeological material.

Fan evolution

Fan morphology and longitudinal profiles

In planform terms, the St. Johns, North Xilocambion and St Saviours fans are characteristically conical (Fig. 2c–e) with their distal margins coalescing to form a narrow bajada. Topographic surveys indicate that each fan possesses a slightly concave-upwards profile, comprising four discrete radial segments (labelled Qf1–Qf4, from oldest to youngest: Fig. 3c–e) delimited by subtle breaks of slope. The highest mean gradients are associated with the proximal segments with the medial and distal segments recording progressively lower gradients (Table 1). By way of contrast, the Kamares and South Parorion fans are telescopically segmented with lower surfaces nested within upper surfaces (labelled Qf1–Qf3, from oldest to youngest: Fig. 2a,b). Both fans are characterized by a stepped profile with successively lower surfaces recording lower gradients (Fig. 3a, b).

Fan surface soils

Erosional processes and long-term cultivation have removed the A-horizons of soils that have developed on proximal and medial fan surfaces, leaving chronosequences that consist of B- and C-horizons. The parent materials for all the fan soils are composed predominantly of reworked crystalline and platy limestones, and, more rarely, dolomitic limestones. All are derived from source areas along the eastern flanks of the Taygetos range.

The Qf1 and Qf2 soils of the St Johns, St Saviours and North Xilocambion fans have partially eroded,

33–43 cm thick, reddish-brown (Munsell colour 5YR 4/3–4/4) Bt-horizons, and a characteristically subangular blocky structure. Orientated clay occurs mainly within pores, and as moderately thin films on grains. The underlying Bk-horizons show stage II+ and, occasionally, stage III carbonate accumulation (nomenclature after Machette's 1985 classification scheme), with carbonate coatings occurring as partial and continuous coverings over the surfaces of clasts. The Qf1 and Qf2 soils of the Kamares and South Parorion fans share the same textural and structural properties as the proximal soils of the other study fans. However, intense localized erosion has reduced the thickness of the reddish-brown (Munsell colour 5YR 4/4) Bt-horizon to 30 cm, and at several points completely removed the soil cover to expose a well-cemented C-horizon.

The Qf3 soils of the St Johns, St Saviours and North Xilocambion fans consist of 22–29 cm thick, brown (7.5YR 5/2–5/4) B-horizons of characteristically subangular blocky structure. Orientated clay occurs within pores and, less commonly, as moderately thin coatings on grains. Stage I+ and stage II carbonate accumulation characteristics are found within the Bk-horizons, occurring as occasional fine discontinuous nodules, or as thin discontinuous coatings on the undersides of clasts and the surrounding matrix. The Qf3 soils of the Kamares and South Parorion fans again share the same textural and structural qualities of the other study fans. However, the brown–strong brown (7.5YR 4/3–4/6) B-horizon displays greater variation in thickness, varying between 26 and 40 cm. Carbonate accumulation within the Bk-horizons of the Kamares and South Parorion fans is characteristic of stage I+ development and takes the form of very thin discontinuous coatings on the undersides of large carbonate clasts.

Soils associated with Qf4 surfaces of the St Saviours fan are characterized by a 20–28 cm-thick, brown (7.5YR 5/3–5/4) granular A/B-horizon. By contrast Qf4 soils of the St Johns and North Xilocambion fans are 14–16 cm thick, subangular, partly blocky, and of a brown (10YR 5/3) to yellowish-brown (10YR 5/4–5/6) colour. The immaturity of the distal soils is reflected by the lack of significant carbonate accumulation. Carbonate occurs only as occasional discontinuous coating on fine grains and, more rarely, on the undersides of larger clasts. These characteristics are indicative of very early stage I carbonate accumulation.

Fan sediments and facies types

The Qf1 and Qf2 segments of the South Parorion, Kamares, St Johns and North Xilocambion fans are composed primarily of debris flow and rare hyperconcentrated flow deposits (Fig. 4a–c,e). The debris

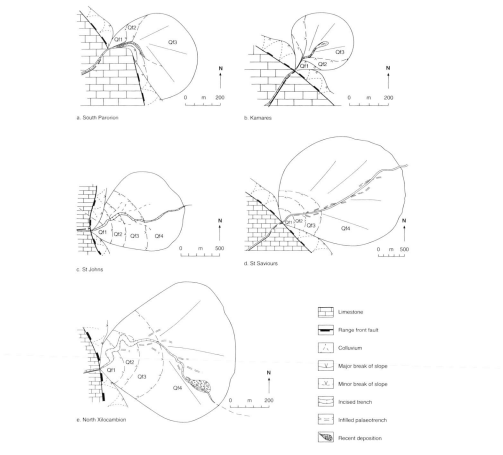

Fig. 3. Surveyed axial-surface and trench profiles of sample fans. The location of each fan system is shown in Figure 1.

flows comprise angular–subrounded pebbles, cobbles and boulders composed predominantly of limestone. The basal debris flows vary between 0.6 and 1.6 m in thickness, and consist of a reddish-brown (5YR 4/3–4/4) clay and fine-silt matrix, derived from the remnants of soil cover or colluvium eroded from hill-slopes. Individual units display variable internal structures, ranging from well-developed local shear fabrics, reflecting internal deformation by pervasive laminar shear (Naylor 1980), to inverse grading or random clast fabrics. The middle and upper debris flows are supported by a brown (10YR 5/4–5/6) coarse-silt–fine-sand (10YR 5/4–5/6) matrix derived from reworked range-front colluvium (Pope 1995). Individual units vary between 0.5 and 1.2 m in thickness, and internally are characterized by rare shear fabrics or crude inverse grading. Hyperconcentrated flow deposits consist of partially cemented, clast-supported gravels and rare cobbles. Individual units reach up to 1.0 m in thickness, and are differentiated from debris flows by their lack of sorting or internal

structure. The bulk of the sediment that forms the Qf1 and Qf2 segments of the St Saviours fan consists of coarse- and fine-grained hyperconcentrated flow deposits (Fig. 4d). The former are typically poorly defined and form 0.4–1.0 m-thick tabular beds, consisting of crystalline limestone gravels and cobbles, supported by fine gravels and coarse brown (10YR 5/3) sand. Within individual units internal structure is lacking and deposits are characterized by random clast fabrics. The fine-grained hyperconcentrated flow deposits form 0.2–0.4 m-thick lenticular beds, comprising structureless fine gravels, supported by a thin brown (10YR 4/3) silt matrix.

The basal units of Qf3 segments comprise rapidly thinning debris-flow deposits, consisting of poorly sorted matrix-supported gravels and cobbles. The overlying units are composed entirely of alternating clast- and matrix-supported hyperconcentrated flow deposits. The former typically form massive (up to 1.6 m thick) tabular beds, which are composed of structureless limestone gravels. However, within

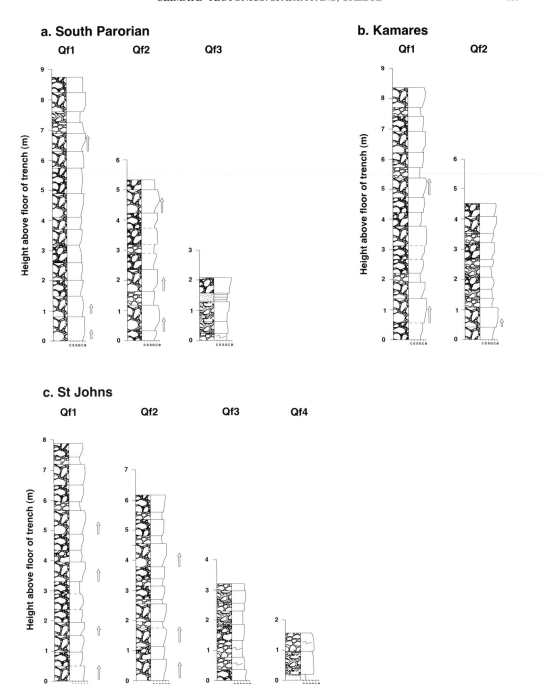

Fig. 4a. Detailed lithostratigraphies of sections associated with Qf1, Qf2, Qf3 and Qf4 segments of the study fans.

the Kamares, South Parorion and St Johns fans clast-supported gravels display a distinctive collapse fabric, whereby following initial deposition the supporting matrix was flushed out and internal support removed (Wells & Harvey 1987). The matrix-supported hyperconcentrated flow deposits are characterized by either a thick brown (10YR 4/3) silt-dominated matrix (in the St Johns, St Saviours and North Xilocambion fans) or a by a thin brown (10YR 4/3–5/3) silty-sand matrix (in the Kamares

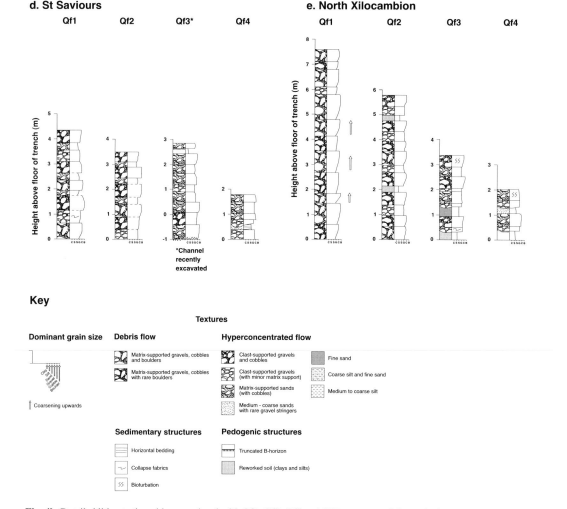

Fig. 4b. Detailed lithostratigraphies associated with Qf1, Qf2, Qf3 and Qf4 segments of the study fans.

and South Parorion fans), which was deposited by downwards percolating sediment-rich water (Pope 1995). Within individual units internal structure is lacking and deposits tend to be characterized by random clast fabrics.

Gravel units constitute the bulk of sediment forming Qf4 segments (Fig. 4c–e). The matrix-supported basal and middle gravels display variable amounts of sorting and partial internal stratification, which strongly indicates that sediment has been transported and deposited by less cohesive hyper-concentrated flows (Smith 1986). In contrast, the upper gravels are predominantly structureless and supported by a clay matrix derived in part from erosion of an underlying palaeosol. Within the St Saviours fans, the upper clast-supported gravels display crude imbrication and crude normal grading,

which strongly indicates that increasingly less cohesive hyperconcentrated flows or stream-flow processes were also responsible for transporting and depositing sediment (Wells & Harvey 1987).

Fan geochronologies

Analysis of soil redness, extractable iron concentration (Fe_d) and mineral magnetic properties of surface soils was undertaken to establish the relative chronology of fan surfaces. For each study fan, a clear and consistent time dependent trend emerges with regard to soil redness ratings and Fe_d characteristics (Table 2). The lowest mean redness ratings and Fe_d values are associated with Qf4 soils. Progressively higher mean redness ratings and Fe_d values are

Table 2. *Average redness ratings and extractable iron (Fe$_d$) concentrations from the B-horizons of soils developed on the surfaces of the sample fans. Note that redness ratings are derived using the approach of Alexander (1985)*

| | Sample size | | | | Redness rating | | | | Fe$_d$(mg g^{-1}) | | | |
	Qf1	Qf2	Qf3	Qf4	Qf1	Qf2	Qf3	Qf4	Qf1	Qf2	Qf3	Qf4
SP	16	20	20	–	11.9	9.9	8.2	–	2.70	2.34	1.78	–
K	25	20	20	–	11.7	9.1	7.2	–	2.56	2.20	1.65	–
SJ	14	9	22	26	11.9	9.0	8.5	5.5	2.41	1.94	1.71	1.43
SS	20	20	20	40	11.7	9.7	8.1	5.3	2.74	2.42	1.66	1.40
NX	10	10	20	25	10.9	9.5	8.5	5.2	2.66	2.48	1.77	1.37

SP, South Parorion; K, Kamares; SJ, St Johns; SS, St Saviour; NX, North Xilocambion.

associated with Qf3 and Qf2 soils, with the highest values recorded for the Qf1 soils. The Fe$_d$ data clearly show that the Bt-horizons of proximal soils contain the highest concentration of secondary iron oxides (and a significant amount of haematite given the high redness rating) and fine-grained secondary magnetite, indicating that such soils formed first. Analysis of variance indicates that for each surface within individual fans, the differences in mean Fe$_d$ values are statistically significant ($p<0.02$).

The results of the mineral magnetic analysis for individual fans are summarized in Table 3. The trend in each of the magnetic parameters very closely follows that of the Fe$_d$ and soil redness data. Measurements of saturation isothermal remanent magnetization (SIRM) progressively increase upfan, indicating that Qf1 soils contain the highest concentrations of remanence carrying minerals. Given the age-related trend exhibited by the Fe$_d$ and soil redness data it is likely that the trend evident in the SIRM data reflect an increased contribution from very fine-grained iron derived from *in situ* weathering or pedogenic processes rather than primary iron oxides (J. Walden pers. comm. 2003). The accumulation of very-fine-grained secondary iron in Qf1 soils, particularly pedogenic magnetite, is further supported by the clear trend in the $\chi_{fd\%}$ data (where $\chi_{fd\%}$ is frequency-dependent susceptibility) (Dearing *et al.* 1996) and the SIRM/χ_{lf} ratio (where χ_{lf} is low field susceptibility) while the HFIRM data suggest that such soils contain significant amounts of haematite or possibly goethite. The progressive upfan increase in both χ_{lf} and low field isothermal remanent magnetization (LFIRM) suggests that the Bt-horizons of progressively older soils also contain increasing concentrations of ferrimagnetic minerals, predominantly ultra-fine-grained magnetite and possibly maghemite (Maher 1988). Analysis of variance indicates that for each surface within individual fans the differences in the mean value of each magnetic parameter are statistically significant ($p<0.02$).

TL and OSL dates (Table 4) enable a calibration of the relative chronology, while at the same time allowing major episodes of aggradation and periods of landscape stability to be placed in a chronological framework. A maximum age for the proximal gravels and Qf1 surfaces is provided by TL dates of between 235 and 137 ka BP on separate beds of coarse silts and fine sands in the Mystras fan (Fig. 5a,b). The partly eroded Bt horizon of the soil developed on the proximal surface of the Mystras fan produces similar magnetic, Fe$_d$ and redness rating values to the proximal soils of the fans discussed in this study, suggesting that the TL chronology can be applied across the piedmont. Whereas the Qf1 can be broadly dated by reference to the Mystras fan, the chronology of Qf2 is only known in relative terms. No luminescence dates have been obtained from Qf2 stratigraphy. Qf2 must post-date 137 ka BP (the youngest date for the Qf1 segment of the Mystras fan), but predate 66 ka BP (the oldest date for the Qf3 segment of the North Xilocambion fan).

The lower gravels of the Qf3 segment in the North Xilocambion fan were deposited at *c.* 66 ka BP (Fig. 5i). The middle gravels of the Qf3 segment in the St Johns fan were deposited in at least two phases at approximately 37 and 29.3 ka BP (Fig.5c, d), while in the St Saviours fan, gravels in a similar stratigraphic position were deposited around 26 ka BP (Fig. 5h). The final depositional events, and hence complete formation of the Qf3 segments of the North Xilocambion and St Saviours fans, occurred at *c.* 25.7 and 17.6 ka BP, respectively (Fig. 5i, g). Given that the mean mineral magnetic and Fe$_d$ values of the Qf3 surfaces of these two fans are similar to those of other Spartan fans discussed in this paper, the luminescence chronology can again be applied to all Qf3 surfaces.

It is uncertain precisely when deposition of the Qf4 segment began, except to state that it post-dated 17.6 ka BP. However, three separate OSL dates from the St Johns and North Xilocambion fans bracket deposition of the last significant Qf4 gravel unit to between *c.* 15–14.6 and 11 ka BP, respectively

Table 3. *Mineral magnetic data for B-horizons of soils developed on the surfaces of the sample fans. Figures given are average values based upon the number of samples given in Table 2*

	χ_{lf} (10^{-8} m^3 kg^{-1})				χ_{fd} (%)				LFIRM (10^{-5} Am^2 kg^{-1})			
	Qf1	Qf2	Qf3	Qf4	Qf1	Qf2	Qf3	Qf4	Qf1	Qf2	Qf3	Qf4
SP	19.08	16.73	12.70	–	9.65	9.40	8.87	–	346	289	194	–
K	18.60	15.24	12.04	–	9.56	9.34	9.18	–	323	242	169	–
SJ	17.59	14.42	11.99	7.67	9.91	9.64	9.28	8.59	258	269	205	152
SS	20.23	15.04	11.02	8.12	9.80	9.46	9.16	8.39	339	223	195	163
NX	19.93	16.66	12.54	7.94	9.53	9.44	8.83	8.35	363	274	204	146

(Fig. 5e, j). On the basis of archaeological evidence, the final deposition events on Qf4 surfaces are dated to no earlier than 3.6 ka BP, and continued intermittently on some of the larger fans until around 1.3 ka BP (Pope *et al.* 2003). Thereafter depositional events were increasingly localized, and culminated in the formation of minor fill terraces within fan trenches (Fig. 5f).

Discussion: climate and tectonism as triggers of fan sedimentation and incision

The role of tectonic activity

Throughout the Pleistocene continuous regional uplift of the order of 300–400 m accentuated and then maintained the gross relief that is necessary for fan development within the Spartan piedmont (Pope 1995). Through its control on gross relief, regional uplift has indirectly influenced fan development in three ways. First, by partially controlling long-term rates of bedrock disintegration and chemical decomposition, which in turn determines potential debris generation within carbonate-dominated sediment-source areas occupying the eastern flanks of the Taygetos range (Pope 1995). Second, through increasing the energy gradients within sediment-source areas, uplift partially provides the potential for high rates of sediment transfer to transverse drainage systems and the piedmont zone (Beaty 1990). Third, given that the Sparta Basin is physically and structurally separated from the coastal plain of the Gulf of Lakonia (IGME 1977, 1983), continuous regional uplift rather than changes in sea level controlled the local base level of erosion throughout the Quaternary (Pope *et al.* 2003). This sustained uplift resulted in a progressive lowering of base level and drainage basin rejuvenation, which during periods of reduced sediment availability resulted in pronounced incision within upland and lowland drainage systems, and a progressive dissection of fan surfaces (Pope 1995).

The culmination of these processes was a distal shift in the locus of deposition, leading to a short-term reduction or complete cessation of deposition on existing fan surfaces.

A clear distinction can be made between the role of Tertiary and Quaternary faulting in respect of fan development. A (now inactive) group of late Miocene WSW–ENE-trending transfer faults has influenced fan development in two ways. First, the faults subdivided sections of the eastern Taygetos range into discrete blocks (Fig. 6). The boundaries of such blocks served to physically constrain the expansion of some of the transverse drainage systems (cf. Eliet & Gawthorpe 1995), and thus partly controlled potential sediment generation rates throughout the Quaternary. Second, the faults served as zones of pronounced weathering and erosion. The principal channels of nine mountainous drainage systems developed preferentially along transfer faults during the Plio-Pleistocene period, and were guided to the range front along the faults. As a consequence, not only did the transfer faults control the spacing of upland drainage basins, but also determined the points where principal channels emerged on to the piedmont zone and, hence, the locations of several fan systems along the range front.

The extensive, partially segmented Sparta fault that bounds the eastern edge of the Taygetos range has played only a limited role in the morphological development of fan systems during the Holocene. Known seismic activity at the start of the Holocene, during the middle Holocene and in 464 BC. culminated in the formation of the fault scarp and vertically displaced Holocene colluvium along the range front, creating pronounced scarps up to 7 m in height and up to 140 m in length (Pope 1995). In comparison, proximal fan surfaces that are in immediate contact with the fault scarp show no obvious vertical displacement or lateral offsetting. However, the Qf1 surfaces of the South Parorion, Kamares and St Johns fans (Fig. 2; Table 1) are anomalously steep even when the steepness of the drainage basins is taken into account, suggesting a possible tectonic

Table 3. (*continued*)

HFIRM (10^{-5} Am^2 kg^{-1})				SIRM (10^{-5} Am^2 kg^{-1})				SIRM/χ_{lf} (10^{-3} Am^{-1})			
Qf1	Qf2	Qf3	Qf4	Qf1	Qf2	Qf3	Qf4	Qf1	Qf2	Qf3	Qf4
15.86	13.06	11.64	–	1375	1243	975	–	72.60	74.29	76.77	–
15.42	12.18	10.11	–	1303	1089	908	–	70.15	72.04	75.47	–
14.55	12.02	10.68	8.47	1270	1007	937	647	72.20	71.31	80.13	85.34
20.23	15.04	11.02	8.12	1405	1270	923	715	69.45	79.17	83.75	89.45
17.83	15.05	12.65	10.46	1290	905	810	590	64.72	54.32	64.59	74.30

Table 4. *TL and OSL determinations on fine-grained sediment from the St Johns, North Xilocambion, St Savours and Mystras fans*

Alluvial fan	Sample number	Technique	Fan surface	Sediment[1]	Age (ka BP)	Stratigraphic position of unit
St Johns	SPA12[2,3]	TL	Qf3	m. silt–c. silt	37 ± 2	Beneath lower gravels
St Johns	Risø22402	OSL	Qf3	c. silt	29.3 ± 4	Beneath uppermost gravels
St Johns	Risø12402	OSL	Qf4	c. silt–f. sand	14.6 ± 1.4	Beneath lower gravels
St Johns	Risø12403	OSL	Qf5[4]	c. silt–f. sand	0.84 ± 0.05	Beneath uppermost gravels
St Johns	Risø12404	OSL	Qf5[4]	c. silt–f. sand	0.54 ± 0.04	Beneath uppermost gravels
St Saviours	Risø32403	OSL	Qf3	c. silt–f. sand	26.1 ± 2	Middle gravels
St Saviours	Risø32404	OSL	Qf3	c. silt	17.6 ± 1.4	Middle gravels
North Xilocambion	Risø22401	OSL	Qf3	f. sand	66.1 ± 6	Upper middle gravels
North Xilocambion	Risø32401	OSL	Qf3	m. silt–c. silt	32.7 ± 5	Middle gravels
North Xilocambion	Risø22403	OSL	Qf3	f. sand	25.7 ± 2	Beneath uppermost gravels
North Xilocambion	Risø22404	OSL	Qf4	c. silt–f. sand	15.2 ± 1.2	Beneath lower gravels
North Xilocambion	Risø22405	OSL	Qf4	m. silt–c. silt	11.2 ± 0.8	Beneath uppermost gravels
Mystras	SPA-2[3]	TL	Qf1	c. silt–f. sand	235 ± 30	Between lower and middle gravels
Mystras	SPA-1[2,3]	TL	Qf1	c. silt–f. sand	137 ± 13.7	Beneath uppermost gravels

[1] f, fine; m, medium; c, coarse.
[2] TL dates from Pope *et al.* (2003).
[3] TL dates from Wilkinson & Pope (2003).
[4] Sample collected from fill terrace within the trench immediately *downfan* of the distal fan margin.

influence (Pope 1995). Although faulting is not observed on Qf1 surfaces of these fans, rare faulting through debris flows exposed in the fanhead trenches has produced slight vertical offsetting and minor lateral distortion of gravels (cf. Gerson *et al.* 1993). By way of comparison, where faulting is absent from coarse-grained proximal sediments, the overlying fan surfaces only record gradients of between 4.9° and 9.0° (Table 1).

The 3–4 m of vertical displacement that accompanied each of the three Holocene range-front faulting events (Armijo *et al.* 1991) produced the effect of an instantaneous change in the base level of erosion.

However, pulses of rejuvenation associated with each seismic event were restricted to the upland drainage systems and the fanhead area. With regard to the former, renewed incision of bedrock channels produced up to three minor strath terraces within the lowermost sections of the principal stream channels close to the range front. The same incision resulted in the formation of up to three U-shaped notches into the upper portion of the fault and deeply dissected the surface of the fault scarp (Pope 1995). At the fanhead, incision initially deepened the existing trench, and then progressively extended downfan causing entrenchment of medial and distal fan surfaces.

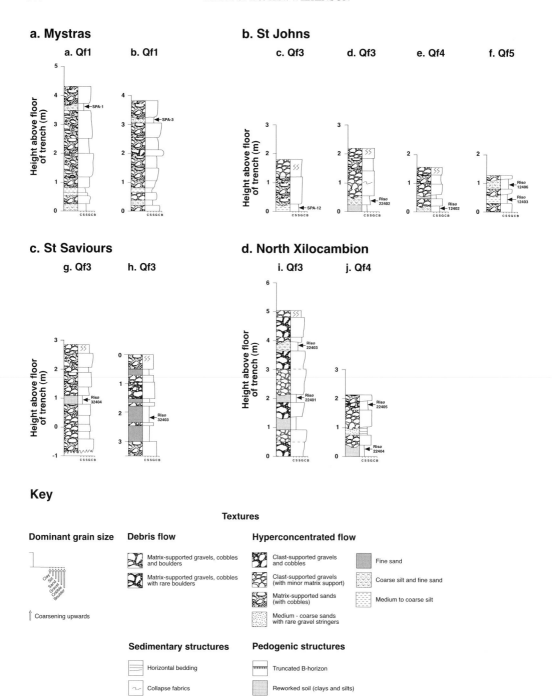

Fig. 5. Detailed lithostratigraphies illustrating the locations of TL and OSL sampling sites within Qf1, Qf2, Qf3 and Qf4 segments of the study fans.

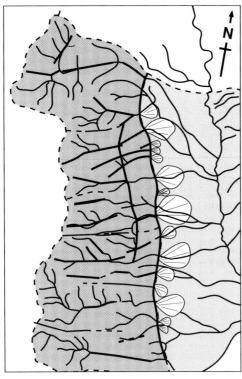

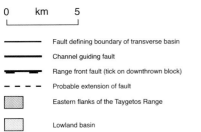

———————	Fault defining boundary of transverse basin
———————	Channel guiding fault
—•——•—	Range front fault (tick on downthrown block)
– – – –	Probable extension of fault
(shaded)	Eastern flanks of the Taygetos Range
(light)	Lowland basin

Fig. 6. The relationship between faulting and the location of transverse drainage systems and piedmont fan systems along the Taygetos range front.

The role of climate

Climate has long been recognized as an important mechanism in the control of sedimentation/erosion on the Sparta piedmont. However, exactly *how* climate acts as a trigger has been a subject to much debate (e.g. Dorn 1994). Bintliff (1977) suggested two phases of deposition distinguished on the basis of sediment calibre and nature of the supporting matrix; the first relating to Würmian 'pluvials' and the second to wet climates at *c.* 1.5 ka BP. It is now known that arid phases characterized by highly seasonal precipitation, rather than pluvials, were the expression of Middle and Upper Pleistocene stadi-

als in the Mediterranean region (Prentice *et al.* 1992), while the wet climatic episode at 1.5 ka BP is not known in southern Europe. In comparison, Hempel's (1987) model of accretion on the Sparta piedmont identified two depositional phases associated with cold and moist phases (*c.* 33–20 and *c.* 16–13 ka BP), and a single depositional phase during a warm and dry phase during the early Holocene (*c.* 10–7 ka BP). Pope & Millington (2000) have proposed that late Würm fan development comprised up to six distinct phases of aggradation, separated by periods of incision. These were tied to alternating cold and dry, and warm and humid conditions, respectively. On the basis of TL dating, Pope *et al.* (2003) were able to place the Sparta fan model into a tentative chronological framework and correlate some of the aggradational phases to known climate events. On a wider scale Macklin *et al.* (2002) have carried out a survey of published fluvial deposits (including alluvial fans in the Peloponnese) with chronometric dates from the Mediterranean region as a whole. They conclude that deposition in the Late Pleistocene is climatically driven and can be correlated with cold stands found during stadials and Heinrich events.

Explaining climate and tectonic influences of the Sparta fans

The recently obtained luminescence chronology from the Sparta Basin enables fan evolution to be considered alongside the palynological and isotopic proxies reviewed earlier, as well as the conclusions reached by Macklin *et al.* (2002).

Initial development phases. Throughout this phase (and all subsequent phases) regional uplift and Miocene-age transfer faults exerted a continuous influence on fan development. During this stage of development net uplift was of the order of 51 m with the volume of sediment delivered to individual fans (South Parorion and Kamares fans excepted) of the order of 2.10×10^5–11.36×10^6 m³ (Fig. 7). TL dating suggests that basal Qf1 debris flow deposits were deposited before *c.* 235 ka BP (i.e. during MIS 6–8: Fig. 8), and may have coincided with major deposition in fan systems in southern Crete (Nemec & Postma 1993). Although the depositional chronology for Qf1 segments is broad, early proximal sedimentation coincided with the extreme aridity of MIS 6, as evidenced by TL dates of *c.* 235 and 137 ka BP for a 2 m-thick gravel unit within the Mystras fan, with a considerably greater thickness of sediments deposited thereafter. Elsewhere in the Mediterranean region MIS 6 fluvial deposition has been identified in Libya and SE Spain (Fuller *et al.* 1998; Rowan *et al.* 2000).

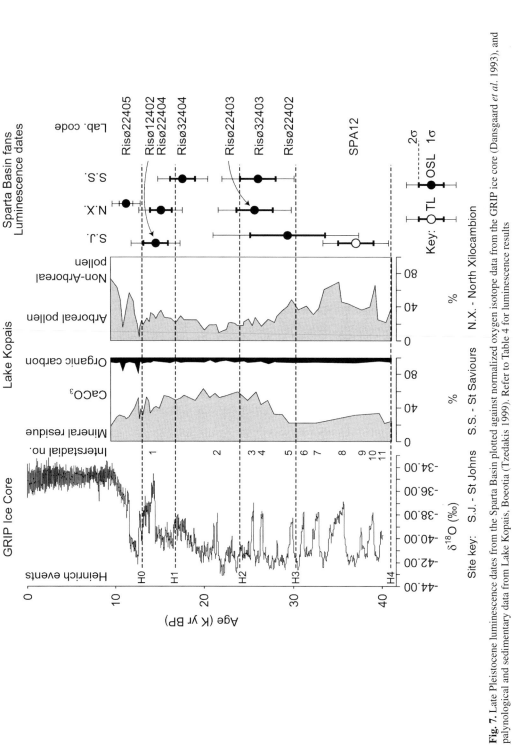

Fig. 7. Late Pleistocene luminescence dates from the Sparta Basin plotted against normalized oxygen isotope data from the GRIP ice core (Dansgaard *et al.* 1993), and palynological and sedimentary data from Lake Kopais, Boeotia (Tzedakis 1999). Refer to Table 4 for luminescence results

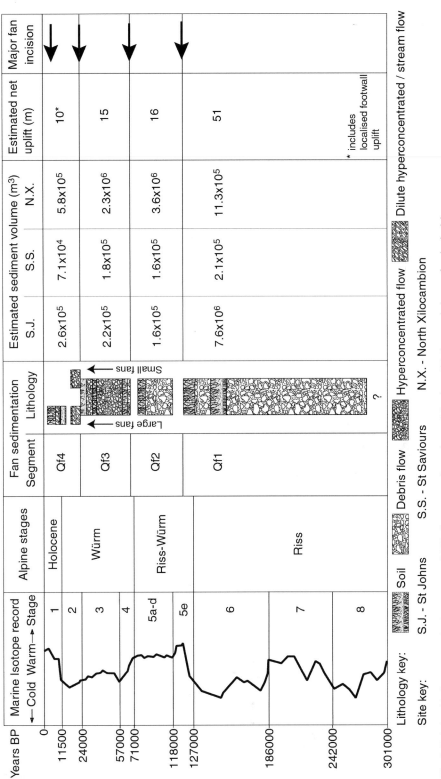

Fig. 8. Relationship between marine isotope stage (MIS), fan segment formation and dominant lithology, and phases of major incision.

Mid-development phases. Between the latter part of MIS 6 and throughout MIS 5, reduced sediment transfer from increasingly vegetated drainage basins coupled with uplift of the order of 17 m initiated fan incision. This phase also marked the abandonment of the Qf1 surfaces and subsequent development of proximal soils. In comparison, records from the eastern Sparta Basin suggest that during MIS 5e minor deposition was occurring on the North Menelaion fan, while the western margins of this fan were being trimmed by the River Evrotas (Wilkinson & Pope 2003). A second phase of fan building, eventually forming Qf2, occurred late in MIS 5 resulting in the deposition of debris flows on the western fans. The timing of individual depositional phases has yet to be fully resolved. However, given the aridity and consequent accelerated input of weathering products into Lake Kopais during MIS 5b (Tzedakis 1999, p. 427, fig. 2), climate change during this substage almost certainly triggered renewed sediment transfer to the Spartan piedmont. Macklin *et al.* (2002) have suggested that the early part of MIS 5e was a period of fan and floodplain incision across the Mediterranean, but note widespread alluvial deposition at the MIS 5b–5a boundary. Climatic and tectonic factors led to between 1.62×10^5 and 3.65×10^6 m^3 of deposition within the study fans, with the complete formation of the Qf2 surfaces occurring before 66 ka BP (Fig. 7).

Late development phases. Morphological evidence strongly suggests that the start of MIS 4 marked a phase of significantly reduced sediment deposition on the piedmont, which against a backdrop of continuing uplift brought about entrenchment of the fanhead and Qf2 surfaces, and a distal shift in the main zone of deposition (Pope unpublished data). There was renewed deposition later in MIS 4 marking the third phase of fan building, and forming the basal sediments of the Qf3 segments. Periods of significantly reduced deposition and concomitant soil development coincided with short-lived periods of woodland expansion during the middle of MIS 4 (i.e. *c.* 66 ka BP) (Tzedakis 1999, p. 429, fig. 3), as well as late interstadials of MIS 3 (Fig. 7 & 8). It is notable that at two standard deviations, only one of the luminescence dates from Qf3 or Qf4 extends to the MIS glacial maximum (*c.* 18–22 ka BP), suggesting that fan gravels were being deposited at this time. It is also apparent that depositional style changed late in MIS 3 or early in MIS 2 to one dominated by hyperconcentrated flows (except in the North Parorion and Kamares fans), as is attested by gravels overlying incipient palaeosols dating to *c.* 26 ka BP in both the North Xilocambion and St Saviours fans. Throughout this phase, tectonic and climate factors culminated in deposition of the order of 1.85×10^5 –2.32×10^6 m^3 of gravels. Dating of Qf4

sediments suggests that deposition on Qf3 surfaces almost certainly ceased before *c.* 15 ka BP. The end of this particular fan-building phase marked the complete formation of the smaller Kamares and North Parorion fan surfaces.

Later development phases. In the Kamares and South Parorion fans, a low discontinuous cut-and-fill terrace provides evidence for the transfer of a small volume of sediment from small range-front drainage basins during MIS 2 (Late Glacial). In contrast, OSL dating from the St Johns and North Xilocambion fans, and field evidence from the St Saviours fan, strongly indicates that the Late Glacial was characterized by a predominantly depositional regime. However, short-lived erosional phases displaced deposition towards the fan margins of Qf3 surfaces, forming the basal and middle gravels of the Qf4 segments. It is very likely that these episodes reflected the well-defined climate changes of the Late Glacial seen at Kopais. Indeed, Late Glacial climate changes have apparently been used by Macklin *et al.* (2002) to explain apparently synchronous deposition of alluvium across the Mediterranean region as a whole at this time. As with previous phases, fan entrenchment can be correlated with a reduction in sediment supply following the expansion of vegetation cover in upland drainage systems, while renewed deposition coincided with decreased vegetation cover (Fig. 8). It is of particular interest in this respect that both the Kopais vegetation record and episodes of stability in the Sparta Basin lag behind amelioration events seen in the oxygen isotope record of the GRIP ice core. These data suggest that it was the impact that climate had on vegetation rather than its association with water availability that was the main determining factor in fan evolution.

Following the deposition of basal and medial gravels, reduced sedimentation coincided with the formation of a Late Glacial soil within several fan systems. During the early and middle stages of MIS 1 regional uplift and two faulting events produced up to 7.5 m of vertical displacement along the front of the Taygetos range, leading to further fanhead entrenchment and the progressive incision of proximal surfaces. Archaeological evidence indicates that incision and soil formation were interrupted by renewed deposition at *c.* 3.6 ka, culminating in the deposition of at least 1.5 m of predominantly fine-grained deposits and the formation of Qf4 surfaces. These events may be associated with population expansion in the Evrotas Valley in the Late Bronze Age and a move to exploit the Taygetos foothills in the Hellenistic period, respectively (Pope *et al.* 2003). The faulting event of 464 BC instigated further incision of Qf4 surfaces, while terraces formed within distal trenches as a result of vertical and lateral incision of backfilled deposits (Pope & Millington 2002). After these events deposition became localized and infrequent, with the final depositional event on Q4 surfaces occurring around 1.3 ka BP (Pope *et al.* 2003). For this developmental stage a mixture of climate, tectonic, and, possibly, human factors resulted in deposition of the order of $7.13 \times 10^4 – 5.8 \times 10^5$ m^3. Around 0.8 ka BP the fan trenches had coupled with the Evrotas river, and at the present day deposition on the fans only occurs as minor terraces inset within the trenches, while all fine-grained material is transferred to the river.

Conclusions

Previous studies of fan evolution in the Sparta Basin have emphasized the importance of processes operating on Holocene timescales (e.g. Pope *et al.* 2003). This is understandable given that: (i) the Holocene surface of the Qf4 segment occupies a greater area than that of any other fan segment; and (ii) Holocene deposits can be relatively easily recognized and dated using artefactual evidence. However, luminescence dating clearly shows that the previous emphasis was misplaced. Fan deposition commenced in the later Middle Pleistocene and subsequently all of the remaining fan segments (except for the uppermost deposits of Qf4) formed during the Upper Pleistocene.

Evidence emerging from further investigations of Spartan fans suggests that tectonic processes exert only a limited influence on patterns of fan development. Long-term tectonic uplift has created the gross relief necessary to partially control sediment generation and sediment transfer rates. In addition, the Tertiary fault system has effectively controlled the positioning of several of the fan systems within the range-front piedmont zone. By comparison, Quaternary faulting along the range front initiated localized uplift, culminating in short phases of renewed fanhead entrenchment and incision of distal-fan surfaces. However, as yet, there is no conclusive evidence to directly link tectonic activity with either the gross morphology or stratigraphy of the Spartan fans. Moreover, the weight of evidence suggests that regional tectonic activity has provided the background against which deposition processes operated rather than being the trigger for such events. It is probably no coincidence that the first major episode of fan sedimentation occurred in MIS 6, the longest and most severe episode of cold and arid climates during the Pleistocene. Thereafter interglacials, possibly the long MIS 5a interstadial, and one or more of the late MIS 3 interstadials are manifested by fan incision and subsequent progradation. Shorter and/or less humid interstadials (e.g. the Late Glacial interstadial) are represented in the fan record by buried palaeosols, but no significant

incision. Major episodes of deposition are represented by lacunae in the luminescence record. These correspond to stadial phases of MIS 2, 3–4, 6 and, possibly, 5b, and also Heinrich events (Fig. 8). The correspondence of the late MIS 2 luminescence dates with increases in arboreal pollen in Lake Kopais, which in turn lag behind climate change seen in the GRIP ice core, suggest that it is the effect of climate change on vegetation in the fan catchment that is the greatest control on fan processes. It is also worth re-emphasizing here that chronometric dates from the Sparta fans define periods of relative fan stability, not deposition. As such if climate change is the main reason for fan evolution, the dates should correspond to warmer intervals. As Figure 8 shows, they broadly do, at least at one standard deviation.

Proxy records from the eastern Mediterranean provided by lake level, sapropel, isotopic and palynological studies suggest that the only significant climate change in the Holocene was from a relatively humid to an arid Mediterranean-type climate at about 6.5–5.5 ka cal BP (e.g. Harrison & Digerfeldt 1993). Although absolute dating evidence is lacking from the Sparta Basin for the Holocene, no fan deposits have yet been attributed to this phase. While it is possible that this trend to arid conditions 'pre-adapted' the landscape to subsequent erosion, as has been suggested by Bintliff (2002), it would nevertheless appear that anthropogenic processes provided a stimulus for later Holocene fan deposition (Pope *et al.* 2003).

The changed views of the chronological development of the Sparta Basin fans that have resulted from luminescence dating provide a salutary lesson. No matter how detailed the geomorphological and stratigraphic examination that is undertaken of alluvial fan systems, no estimate of age obtained by these methods can ever be deemed reliable except in the grossest possible terms. It is only with the application of chronometric dating that a reliable temporal framework can be constructed, and only with such a framework can the triggers of fan-forming processes be independently assessed. Similarly, without an absolute chronology it is impossible to determine rates at which deposition and incision occurred, or indeed how long periods of stasis lasted. Although a detailed chronology of events in the fans of the western part of the Sparta Basin is still lacking, it has nevertheless proven possible in this paper to address some of these issues.

The authors would like to thank J. Walden (University of St Andrews) for allowing access to mineral magnetic facilities; M. Frechen (formerly of the University of Gloucestershire) and A. Murray (University of Aarhus) for undertaking the TL and OSL dating, respectively; A. Mistry (University of Sunderland) for carrying out the iron analysis; and S. Hodgson (University of Derby) for drawing Figures 1–6. Permission to undertake fieldwork in the Sparta Basin was granted by the Institute for Geological and Mineral Exploration (Athens). Fieldwork was supported by grants from the University of Derby, University College Winchester and the British School at Athens.

References

ALEXANDER, E.B. 1985. Estimating relative ages from iron-oxide/total–iron ratios of soils in the Western Po valley, Italy - a discussion. *Geoderma*, **35**, 257–259.

ALLEN, H.D. 1990. A postglacial record from the Kopais Basin, Greece. *In*: BOTTEMA, S., ENTJES-NIEBORG, G. & VAN ZEIST, W. (eds) *Man's Role in the shaping of the Eastern Mediterranean*. Balkema, Rotterdam, 173–182.

ANGELIER, J., CYBERIS, N., LE PICHON, X., BARNER, E. & HUCHON, P. 1982. The tectonic development of the Hellenic Arc and the Sea of Crete: a synthesis. *Tectonophysics*, **86**, 159–196.

ARMIJO, R., LYON-CAEN, H. & PAPANASTASSIOU, D. 1991. A possible normal-fault rupture for the 464 BC Sparta earthquake. *Nature*, **351**, 137–139.

AUBOUIN, J., BONNEAU, M., DAVIDSON, J., LEBOULENGER, P., MATESCO, S. & ZAMBETAKIS, A. 1976. Esquisse structurale de l'arc égéen externe: des Dinarides aux Taurides. *Bulletin of the Society Geologique Française*, **18**, 327–336.

BEATY, C.B. 1961. Topographic effects of faulting: Death Valley California. *Annals of the Association of American Geographers*, **51**, 234–240.

BEATY, C.B. 1990. Anatomy of a White Mountains debris-flow – The making of an alluvial fan. *In*: RACHOCKI, A.H. & CHURCH, M. (eds) *Alluvial Fans: A Field Approach*. Wiley, Chichester, 69–89.

BINTLIFF, J.L. 1977. *Natural Environment and Human Settlement in Prehistoric Greece*. BAR International Series 28 (i and ii). British Archaeological Reports, Oxford.

BINTLIFF, J.L. 2002. Time, process and catastrophism in the study of Mediterranean alluvial history: a review. *World Archaeology*, **33**, 417–435.

BLAIR, T.C. & MCPHERSON, J.G. 1994. Alluvial fans and processes. *In*: ABRAHAMS, A.D. & PARSONS, A.J. (eds) *Geomorphology of Desert Environments*. Chapman & Hall, London, 354–402.

BULL, W.B. 1964. *Geomorphology of Segmented Alluvial Fans in Western Fresno County, California. Erosion and Sedimentation in Semi Arid Environments*. US Geological Survey Professional Paper, **352-E**, 89–129.

CALVACHE, M., VISERAS, C. & FERNANDEZ, J. 1997. Controls on alluvial fan development – evidence from fan morphometry and sedimentology: Sierra Nevada, SE Spain. *Geomorphology*, **21**, 69–84.

DANSGAARD, W., JOHNSEN, S.J. *ET AL.* 1993. Evidence for general instability of past climate from a 250-kyr ice-core record. *Nature*, **364**, 218–220.

DEARING, J.A., HAY, K., BABAN, S., HUDDLESTON, A.S., WELLINGTON, C.M.H. & LOVEBOND, P.J. 1996. Magnetic susceptibility of top soils: a test of conflicting theories using a national database. *Geophysics Journal International*, **127**, 728–734.

DEMITRACK, A. 1986. *The late Quaternary geologic history of the Larissa Plain Thessaly, Greece: Tectonic, climatic, and human impact on the landscape*. PhD Thesis Stanford University, CA.

DENNY, C.S. 1967. Fans and sediments. *American Journal of Science*, **265**, 81–105.

DORN, R.I. 1994. The role of climate change in alluvial fan development. *In*: ABRAHAMS, A.D. & PARSONS, A.J. (eds) *Geomorphology of Desert Environments*. Chapman and Hall, London, 593–615.

DOUTSOS, T. & PIPER, D.J.W. 1990. Listric faulting, sedimentation and morphological evolution of the Quaternary eastern Corinth rift, Greece: First stages of continental rifting. *Bulletin of the Geological Society of America*, **102**, 812–829.

DUCAT, J. 1983. *Actes IV Rencontres Internationales d'Archéologie et d'Histoire*. Centre National de la Recherche Scientifique, Paris, 73–85.

DUFAURE, J.J. 1977. *Carte Geologique Le Peloponnese*. Institut de Geographie Nationale, Paris.

ELIET, P.P. & GAWTHORPE, R.L. 1995. Drainage development and sediment supply within rifts, examples from the Sperchios basin, central Greece. *Journal of the Geological Society, London*, **152**, 883–893.

FRENZEL, B., PÉCSI, M. & VELICHKO, A.A. 1992. *Atlas of Palaeoclimates and Palaeoenvironments of the Northern Hemisphere.*, Gustav Fischer, Stuttgart.

FULLER, I.C., MACKLIN, M.G., LEWIN, J., PASSMORE, D. & WINTLE, A.G. 1998. River response to high-frequency climate oscillations in southern Europe over the past 200 k.y. *Geology*, **26**, 275–278.

GERSON, R., GROSSMAN, S, AMIT, R. & GREENBAUM, N. 1993. Indicators of faulting events and periods of quiescence in desert alluvial fans. *Earth Surface Processes and Landforms*, **18**, 181–202.

GREIG, J.R.A. & TURNER, J. 1974. Some pollen diagrams from Greece and their archaeological significance. *Journal of Archaeological Science*, **1**, 177–194.

HARRISON, S.P. & DIGERFELDT, G. 1993. European lakes as palaeohydrological and palaeoclimatic indicators. *Quaternary Science Reviews*, **12**, 233–248.

HARVEY, A.M. 1990. Factors influencing Quaternary alluvial fan development in southeast Spain. *In*: RACHOCKI, A.H. & CHURCH, M. (eds) *Alluvial Fans: A field Approach*. Wiley, Chichester, 247–269.

HARVEY, A.M. 1997. The role of alluvial fans in arid zone fluvial systems. *In*: THOMAS, D.S.G. (ed.) *Arid Zone Geomorphology: Form and Change in Drylands*, 2nd ed. Wiley, Chichester, 231–259.

HARVEY, A.M. 2002. The role of base-level change in the dissection of alluvial fans: case studies from southeast Spain and Nevada. *Geomorphology*, **45**, 67–87.

HARVEY, A.M. & WELLS, S.G. 1994. Late Pleistocene/ Holocene changes in hillslope sediment supply to alluvial fan systems: California. *In*: MILLINGTON, A.C. & PYE, K. (eds) *Environmental Change in Drylands: Biogeographical and Geomorphological Perspectives*. Wiley, Chichester, 67–84.

HARVEY, A.M., FOSTER, G., HANNAM, J. & MATHER, A.E. 2003. The Tabernas alluvial fan and lake system, southeast Spain: applications of mineral magnetic and pedogenic iron oxide analyses towards clarifying the Quaternary sediment sequences. *Geomorphology*, **50**, 151–171.

HARVEY, A.M., SILVA, P.G., MATHER, A.E., GOY, J.L., STOKES, M. & ZAZO, C. 1999. The impact of Quaternary sea-level and climate change on coastal alluvial fans in the Cabo de Gato ranges, southeast Spain. *Geomorphology*, **28**, 1–22.

HEMPEL, L. 1984. Geoökodynamic im Mittlemeerraum während des Jungquartärs. Beobachtungen zur Frage 'Mensch und/oder Klima?' in Sübgriechenland und uf Kreta. *Geoökodynamic*, **5**, 99–140.

HEMPEL, L. 1987. The 'Mediterraneanization' of the climate in Mediterranean countries – a cause of the unstable ecobudget. *Geojournal*, **14**, 163–173.

HOOKE, R. LeB. 1967. Processes on arid-region alluvial fans. *Journal of Geology*, **75**, 438–460.

HOOKE, R. LeB. 1972. Geomorphic evidence for Late-Wisconsin and Holocene deformation, Death Valley, California. *Bulletin of the Geological Society of America*, **83**, 2073–2098.

IGME. 1977. *Geological Map of Gythion* 1:50 000 Scale. Institute of Geological and Mineral Exploration, Athens.

IGME. 1983. *Geological Map of Xilocambion*, 1:50 000 Scale. Institute of Geological and Mineral Exploration, Athens.

JACOBSHAGEN, V., ST DURR, F., KOCKEL, F., KOPP, K.O., KOWALCZKK, G., BERKHEMER, H. & BUTTNER, D. 1978. Structure and geodynamic evolution of the Aegean region. *In*: CLOSS, H., ROEDER, D. & SCHMIDT, K. (eds) *Alps, Alpennines and Hellenides*. E.Schweizerbart'sche Verlagsbuchhandlung, Stuttgart, 536–564.

KELLETAT, D., KOWAKZYK, G., SCHRODER, B. & WINTER, K.P. 1976. A synoptic view on the Neotectonic development of the Peloponnesian coastal regions. *Zeitschrift der Deutschen Geologischen Gesellschaft*, **127**, 447–465.

KERAUDREN, B. & SOREL, D. 1987. The terraces of Corinth (Greece) – A detailed record of eustatic sea-level variations during the last 500 000 years. *Marine Geology*, **77**, 99–107.

LANGBEIN, W.B. & SCHUMM, S.A. 1958. Yield of sediment in relation to mean annual precipitation. *American Geophysical Union Transactions*, **39**, 1076–1084.

LEEDER, M.R., ORD, D.M. COLLIER, R. 1988. Development of alluvial fans and fan deltas in neotectonic extensional settings: implications for the interpretation of basin-fills. *In*: NEMEC, W. & STEEL, R.J. (eds) *Fan Deltas: Sedimentology and Tectonic Settings*. Blackie, London, 173–185.

LUSTIG, L.K. 1965. *Clastic Sedimentation in Deep Springs Valley, California*. US Geological Survey, Professional Paper, **352-F**, 131–192.

LYON-CAEN, H., ARMIJO, R. ET AL. 1988. The 1986 Kalamata (S. Peloponnesus) earthquake: Detailed study of a normal fault, evidence for East–West extension in the Hellenic Arc. *Journal of Geophysical Research*, **930**, 14 967–15 000.

MACHETTE, M.N. 1985. Calcic soils of the southwestern United States. *In*: WEIDE, D.L. & FABER, M.L. (eds) *Quaternary Soils and Geomorphology of the American Southwest*. Geological Society of America, Special Paper, **203**, 1–21.

MACKLIN, M.G., FULLER, I.C. ET AL. 2002. Correlation of fluvial sequences in the Mediterranean basin over the last 200 ka and their relationship to climate change. *Quaternary Science Reviews*, **21**, 1633–1641.

MAHER, B.A. 1988. Magnetic properties of some synthetic sub-micron magnetites. *Journal of Geophysical Research*, **94**, 83–96.

MILLINGTON, A.C., HARDY, J.R., STYLES, P.J., McEACHRAN, K.M., PENROSE, J.P. & TRIBE, A. 1990. *Establishment of Land Information Systems in the Nomos of Messinia, Greece*. Final Report (Contract No.3591–88–12 ED ISP GB) Department of Geography, University of Reading.

NAYLOR, M.A. 1980. The origin of inverse grading in muddy debris flow deposits: A review. *Journal of Sedimentary Petrology*, **50**, 1111–1116.

NEMEC, W. & POSTMA, G. 1993. Quaternary alluvial fans in southwestern Crete: sedimentation processes and geomorphic evolution. *In*: MARZO, M. & PUIGDEFÁBREGAS, C. (eds) Alluvial Sedimentation. International Association of Sedimentologists, Special Publication, **17**, 235–276.

LE PICHON, X. & ANGELIER, J. 1981. The Aegean Sea. *Philosophical Transactions of the Royal Society of London*, **A300**, 357–372.

PE-PIPER, G.G. & PIPER, D.J.W. 1985. Late-Cenozoic clays and climatic change in the post-orogenic Lakonia graben, southern Greece. *Neues Jahrbuch für Mineralogie Abhandlungen*, **151**, 301–313.

PIPER, D.J.W., PE-PIPER, G., KONTOPOULOS, N. & PANAGOS, A.G. 1982. Plio-Pleistocene sedimentation in the western Lakonia, graben, Greece. *Neues Jahrbuch für Geologie und Paläontologie Monastshefte*, **11**, 679–691.

POPE, R. J. J. 1995. *Late Pleistocene to Late Holocene alluvial fan development, the Sparta Basin, Greece*. PhD Thesis, University of Reading.

POPE, R.J.J. 2000. The application of mineral magnetic and extractable iron (Fe_d) analysis for differentiating and relatively dating fan surfaces in central Greece. *Geomorphology*, **32**, 57–67.

POPE, R.J.J. & MILLINGTON, A.C. 2000. Unravelling the patterns of alluvial fan development using mineral magnetic analysis: Examples from the Sparta Basin, Lakonia, southern Greece. *Earth Surface Processes and Landforms*, **25**, 601–615.

POPE, R.J.J. & MILLINGTON, A.C. 2002. The role of alluvial fans in mountainous and lowland drainage systems: Examples from the Sparta Basin, Lakonia, southern Greece. *Zeitschrift für Geomorphologie Neue Folge*, **46**, 109–136.

POPE, R.J.J., WILKINSON, K.N. & MILLINGTON, A.C. 2003. Human and climatic impact on Late Quaternary deposition in the Sparta Basin piedmont: evidence from alluvial fan systems. *Georachaeology: an International Journal*, **18**, 685–724.

POULIMENOUS, G. & DOUTSOS, T. 1997. Flexural uplift of rift flanks in central Greece. *Tectonics*, **16**, 912–923.

PRENTICE, I.C., GUIOT, J. & HARRISON, S.P. 1992. Mediterranean vegetation, lake levels and palaeoclimate at the last glacial maximum. *Nature*, **360**, 658–600.

RITTER, J.B., MILLER, J.R., ENZEL. Y. & WELLS, S.G. 1995. Reconciling the roles of tectonism and climate in Quaternary alluvial fan evolution. *Geology*, **23**, 245–248.

ROCKWELL, T.K., KELLER, E.A. & JOHNSON, D.L. 1985. Tectonic geomorphology of alluvial fans and mountain fronts near Ventura, California. *In*: MORISAWA, M. & HACK, J.T. (eds) *Tectonic Geomorphology*. Allen & Unwin, Boston, MA, 183–207.

ROWAN, J.S., BLACK, S., MACKLIN, M.G., TABNER, B.J. & DORE, J. 2000. Quaternary environmental change in Cyrenaica evidenced by U–Th, ESR and OSL dating of coastal alluvial fan sequences. *Libyan Studies*, **31**, 5–16.

SCHNEIDER, C. 1986. *Untertsuchungen zur jungQuartären morphogense im becken von Sparta (Peloponnes)*. PhD Thesis, University of Münster.

SMITH, G.A. 1986. Coarse-grained nonmarine volcaniclastic sediment: Terminology and depositional process. *Bulletin of the Geological Society of America*, **97**, 1–10.

TURNER, J. & GREIG, J.R.A. 1975. Some Holocene pollen diagrams from Greece. *Review of Palaeobotany and Palynology*, **20**, 171–204.

TZEDAKIS, P.C. 1999. The last climatic cycle at Kopais, central Greece. *Journal of the Geological Society*, **156**, 425–434.

WATTS, W.A., ALLEN, J.R.M. & HUNTLEY, B.A. 1996. Vegetation history and palaeoclimate of the last glacial period at Lago Grande di Monticchio, southern Italy. *Quaternary Science Reviews*, **15**, 133–153.

WELLS, S.G. & HARVEY, A.M. 1987. Sedimentologic and geomorphic variations in storm-generated alluvial fans, Howgill Fells, northwest England. *Bulletin of the Geological Society of America*, **98**, 182–198.

WELLS, S.G., MCFADDEN, L.D. & DOHRENWEND, J.C. 1987. Influence of late Quaternary climatic changes on geomorphic and pedogenic processes on a desert piedmont, eastern Mojave Desert, California. *Quaternary Research*, **27**, 130–146.

WHITE, K.H. & J. WALDEN, J. 1994. Mineral magnetic analysis of iron oxides in arid zone soils, Tunisian Southern Atlas. *In*: MILLINGTON, A.C. & PYE, K. (eds) *Environmental Change in Drylands: Biogeographical and Geomorphological Perspectives*. Wiley, Chichester, 44–65.

WHITE, K.H., DRAKE, N., MILLINGTON, A.C. & STOKES, S. 1996. Constraining the timing of alluvial fan response to Late Quaternary climatic changes, southern Tunisia. *Geomorphology*, **17**, 295–304.

WILKINSON, K.N. & POPE, R.J.J. 2003. Quaternary alluviation and archaeology in the Evrotas Valley, southern Greece. *In*: HOWARD, A.J., MACKLIN, M.G. & PASSMORE, D. (eds) *Alluvial Archaeology in Europe*. Balkema, Lisse, 187–202.

WINTLE, A. 1991. Luminescence dating. *In*: SMART, P.L. & FRANCES, P.D. (eds) *Quaternary Dating Methods – A Users Guide. Technical Guide 4*. Quaternary Research Association, Cambridge, 108–127.

Luminescence dating of alluvial fans in intramontane basins of NW Argentina

R.A.J. ROBINSON[1], J.Q.G. SPENCER[1], M.R. STRECKER[2],
A. RICHTER[2] & R.N. ALONSO[3]

[1] *School of Geography & Geosciences, University of St Andrews, St Andrews, Fife KY16 9AL,
UK (e-mail: rajr@st-andrews.ac.uk)*
[2] *Institut für Geowissenschaften, Universität Potsdam, P.O. Box 601553, Potsdam, Germany*
[3] *Facultad de Ciencias Naturales-Geologia, Universidad Nacional de Salta, Salta, Argentina*

Abstract: Alluvial fans are sensitive recorders of both climatic change and tectonic activity. The ability to constrain the age of alluvial-fan sequences, individual sedimentary events and the rates of sediment accumulation are key for constraining which mechanisms most control their formation. Recent advances in optically stimulated luminescence (OSL) measurement and analysis have resulted in vast improvements in the dating technique and reliability of age determinations, particularly for OSL dating of quartz grains, and routine application to a wide variety of depositional environments is now possible. Here we apply OSL methods to date a variety of deposits within Late Pleistocene conglomeratic alluvial sequences in NW Argentina. The ages obtained range from 39 to 83 ka and were determined from debris-flow- and fluvial-dominated deposits and lacustrine sequences in intramontane basins bounded by tectonically active mountain ranges with as much as 2 km of relief. With careful choice of facies and sample collection, OSL techniques can be used to date Late Pleistocene, predominately matrix-supported, cobble–conglomerate alluvial deposits.

Alluvial fans are ubiquitous along mountain fronts, and their evolution is tied to tectonic and climatic conditions (Bull 1977; Blair & McPherson 1994, 1998; Harvey 1997). The dating techniques applied to such sediments in the past include radiocarbon methods, radiometric dating of volcanic ash beds, U–Th disequilibria, cosmogenic-ray nuclides, and thermoluminescence (TL) and optically stimulated luminescence (OSL) of feldspar and quartz (e.g. Hooke & Dorn 1992; Clarke 1994; Meyer *et al.* 1995; Tandon *et al.* 1997; Cerling *et al.* 1999; Owen *et al.* 1999; Nott *et al.* 2001; Singh *et al.* 2001; Brown *et al.* 2003; Keefer *et al.* 2003; Tatumi *et al.* 2003). Most previous alluvial-fan luminescence studies have been in material younger than about 10 ka and the oldest independently confirmed date of alluvial fan-related material is 55 ka for fluvial terraces (Tanaka *et al.* 2001; Jain *et al.* 2004). Recent advances in OSL methodology, including improvements to the single-aliquot regeneration-dose (SAR) protocol (e.g. Murray & Wintle 2003), have enhanced the reliability of OSL dating and reduced the error associated with dating quartz grains. At the same time techniques for analysing complex luminescence distributions have advanced, permitting a broader range of environments to be dated with improved accuracy and precision (e.g. Galbraith 1990; Olley *et al.* 1999; Lepper *et al.* 2000; Spencer *et al.* 2003).

In many alluvial-fan environments with sediments older than approximately 50 ka luminescence is the only method available for routine dating of clastic sediments. In this paper we demonstrate how careful sample selection of suitable facies associated with debris- and mud-flow, fluvial and sheetflood deposits can produce OSL ages that are stratigraphically consistent and supported by independent age control. The facies representing unconfined flows at the interface between distal alluvial-fan and lacustrine environments display a normal distribution of equivalent doses, but one of the ages may be overestimated when compared to independent age control. The superbly exposed alluvial fans and intercalated mass-flow, fluvial, sheetflood and lacustrine deposits in the intramontane Quebrada de Humahuaca and Quebrada del Toro of the Eastern Cordillera of NW Argentina constitute a well-suited area to assess the suitability of different facies for OSL dating. The results and OSL techniques are widely applicable to similar environments, and such a dating technique can be used to unravel how surface processes and landscape behaviour evolve during alternating climatic phases in tectonically active environments over 10^2–10^5 year timescales. In particular, this paper demonstrates the suitability of certain facies within alluvial-fan sequences for OSL dating and concludes that, through careful selection of sample sites, successful results from generally perceived problematic environments are possible. The problems that can be encountered in applying the technique to facies with a complex sedimentary history are illustrated and discussed.

From: HARVEY, A.M., MATHER, A.E. & STOKES, M. (eds) 2005. *Alluvial Fans: Geomorphology, Sedimentology, Dynamics*. Geological Society, London, Special Publications, **251**, 153–168. 0305–8719/05/$15 © The Geological Society of London 2005.

Geological, geomorphic and climatic setting

The intramontane basins of the Eastern Cordillera (*c.* 22°–25°S) in the southern central Andes of NW Argentina (Fig. 1) are characterized by tiered Quaternary terraces of alluvial fans, pediments and the aggradational surfaces of lacustrine basins (Fig. 2). Previous work has linked some of the Late Quaternary alluviation to increased frequency of landsliding during humid climatic periods with subsequent bedrock and sediment damming in valleys (Hermanns & Strecker 1999; Trauth *et al.* 2000). Recently published palaeoclimate records from Pleistocene and Holocene lacustrine deposits of NE Chile, southern Bolivia and NW Argentina (Bobst *et al.* 2001; Godfrey *et al.* 2003, Lowenstein *et al.* 2003; Fritz *et al.* 2004) document alternating humid and dry phases with approximately 20 ka periodicity in the southern central Andes during the last 180 ka.

The Eastern Cordillera fold and thrust belt is located east of the intra-Andean Puna Plateau (Fig. 1) and comprises intramontane basins at 2–3 km elevation, which are bound by high-angle reverse-faults uplifting ranges that are 4–5 km in elevation. These NNE- to NNW-oriented ranges are mainly composed of Precambrian–early Paleozoic rocks, whereas the basins are characterized by deformed Late Miocene–Pliocene sedimentary rocks. The sedimentary units are eroded and overlain by predominantly flat-lying Quaternary gravel and boulder conglomerate deposits that constitute terrace, alluvial-fan and pediment cover units. The predominate provenance of the Quaternary gravels in both basins are quartz and lithic arenites and shales of the Precambrian Puncoviscaña Formation, quartz arenites and subfeldspathic arenites of the Cambrian Meson Group, lithic arenites, greywackes

and shales of the Ordovician Santa Victoria Formation, and lithic and quartz arenites and limestones of the Cretaceous Pirgua, Lecho and Yacoraite formations, respectively. The Palaeocene Santa Barbara Group and Tertiary volcanic and granite clasts do contribute a minor component to the clast lithology but their distribution is more localized. In many sequences, all of the above-mentioned clasts may be recycled from within the Miocene–Pliocene formations and the Quaternary cover units.

Recent research on these intramontane basins has resulted in a detailed geodynamic and geological history that includes palaeoclimate proxy studies (Marrett *et al.* 1994; Hermanns *et al.* 2000; Marrett & Strecker 2000; Trauth *et al.* 2000, 2003). These studies show that ongoing local intrabasin faulting and folding, and basement-margin thrusting, lead to base-level changes that are compensated by fluvial incision (Strecker & Hilley 2003). During humid climate phases, critical erosion thresholds are overcome and slopes oversteepen by lateral scouring of rivers, resulting in a higher probability of slope failure. This leads to increased frequencies of landslides and valley impoundment, ultimately leading to aggradation and increased rates of sediment storage (Allen & Hovius 1998; Hermanns & Strecker 1999; Hermanns *et al.* 2000; Hovius *et al.* 2000; Trauth *et al.* 2000). In addition, Strecker & Hilley (2003) have suggested that the interplay between tectonic uplift and fluvial processes in the low-elevation outlets of these intramontane basins may cause aggadation and erosion cycles. So although active tectonics is influencing landscape behaviour, Quaternary deposition recording multiple episodes of alluvial aggradation on timescales of 10^4 years suggests that the most likely driver of aggradation (and incision) is climate change. Ar–Ar dating of tephra, radiocarbon dating and a chronology based on *in situ*-produced cosmogenic-ray nuclides has constrained the timing

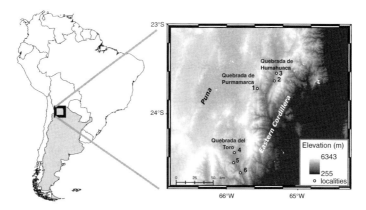

Fig. 1. Study area and sample locations in NW Argentina, with grey shaded digital elevation model created from the 90 m shuttle radar topography mission (SRTM) of South America from JPL/NASA.

of landslide events in the Santa Maria Basin and, to a lesser extent, in the Quebrada del Toro (Fig. 1) about 100 km further north. This synopsis shows that landsliding, damming and lake formation coincide with two regionally documented humid periods at about 35–25 ka ^{14}C calibrated (cal) BP and < 5000 years ^{14}C cal BP (Trauth *et al.* 2000).

The modern climate system is controlled by three competing precipitation sources: the tropical easterlies, the South Atlantic Convergence Zone (SACZ)

Fig. 2. Photographs of sample sites. (**a**) Lower terrace in Quebrada de Purmamarca showing equivalent units to sample PMH200301. (**b**) Inset terraces in the Quebrada de Humahuaca, Río de la Huerta section. (**c**) Purmamarca sample location 1 (see arrow) at the interface of the alluvial fan (base of photograph) and the overlying lacustrine units. Debris-flow deposits overlie the lacustrine units. (**d**) Tilcara section (2) for sample TILC-040303–21 in the Quebrada de Humahuaca (arrow shows person for scale and sample site). Overlying conglomeratic units are debris-flow- and fluvial-dominated. (**e**) Sample horizon for TILC-040303–21. (**f**) Río de la Huerta section (3) for sample HUE060303–30 in the Quebrada de Humahuaca (arrow shows person for scale above sample site). Note debris-flow unit immediately below the sample horizon and the truncation of the sample unit by the erosional terrace surface. (**g**) Location 4 for sample O-080301-05 in the Quebrada del Toro showing the overall coarsening-upwards sequence of a stream-dominated alluvial fan (arrow shows site with gamma spectrometer in the sample hole). (**h**) Sample horizon for O-080301-05 showing empty sample hole and hammer for scale. Horizon below is a small chute-channel infilled with a currently undated volcanic ash. (**i**) Quebrada del Toro north of Sola. An overall fining-upwards sequence illustrating the interface between alluvial-fan and lacustrine deposits. Samples O-080301-04 and -03 come from the base of this sequence. (**j**) Sample location 5 for O-080301-04 by the base of the hammer. (**k**) Sample location 6 for O-140301-08 at Golgota with sample pipe for scale.

and the westerlies (Schwerdtfeger 1976; Godfrey *et al*. 2003). Interannual through to decadal climate variability is influenced by the sea-surface temperatures of the equatorial Atlantic and Pacific oceans, the El Niño Southern Oscillation (ENSO) and the tropical Atlantic sea-surface temperature dipole (TAD) (Lenters & Cook 1999). The position of the intense moisture regions, the Bolivian High and SACZ are affected by ENSO events, extratropical cyclones and the strength of the westerlies (Vuille 1999; Garreaud *et al*. 2003; Trauth *et al*. 2003), and the changing position and intensity of these moisture zones influences the magnitude and timing of precipitation events in the basins of the southern central Andes.

Facies description

The Quaternary deposits in the Quebrada de Humahuaca and the Quebrada del Toro are predominately debris- and mud-flow-, stream-flow- or sheet-flood-dominated, and many vertical sequences of decametre thickness contain all of these types of deposits (Fig. 3). The Quebrada de Purmamarca, a major tributary to the Quebrada de Humahuaca, and the Quebrada del Toro also contain sequences that are characterized by rapid lateral facies transitions from alluvial-fan into lacustrine deposits (Schwab & Schaefer 1976; Trauth & Strecker 1999). The top 20–40 cm of most terraces in both basins have palaeosols that can be developed within debris-flow, hillslope-wash and sheetflood deposits, and thin (<15 cm) horizons of windblown sediment (e.g. at Río de La Huerta, Figs 2f & 3c).

Although conglomerates dominate the sedimentary sequences, many facies contain the fine–medium grain sizes required for standard OSL procedures, although the degree of sorting observed varies. The tops of coarse-grained fluvial bars are commonly marked by thin (20–30 cm), laterally discontinuous (50–100 cm) tabular deposits containing well-sorted, silt–fine sands with thin laminations and local ripple cross-lamination (Fig. 2d, e). These are interpreted as windblown deposits and, in places, laterally discontinuous palaeosols are developed within them. The ripple cross-lamination can display very different flow directions to those measured from imbricated pebbles in the fluvial bar, and in some cases this is similar to the modern up-valley wind direction in austral summer afternoons. Windblown deposits are a common feature in the modern gullies, where sand-sized sediment that is readily available from friable bedrock and from winnowing of bar tops is protected in the lee of bedforms formed on the tops of gravel bars or in chute channels. Other thin (20–40 cm) laterally continuous tabular layers are also developed and consist of very well–moderately well sorted, very

fine–medium-grained sands. These layers show no lamination, display various degrees of weathering and locally contain pebbles, some of which are carbonate-coated. These represent different stages of calcic palaeosol development formed within windblown (and loess) deposits. Other suitable facies present include laterally discontinuous (100–200 cm), dm-thick tabular units that are moderately sorted, medium- to coarse-grained with thin laminations or ripple cross-laminations, and display both fining- and coarsening-upward trends (cm scale); these are interpreted as fluvial bar-top sequences. Similar deposits, except with a lenticular geometry (50–100 cm wide), represent small chute-channel-fill deposits on the tops of fluvial bars (Fig. 2g, h). Thin (10–20 cm), laterally continuous, poorly sorted, medium-grained – granular-sized layers and moderately sorted, medium-grained horizons associated with fine-grained tabular units (dm-thick) are unconfined sheetflood deposits formed at the interface between distal alluvial-fan and lacustrine environments (Fig. 2i–k).

Sampling protocol

Optical dating is now recognized as a valuable tool for dating a range of sediments and environments (Stokes 1999). However, the accuracy and precision of OSL dates is partially controlled by the contribution of unbleached grains within the sample (e.g. Olley *et al*. 1999; Murray & Olley 2002) and this influences which environments will give the most easily interpreted results. Deposits whose sediment-transport processes, prior to final deposition and burial, are likely to give adequate solar exposure that bleaches remnant OSL to a negligible level (such as aeolian deposits) are preferred targets. In intramontane basin settings, the potentially low cumulative transport time of sediment grains may influence the efficiency of bleaching. The grain-size ranges suitable for OSL dating are 4–11 and 90–250 μm; in this project we aimed to collect samples with sufficient material in the 180–212 μm range. Olley *et al*. (1998) studied modern and recent fluvial samples from Australia and determined that this size range contains the best bleached material. Five main facies were sampled: (1) thin, laterally discontinuous windblown deposits on the top of coarse-grained fluvial bars; (2) laterally continuous or locally pervasive palaeosols developed within fine-grained wind-blown (and loess) sediments; (3) medium-grained, medium- to well-sorted fluvial bar tops; (4) chute-channel-fill deposits; and (5) medium- to coarse-grained unconfined flow deposits from the interface of alluvial-fan and lacustrine deposits (Fig. 2c–k). Windblown deposits are interpreted to have a high probability of containing well-bleached grains.

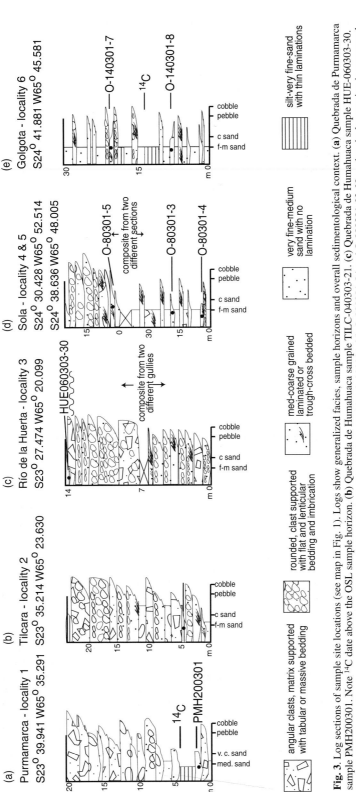

Fig. 3. Log sections of sample site locations (see map in Fig. 1). Logs show generalized facies, sample horizons and overall sedimentological context. (**a**) Quebrada de Purmamarca sample PMH200301. Note ^{14}C date above the OSL sample horizon. (**b**) Quebrada de Humahuaca sample TILC-040303-21. (**c**) Quebrada de Humahuaca sample HUE-060303-30. Note that the log section is composed of two adjacent gullies. (**d**) Quebrada del Toro samples O-080301-05, O-080301-04 and O-080301-03. Note that the log section is composed of two stratigraphically related sequences from different locations. (**e**) Quebrada del Toro samples O-140301-08. Note the ^{14}C date above the OSL sample horizon.

The laterally discontinuous palaeosols are developed within a variety of deposits, including hillslope-wash and reworked sheetflood deposits, and could contain partially and heterogeneously bleached grains. Modern fluvial bar-top and chute-channel-fill sediments can also contain a mixture of bleached and unbleached grains (Rhodes & Pownell 1994; Murray et al. 1995; Olley et al. 1998, 1999; Jain et al. 2004) depending on the sediment-transport mechanisms and the turbidity of the flow. Depending on the turbidity and concentration of the flow, unconfined (sheetflood) flows at the interface between alluvial-fan and lacustrine environments could produce a range of bleaching efficiencies during transport.

Although some of the aggradational fill units in the Quebrada de Humahuaca are dominated by conglomerates and mass flow deposits, we have been able to sample suitable horizons in most units at several localities throughout the basin. Alluvial-fan and lacustrine deposits were sampled in Quebrada del Toro, which is, relatively speaking, less conglomeratic than the Quebrada de Humahuaca (Fig. 2). Five representative log sections (Fig. 3) illustrate the vertical facies variability and the overall context of the depositional environment sampled.

Samples were collected in 4 cm-diameter, 16 cm-long, black PVC tubes. One end was capped, the open end was tapped into the sediment until the tube was filled, and then the tube was extracted and immediately capped. For two of the samples (Table 1) the sample hole was then widened with a 6 cm-diameter auger and a portable EG&G ORTEC MicroNOMAD gamma spectrometer, comprising a 2-inch NaI crystal, photomultiplier base and MCA, was inserted for 30 min in order to measure the spectrum of environmental radiation due to gamma rays, and, hence, to calculate U, Th and K concentrations. The dosimetry for the remaining samples (Table 1) was determined by neutron activation analysis (NAA) at Becquerel Laboratories in Australia. A detailed record comprising sedimentary logs, photographs and location information (longitude, latitude, altitude, depth of overburden) was made at each sampling site. Altitude measurements were taken using a Garmin Vista Etrek GPS and a Suunto barometric digital altimeter. The digital altimeter was reset to known benchmarks twice per day. Depths of overburden and terrace heights were measured using a laser Impulse 200 rangefinder or the digital altimeter.

OSL methodology

In the luminescence laboratories at St Andrews, standard preparation techniques (e.g. Spencer & Owen 2004) were employed. About 2–3 cm of sediment, which may have been exposed to daylight during sampling, was removed from each end of the tube before the sediment used for dating was extracted. Estimates of in situ water content (mass of moisture/dry mass) were calculated by recording the mass of sediment before and after weight stabilization in a 50 °C oven. A quartz fraction of the 180–212 μm-size range was obtained by sieving, HCl and H_2O_2 treatments, heavy liquid density separation and 40% HF etching for 40 min. The density separation was used to separate a <2.70 g cm^{-3} density fraction, and this fraction was treated with HF to dissolve feldspars and remove the surface few microns from the quartz grains to minimize the luminescence due to ionization by alpha particles external to the grains. All quartz samples were screened for feldspar contamination using infrared stimulated luminescence (IRSL). Samples with IRSL signals were given further acid treatments (HF and/or H_2SiF_6) and the use of infrared bleaching before luminescence measurements (see below) was investigated in order to remove any remaining contaminant feldspar component from the quartz signals (cf. Wallinga et al. 2002). Standard quartz aliquots comprising of an approximate 5 mm-diameter circle of grains were dispensed onto stainless steel discs, lightly coated with silicon spray, for luminescence measurements.

OSL measurements were made using Risø TL/OSL-DA-15 readers. Luminescence from the quartz grains was stimulated with either blue–green light (420–550 nm) from a filtered halogen lamp and liquid light guide (Bøtter-Jensen 1997) or blue light (470 ± 30 nm) from a blue diode array, with detection using Hoya U-340 filters and a 9635QA photomultiplier tube. Presence of contaminant feldspar grains was confirmed if samples emitted luminescence when stimulated using infrared (IR) diodes. For certain samples IR signals were present even after repeated acid etching and/or leaching. For these samples, a pretreatment of IRSL was used before stimulation with blue–green light. The equivalent dose (D_e) for each quartz aliquot was measured using the SAR protocol (Murray & Wintle 2000, 2003). Internal assessment of the reliability of measured SAR data was monitored using the following tests: ability to recover a known laboratory radiation dose; success of sensitivity correction by replication of regenerated OSL at low and high doses; absence of thermal transfer; and stability of D_e with preheat variation. Several (18–43) aliquots of quartz were measured for each sample and D_e distributions were analysed using histograms, probability density functions (PDFs), radial plots and statistical tests of normality. The data are presented using several graphical methods and statistical tests in order to highlight the distribution behaviour for each facies.

Table 1. *Dosimetry data*

Sample	U (ppm)	Th (ppm)	K (%)	Rb (ppm)	H$_2$O content* (%)	Cosmic dose rate[†] (mGy a^{-1})	Total dose rate (mGy a^{-1})
PMH200301	2.04±0.08	12.4±0.10	2.72±0.04	130±3.79	1.38	0.003	3.93±0.18
TILC040303–21‡	2.48±0.12	9.52±0.29	2.06±0.03	100±10	0.82	0.022	3.27±0.15
HUE060303–30‡	2.25±0.12	9.10±0.28	1.53±0.03	100±10	2.31	0.259	2.92±0.14
O-080301–05	1.61±0.10	11.4±0.09	2.89±0.04	138±3.77	1.04	0.138	4.08±0.18
O-080301–04	1.14±0.15	10.7±0.09	1.95±0.04	91.2±4.01	3.72	0.013	2.80±0.13
O-080301–03	2.36±0.07	9.23±0.07	2.76±0.04	131±3.79	2.25	0.038	3.82±0.17
O-140301–08	1.81±0.09	9.43±0.08	2.29±0.04	106±3.89	1.00	0.009	3.28±0.15

* Per cent moisture compared to dry weight. Uncertainty taken as 5%.

[†] Cosmic dose rate calculated assuming constant burial depth using the method described in Prescott & Hutton (1994). Uncertainty taken as 10%.

‡ For these samples U, Th and K were calculated from on-site gamma spectra, and Rb was taken as 100 ± 10 ppm. U, Th, K and Rb for all other samples measured using NAA.

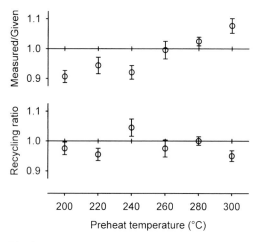

Fig. 4. Dose recovery tests for sample O-080301-05 show dose recovery (measured/given dose) and recycling ratios within 0.9–1.1 of the test thresholds over a 10-s preheat range of 200–300 °C. For this sample a preheat of 260 °C for 10 s was chosen. The dose recovery and recycling data are typical of the behaviour of all samples.

Results

The OSL samples presented have all passed the luminescence behaviour tests mentioned above. All samples have measured/given ratios from dose recovery tests of between 0.9 and 1.1, all the accepted aliquots for a sample have recycling ratios from repeat regenerative doses of between 0.9 and 1.1, and recuperation values are less than 5% of the sensitivity corrected natural signal (Fig. 4). The scatter in D_e distribution evident in many samples (see Figs 5 & 6) masks indication of preheat plateau conditions, and therefore dose recovery tests with preheat variation were employed to support selection of appropriate preheat values. Figure 5a shows the D_e distribution plotted as a histogram and a PDF for a sample from the Quebrada de Purmamarca (Fig. 1). Sample PMH200301 is from a dm-thick, medium-grained, well-sorted sand that is overlain by a fine-grained, organic lacustrine unit. The depositional environment is interpreted as the edge of an alluvial fan at the interface of a small lake (Figs 2a, c & 3a). All histograms display the binned D_e distributions with a bin width equal to the averaged standard error of all the measurements, while the PDF is independent of bin width and, in addition, reflects the precision attributed to each of the D_e measurements (Fig. 5a). The luminescence decay curves for the natural dose, a regeneration dose of 168 Gy and associated test doses of 11 Gy (T1 after the natural and T2 after 168 Gy, respectively) illustrate the typical shape of the luminescence responses from a single aliquot of PMH200301 (Fig. 5b). All the SAR

data for this aliquot demonstrate the non-linearity of growth with increasing regeneration dose, as well as typical recycling and recuperation behaviour (Fig. 5c). Figure 5d shows the radial plot of the D_e data and the two *x*-axis scales represent increasing precision and decreasing relative error of each D_e value. The histogram, PDF and radial plot (Fig. 5) graphically show that the D_e values tend towards a normal distribution with a small positively skewed tail. Table 2 presents values for the mean, median, 1σ, Shapiro Wilk test of normal behaviour (Shapiro & Wilk 1965), skewness and kurtosis for all distributions. For a 95% confidence interval, this sample is normally distributed and tends towards mesokurtic behaviour (the size of the tails expected in a normal distribution). For consistency, all the sample ages are determined using the median D_e value (cf. Murray & Funder 2003), as two samples fail the normality test with a 95% confidence interval. For all our results, this produces median values that are less than, and greater than, the mean value for positively and negatively skewed distributions, respectively. The errors associated with the two skewed distributions are reported as standard deviations (SDs), while the normally distributed samples are reported as standard errors (SEs). We present our median age values and errors in Table 2. The Purmamarca sample (PMH200301) is associated with an accelerator mass spectrometry (AMS) ^{14}C date of charcoal (May 2003) from an overlying (within 5 m) silt- to clay-sized lacustrine deposit rich in organic remains; the OSL age of 47.6 ± 2.8 ka (Table 2) and the AMS ^{14}C age of $49\,550 \pm 1700$ BP compare very well.

Figure 6 shows the D_e distributions from windblown, fluvial and sheetflood samples (see Fig. 3 for their stratigraphic and sedimentological context). Figure 6a represents a lens of windblown silt- to coarse-sand-sized material overlying a fluvial bar deposit from the Quebrada de Humahuaca (Figs 2d, e & 3b: Tilcara TILC040303-21); the range of D_e values is large but the histogram, PDF, radial plot and statistical tests (Table 2) confirm that the D_e values are normally distributed but with a slightly larger tail (leptokurtic) than expected in a normal distribution. A palaeosol (Fig. 6b) from the same quebrada (Figs 2f & 3c: Río de La Huerta HUE060303-30) that developed within a laterally continuous well-sorted, fine- to medium-grained windblown material (loess) has a normal, but negatively skewed, distribution with a slightly smaller tail (platykurtic) than expected in a normal distribution (Table 2). The resulting ages from these deposits are 83.1 ± 4.7 and 39.4 ± 2.3 ka, respectively (Table 2).

The remaining four samples in Figure 6 are from the Quebrada del Toro (Fig. 1). Three are stratigraphically related and consistent. A small chute-channel-fill deposit in a distal alluvial-fan setting

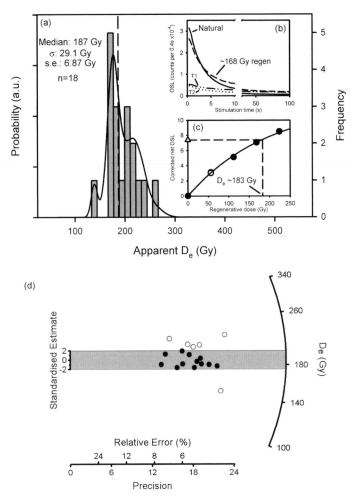

Fig. 5. Luminescence data for sample PMH200301 from the Quebrada de Purmamarca, location 1. (**a**) Histogram (grey bars) and probability density function (PDF: solid line) for 18 estimates of equivalent dose (D_e). Dashed line indicates the median value of 187 Gy. (**b**) Typical decay curves from a quartz aliquot for the natural (solid curve) and *c.* 168 Gy regenerated (dashed curve) luminescence and associated *c.* 11 Gy test dose data (T1 and T2). T1 (dashed–dotted curve) subsequent to natural measurement and T2 (dotted curve) after approximately 168 Gy regenerated dose data. (**c**) Net OSL data corrected for test dose response from the complete SAR measurement for the same aliquot as in (b), showing the luminescence from the natural dose (open triangle), regeneration doses of 56, 112, 168, 224 and 0 Gy (solid circles), and a recycled regeneration point at 56 Gy (open circle). The regenerated points are fitted with a saturating exponential function (solid line) and the D_e is estimated by interpolation (dashed line). The D_e is similar to the median value. (**d**) Radial plot of the data. Horizontal lines represent 2σ and solid circles represent aliquots that fall within 2σ of the median. The right-hand curve represents a logarithmic scale of D_e (Gy) and the lower scale bars illustrate the error associated with each aliquot in relative error (%) and precision ($1/\sigma$).

(Figs 2g, f & 3d: O-080301-05) produces an asymmetrical distribution, and the PDF and radial plot both show that some of the lower D_e values have slightly higher precision (Fig. 6c); the distribution fails the normality test (Table 2) and is positively skewed and leptokurtic. The age of this sample is 41.9 ± 13.3 ka (using SD). If we compare this to the age that would result from choosing the leading

peak as the realistic burial D_e, the age would be approximately 35 ka. Two samples were collected from deposits stratigraphically underlying O-080301-05 at a location about 10 km away (Fig. 3d); the approximately 30 m-thick fining-upwards sequence of tabular cobble–pebble conglomerate alluvial-fan deposits grades upwards into fine-grained lacustrine deposits (Fig. 2i). O-080301-04

Table 2. *Equivalent dose (D_e) data, total dose-rates and OSL ages for 1802–12 µm quartz*

Sample	n*	Mean D_e (Gy)	Median D_e (Gy)	1-sigma (Gy)	Shapiro–Wilk, W[†]	Skewness[‡]	Kurtosis[‡]	Total dose rate (mGy a^{-1})	Age[§] (ka)
PMH200301	18	195	187	29.1	0.96	0.33	−0.10	3.93±0.18	47.6±2.8
TILC040303–1	26	276	272	45.0	0.96	0.61	0.12	3.27±0.15	83.1±4.7
HUE060303–0	28	114	115	19.7	0.95	−0.61	−0.14	2.92±0.14	39.4±2.3
O-080301–5	19	178	171	53.7	0.87	1.30	1.20	4.08±0.18	41.9±13.3
O-080301–4	39	204	189	58.3	0.93	0.88	0.40	2.80±0.13	67.5±21.0
O-080301–3	37	231	231	45.4	0.94	0.92	1.66	3.82±0.17	60.3±3.3
O-140301–8	43	200	195	42.4	0.97	0.54	0.04	3.28±0.15	59.4±3.3

* Number of replicated D_e estimates.

[†] Shapiro–Wilk test of normality. For normality W should be greater than 0 and approach unity. Samples O-080301–4 and O-080301–5 have p values <0.05, indicating non-normal distributions with a 95% confidence interval.

[‡] Skewness values >0 indicate positive skewness, whereas values <0 indicate negative skewness. Low p values (<0.05) for samples O-080301–03, O-080301–4 and O-080301–05 indicate skewness is significantly non-normal. Kurtosis values >0 indicates distribution is more platykurtic than normal, and values <0 are more leptokurtic.

[§] Ages calculated using median D_e. The errors for the two samples that fail the normality test are presented as standard deviations.

is the lower sample collected from a dm-scale fining-upwards sequence (imbricated pebble conglomerate–moderately sorted medium-coarse-grained sand) that represents a bar-top deposit (Fig. 2j). The D_e distribution displays some asymmetry and just fails the normality test; it has a positive, but moderate, skewness and is slightly leptokurtic (Fig. 6d; Table 2). High precision associated with the lowest D_e value produces the small peak on the PDF and corresponds with the highest precision outlier on the radial plot (Fig. 6d). A cluster of D_e values in the radial plot (equal to the second mode in the histogram) is interpreted as a possible older, unbleached population of D_e. The overlying sample, O-080301-03 (Fig. 3d), is from a moderately sorted, medium-grained, laterally continuous layer associated with thin fine-grained, lacustrine deposits. The geometry of the horizon and its facies association with fine-grained lacustrine horizons suggests that it is deposit resulting from an unconfined flow. It passes the normal distribution test, has a moderate, positive skewness and also tends towards leptokurtic behaviour (Table 2; Fig. 6e). Although the median D_e values are very different for these two samples, the ages are consistent (67.5 ± 21.1 ka overlain by 60.3 ± 3.3 ka) because sample O-080301-04 has a lower dose rate compared to O-080301-03. Samples O-080301-04, O-080301-03 and O-080301-05 from different locations are stratigraphically consistent, producing ages of 67.5 ± 21.1, 60.3 ± 3.3 and 41.9 ± 13.3 ka, respectively (Table 2; Fig. 3d). The vertical separation between samples -03 and -05 is not known exactly, but is less than about 80 m.

The final sample (Golgota O-140301-08) is from a similar environment to O-080301-03; it is from a poorly-sorted, coarse-grained, fining-upwards horizon associated with m-scale fine-grained units, and also represents an unconfined flow deposit at the interface between an alluvial-fan and lacustrine environment (Figs 2k, 3e & 6f). It has a normal distribution and yields a median OSL age of 59.4 ± 3.3 ka. AMS ^{14}C dating of freshwater snails (*Heleobia* sp.), sampled from an overlying horizon within the same section (Fig. 3e), produces an age of 30 050 ± 190 years BP (Trauth & Strecker 1999). Although this seems stratigraphically consistent, there is no obvious unconformity within this section that could explain the very slow accumulation rates arising from these ages (*c.* 0.02 mm year⁻¹). Our OSL median age would appear to be an overestimate and an inappropriate parameter as an estimator of burial age for this sample. The first leading peak of the PDF, or a trend line through the lowest D_e values on the radial plot (Fig. 6f), produces an D_e of approximately 130 Gy, which gives an OSL age of about 40 ka (*c.* 0.03 mm year⁻¹). An overlying sample in similar facies (O-140301-07) had very poor luminescence characteristics and no reliable D_e values could be produced.

Discussion

This paper presents D_e distributions using OSL procedures for different facies collected from three quebradas in NW Argentina, and the climatic and tectonic significance of these ages will be presented in a separate article. The results demonstrate that a variety of facies associated with alluvial-fan and fluvial-terrace environments are suitable for OSL dating. Although asymmetry is shown in some of the histograms, PDFs and radial plots, the contribution of the tails is small in most cases and the distributions pass the Shapiro–Wilk normality test. Therefore, partial bleaching is not thought to have a dominant effect. Only O-080301-04 and O-080301-05 have non-normal distributions at the 95% confidence interval, and the latter sample has the highest skewness. Sample O-080301-05 represents a chute-channel-fill deposit. The fluvial bar-top deposit, sample 080301-04, may contain a mixed population of older and younger sediments (Fig. 6d). The sedimentary processes operating within the depositional environment of samples PMH200301, O-080301-03 and O140301-08 (the interface between an alluvial-fan and a lacustrine setting) are interpreted to be dominated by shallow unconfined flows that provide the opportunity for efficient bleaching of grains to occur. The two samples from windblown deposits (TILC040303-21 and HUE-060303-30) are normally distributed; they also represent the oldest (83.1 ± 4.7 ka) and youngest samples (39.4 ± 2.3 ka), respectively. These subenvironments seem very suitable for OSL dating and both of these facies are fairly common deposits in the quebradas of NW Argentina. It is possible that sample O-140301-08 may represent a turbidite deposit formed by subaqueous (unconfined) flow on the margin of a lake. However, it has a normal D_e distribution and does not appear to show evidence of poorly bleached grains within the sample. Our leading edge OSL age is consistent with the ^{14}C date of freshwater snails measured from sediments deposited above our horizon if accumulation rates were very low in these lakes or an unrecognized unconformity exists between these two horizons. We discuss future investigations of the distributions below.

In young fluvial-related samples, the asymmetry detected in two of the seven samples presented would usually be interpreted to reflect insufficient solar exposure of some of the grains before burial (e.g. Olley *et al.* 1999; Murray & Olley 2002; Jain *et al.* 2004), although heterogeneous beta dose received by individual grains during burial (Murray & Roberts 1997) can also cause skewed D_e

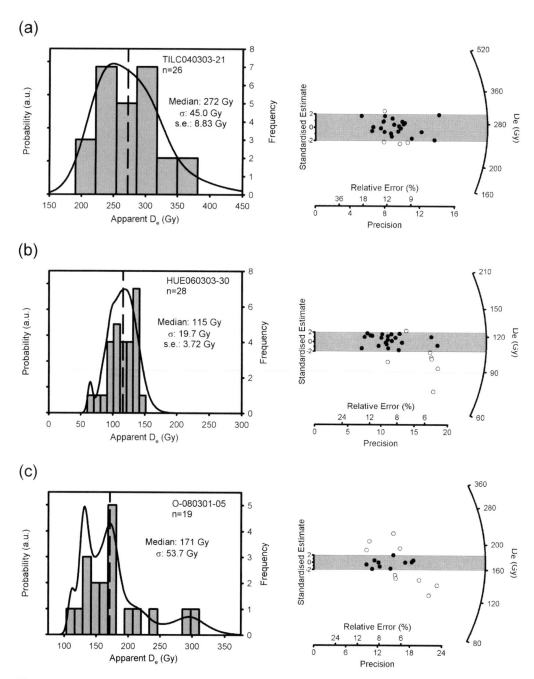

Fig. 6. D_e distribution data for six samples from the Quebrada de Humahuaca and Quebrada del Toro. Combined histogram and PDF plots with median D_e and error are shown on the left, and radial plots are shown on the right. (**a**) Sample TILC-040303-21 from Tilcara, location 2. (**b**) Sample HUE060303-30 from Río de la Huerta, location 3. (**c**) Sample O-080301-05 from approximately 16 km north of Sola, location 4.

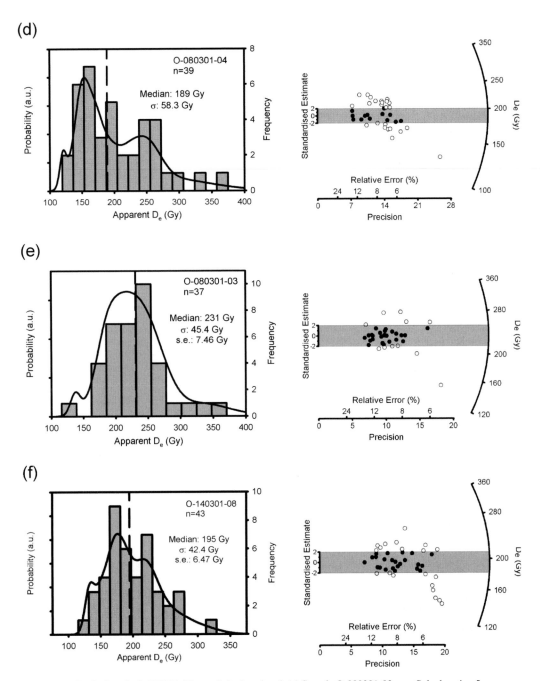

Fig. 6. *(contd.)* (**d**) Sample O-080301-04 near Sola, location 5. (**e**) Sample O-080301-03 near Sola, location 5. (**f**) Sample O-140301-08 from Golgota, location 6.

distributions. Asymmetry is currently not thought to be caused by partial bleaching in older deposits, but can occur as a result of interpolating a normal distribution from OSL measurements onto a saturating exponential function (Murray & Funder 2003). Further work on these samples is needed to test whether the dominant peaks in the distributions represent the dose accumulated during final burial with the use of small aliquot techniques of 50–100 grains (and some single-grain analyses) and statistical analyses that deconvolve complex distributions (e.g. Galbraith 1990; Roberts *et al.* 2000; Spencer *et al.* 2003). If multiple populations, due to samples being composed of a mixture of bleached and unbleached grains (i.e. O-080301-04), can be deconvolved, then objective assignment of a mean and standard error can be made, enabling us to considerably improve both precision and accuracy in older sediments with a short cumulative transport time or a complex depositional history.

Conclusions

In this paper we have presented the results of standard-sized single-aliquot OSL dating of quartz grains collected from windblown, fluvial, sheetflood and lacustrine-related facies associated with alluvial fans that are dominated by cobble–boulder conglomerates, and have demonstrated that they are appropriate facies for OSL dating. Four of the five facies chosen produce a normal distribution of D_e. Our data have yielded OSL ages of 83.1 ± 3.2–37.7 ± 8.1 ka. Sample PMH200301 has excellent concordance with the AMS ^{14}C date and three samples from two different localities show stratigraphic consistency. Our median age of a sample collected from a lacustrine-related unconfined flow deposit appears to overestimate an AMS ^{14}C date available from an overlying deposit, although the leading peak age calculated for this sample is more consistent with the AMS ^{14}C date. Further small aliquot measurements and statistical deconvolution methods to analyse multiple population data are in progress.

Acknowledgment is made to the donors of the Petroleum Research Fund, administered by the American Chemical Society, and The Carnegie Trust for the Universities of Scotland for support of this research. R.N. Alonso, A. Richter and M.R. Strecker thank the German Research Council Deutsch Forschungsemensdaft for financial support. We are very grateful to A. Gemmell for supplementary machine time and to J.-H. May for making his Master's thesis on the Quebrada de Purmamarca, including the AMS ^{14}C date of charcoal by Prof. Dr P.M. Grootes of the Christian-Albrechts-Universität, Kiel, available. G. Duller and A. Murray are acknowledged for all their helpful suggestions and discussions. Many thanks go to E. Rhodes for help in calibrating the gamma spectrometer.

References

ALLEN, P.A. & HOVIUS, N. 1998. Sediment supply from landslide-dominated catchments: implications for basin margin fans. *Basin Research*, **10**, 19–35.

BLAIR, T.C. & MCPHERSON, J. 1994. Alluvial fans and their natural distinction from rivers based on morphology, hydraulic processes, sedimentary processes, and facies assemblages. *Journal of Sedimentary Research*, **A64**, 450–489.

BLAIR, T.C. & MCPHERSON, J. 1998. Recent debris-flow processes and resultant form and facies of the dolomite alluvial fan, Owens Valley, California. *Journal of Sedimentary Research*, **68**, 800–818.

BOBST, A.L., LOWENSTEIN, T.K., JORDAN, T.E., GODFREY, L.V., KU, T.-L. & LUO, S. 2001. A 106 ka paleoclimate record from drill core of the Salar de Atacama, northern Chile. *Palaeogeography, Palaeoclimatology, Palaeoecology*, **173**, 21–42.

BØTTER-JENSEN, L. 1997. Luminescence techniques: instrumentation and methods. *Radiation Measurements*, **27**, 749–768.

BROWN, E.T., BENDICK, R., BOURLES, D.L., GAUR, V., MOLNAR, P., RAISBECK, G.M. & YIOU, F. 2003. Early Holocene climate recorded in geomorphological features in Western Tibet. *Palaeogeography, Palaeoclimatology, Palaeoecology*, **199**, 141–151.

BULL, W.B. 1977. The alluvial-fan environment. *Progress in Physical Geography*, **1**, 222–270.

CERLING, T.E., WEBB, R.H., POREDA, R.J., RIGBY, A.D. & MELIS, T.S. 1999. Cosmogenic He-3 ages and frequency of late Holocene debris flows from Prospect Canyon, Grand Canyon, USA. *Geomorphology*, **27**, 93–111.

CLARKE, M.L. 1994. Infrared stimulated luminescence ages from aeolian sand and alluvial-fan deposits from the Eastern Mojave Desert, California. *Quaternary Science Reviews*, **13**, 5–7.

FRITZ, S.C., BAKER, P.A. *ET AL.* 2004. Hydrologic variation during the last 170,000 years in the southern hemisphere tropics of South America. *Quaternary Research*, **61**, 95–104.

GALBRAITH, R.F. 1990. The radial plot: graphical assessment of spread in ages. *Nuclear Tracks and Radiation Measurements*, **17**, 207–214.

GARREAUD, R., VUILLE, M. & CLEMENT, A.C. 2003. The climate of the Altiplano: observed current conditions and mechanisms of past changes. *Palaeogeography, Palaeoclimatology, Palaeoecology*, **194**, 5–22.

GODFREY, L.V., JORDAN, T.E., LOWENSTEIN, T.K. & ALONSO, R.A. 2003. Stable isotope constraints on the transport of water to the Andes between 22° and 26° S during the last glacial cycle. *Palaeogeography, Palaeoclimatology, Palaeoecology*, **194**, 299–317.

HARVEY, A.M. 1997. The role of alluvial fans in arid zone fluvial systems. *In*: THOMAS, D.S.G. (ed.) *Arid Zone Geomorphology: Process, Form and Change in Drylands*, 2nd edn. Wiley, Chichester, 233–259.

HERMANNS, R. & STRECKER, M.R. 1999. Structural and lithological controls on large Quaternary bedrock landslides in NW-Argentina. *Bulletin of the Geological Society of America*, **111**, 934–948.

HERMANNS, R.L., TRAUTH, M.H., NIEDERMANN, S., MCWILLIAMS, M. & STRECKER, M.R. 2000.

Tephrochronological constraints on temporal distribution of large landslide in Northwest Argentina. *Journal of Geology*, **108**, 35–52.

HOOKE, R.LeB. & DORN, R.I. 1992. Segmentation of alluvial fans in Death Valley, California – new insights from surface exposure dating and laboratory modelling. *Earth Surface Processes and Landforms*, **17**, 557–574.

HOVIUS, N., STARK, C., HAO-TSU, C. & JIUN-CHUAN, L. 2000. Supply and removal of sediment in a landslide-dominated mountain belt. *Journal of Geology*, **108**, 73–89.

JAIN, M., MURRAY, A.S. & BØTTER-JENSEN, L. 2004. Optically stimulated luminescence dating: how significant is incomplete light exposure in fluvial environments? *Quaternaire*, **15**, 143–157.

KEEFER, D.K., MOSELEY, M.E. & DEFRANCE, S.D. 2003. A 38000-year record of floods and debris flows in the Ilo region of southern Peru and its relation to El Nino events and great earthquakes. *Palaeogeography, Palaeoclimatology, Palaeoecology*, **194**, 41–77.

LEPPER, K., AGERSNAP LARSEN, N. & McKEEVER, S.W.S. 2000. Equivalent dose distribution analysis of Holocene aeolian and fluvial quartz sands from Central Oklahoma. *Radiation Measurements*, **32**, 603–608.

LENTERS, J.D. & COOK, K.H. 1999. Summertime precipitation variability over South America: Role of the large-scale circulation. *Monthly Weather Review*, **127**, 409–431.

LOWENSTEIN, T.K., HEIN, M.C., BOBST, A.L., JORDAN, T.E., KU, T.L. & LUO, S. 2003. An assessment of stratigraphic completeness in climate-sensitive closed-basin lake sediments: Salar de Atacama, Chile. *Journal of Sedimentary Research*, **73**, 91–104.

MARRETT, R.A & STRECKER, M.R. 2000. Response of intracontinental deformation in the Central Andes to late Cenozoic reorganization of South American Plate motions. *Tectonics*, **19**, 452–467.

MARRETT, R.A., ALLMENDINGER, R.W., ALONSO, R.N. & DRAKE, R.E. 1994. Late Cenozoic tectonic evolution of the Puna Plateau and adjacent foreland, northwestern Argentine Andes. *Journal of South American Earth Sciences*, **7**, 179–207.

MAY, J. 2003. *The Quebrada de Purmamarca, Jujuy, NW-Argentina: landscape evolution and morphodynamics in the semi-arid Andes*. Diplomarbeit, Julius-Maximilians-Universität, Würzburg.

MEYER, G.A., WELLS, S.G. & JULL, A.J.T. 1995 Fire and alluvial chronology in Yellowstone National Park-Climatic and intrinsic controls on Holocene geomorphic processes. *Bulletin of the Geological Society of America*, **107**, 1211–1230.

Murray, A.S. & FUNDER, S. 2003. Optically stimulated luminescence dating of a Danish Eemian coastal marine deposits: a test of accuracy. *Quaternary Science Reviews*, **22**, 1177–1183.

MURRAY, A.S. & OLLEY, J.M. 2002. Precision and accuracy in the optically stimulated luminescence dating of sedimentary quartz: a status review. *Geochronometria*, **21**, 1–16.

MURRAY, A.S. & ROBERTS, R.G. 1997. Determining the burial time of single grains of quartz using optically stimulated luminescence. *Earth and Planetary Science Letters*, **152**, 163–180.

MURRAY, A.S. & WINTLE, A.G. 2000. Luminescence dating of quartz using an improved single-aliquot regenerative-dose protocol. *Radiation Measurements*, **32**, 57–73.

MURRAY, A.S. & WINTLE, A.G. 2003. The single aliquot regenerative dose protocol: potential for improvements in reliability. *Radiation Measurements*, **37**, 377–381.

MURRAY, A.S., OLLEY, J.M. & CAITCHEON, G. 1995. Measurement of equivalent doses in quartz from contemporary water-lain sediments using optically stimulated luminescence. *Quaternary Science Reviews*, **14**, 365–371.

NOTT, J.F., THOMAS, M.F. & PRICE, D.M. 2001. Alluvial fans, landslides and Late Quaternary climatic change in the wet tropics of northeast Queensland. *Australian Journal of Earth Sciences*, **48**, 875–882.

OLLEY, J.M., CAITCHEON, G. & MURRAY, A.S. 1998. The distribution of apparent dose as determined by optically stimulated luminescence in small aliquots of fluvial quartz: implications for dating young sediments. *Quaternary Science Reviews*, **17**, 1033–1040.

OLLEY, J.M., CAITCHEON, G. & ROBERTS, R.G. 1999. The origin of dose distributions in fluvial sediments, and the prospect of dating single grains from fluvial deposits using optically stimulated luminescence. *Radiation Measurements*, **30**, 207–217.

OWEN, L.A., CUNNINGHAM, D., RICHARDS, B.W.M., RHODES, E., WINDLEY, B.F., DORJNAMJAA, D. & BADAMGARAV, J. 1999. Timing of formation of forebergs in the northeastern Gobi Altai, Mongolia: implications for estimating mountain uplift rates and earthquake recurrence intervals. *Journal of the Geological Society, London*, **156**, 457–464.

PRESCOTT, J.R. & HUTTON, J.T. 1994. Cosmic ray contributions to dose rates for luminescence and ESR dating: large depths and long-term time variations. *Radiation Measurements*, **23**, 497–500.

RHODES, E.J. & POWNELL, L. 1994. Zeroing of the OSL signal in quartz from young glacio-fluvial sediments. *Radiation Measurements*, **23**, 581–586.

ROBERTS, R.G., GALBRAITH, R.F., YOSHIDA, H, LASLETT, G.M. & OLLEY, J.M. 2000. Distinguishing dose populations in sediment mixtures; a test of single-grain optical dating procedures using mixtures of laboratory-dosed quartz. *Radiation Measurements*, **32**, 459–465.

SCHWAB, K. & SCHAEFER, A. 1976. Sedimenation und Tektonik im mittleren Abschnitt des Rio Toro in der Ostkordillere NW-Argentiniens. [Sedimentation and tectonics in the central section of the Rio Toro in the eastern cordillera of northwestern Argentina.] *Geologische Rundschau*, **65**, 175–194.

SCHWERDTFEGER, W. (ed.). 1976. *Climates of Central and South America*. Elsevier, New York.

SHAPIRO, S.S. & WILK, M.B. 1965. An analysis of variance test for normality (complete samples). *Biometrika*, **52**, 591–611.

SINGH, A.K., PARKASH, B., MOHINDRA, R., THOMAS, J.V. & SINGHVI, A.K. 2001. Quaternary alluvial fan sedimentation in the Dehradun Valley Piggyback Basin, NW Himalaya: tectonic and palaeoclimate implications. *Basin Research*, **13**, 449–471.

SPENCER, J.Q. & OWEN, L.A. 2004. Optically stimulated luminescence dating of Late Quaternary glaciogenic sediments in the upper Hunza valley: validating the timing of glaciation and assessing dating methods. *Quaternary Science Reviews*, **23**, 175–191.

SPENCER, J.Q., SANDERSON, D.C.W., DECKERS, K. & SOMMERVILLE, A.A. 2003. Assessing mixed dose distributions in young sediments identified using small aliquots and a simple two-step SAR procedure: the F-statistic as a diagnostic tool. *Radiation Measurements*, **37**, 425–431.

STOKES, S. 1999. Luminescence dating applications in geomorphological research. *Geomorphology*, **29**, 153–171.

STRECKER, M.R. & HILLEY, G.E. 2003. Ephemeral Sedimentary Basins and Limits of Lateral Plateau Growth at the Eastern Puna Margin, NW Argentina. *Eos, Transactions of the American Geophysical Union*, **84**, Fall Meeting Supplement, Abstract T32E-04.

TANAKA, K., HATAYA, R., SPOONER, N.A. & QUESTIAUX, D.G. 2001. Optical dating of river terrace sediments from Kanto plains, Japan. *Quaternary Science Reviews*, **20**, 825–828.

TANDON, S.K., SAREEN, B.K., SOMESHWAR RAO, M. & SINGHVI, A.K. 1997. Aggradation history and luminescence chronology of Late Quaternary semi-arid sequences of the Sabarmati basin, Gujarat, Western India. *Palaeogeography, Palaeoclimatology, Palaeoecology*, **128**, 339–357.

TATUMI, S.H., PEIXOTO, M.N.O. *ET AL*. 2003. Optical dating using feldspar from Quaternary alluvial and colluvial sediments from SE Brazilian Plateau, Brazil. *Journal of Luminescence*, **102–103**, 566–570.

TRAUTH, M.H. & STRECKER, M.R. 1999. Formation of landslide-dammed lakes during a wet period between 40 000 and 25 000 yr B.P. in northwestern Argentina. *Palaeogeography, Palaeoclimatology, Palaeoecology*, **153**, 277–287.

TRAUTH, M.H., ALONSO, R.A., HASELTON, K.R., HERMANNS, R.L. & Strecker, M.R. 2000. Climate change and mass movements in the northwest Argentine Andes. *Earth and Planetary Science Letters*, **179**, 243–256.

TRAUTH, M.H., BOOKHAGEN, B., MULLER, A.D. & STRECKER, M.R. 2003. Late Pleistocene climate change and erosion in the Santa Maria Basin, NW Argentina. *Journal of Sedimentary Research*, **73**, 82–90.

VUILLE, M. 1999. Atmospheric circulation over the Bolivian Altiplano during dry and wet periods and extreme phases of the Southern Oscillation. *International Journal of Climatology*, **36**, 413–423.

WALLINGA, J., MURRAY, A.S. & BØTTER-JENSEN, L. 2002. Measurement of the dose in quartz in the presence of feldspar contamination. *Radiation Protection Dosimetry*, **101**, 367–370.

Factors controlling sequence development on Quaternary fluvial fans, San Joaquin Basin, California, USA

G.S. WEISSMANN*, G.L. BENNETT & A.L. LANSDALE

Department of Geological Sciences, Michigan State University, 206 Natural Science Building, East Lansing, MI 48823-1115, USA (e-mail: weissman@msu.edu)
Present address: Department of Earth Planetary Sciences, MSC03-2040, 1 University of New Mexico, Albuquerque, NM 87131-0001, USA (e-mail: weissman@unm.edu)

Abstract: Variable geometry and distribution of stratigraphic sequences of fluvial fans in the eastern San Joaquin Basin, California, were controlled by tectonics, through basin subsidence and basin width, and response to Quaternary climate change, related to the degree of change in sediment supply to stream discharge ratios and local base-level elevation changes. Three fluvial fans – the Kings River, Tuolumne River and Chowchilla River fans – illustrate the influence of these factors on ultimate sequence geometry. In areas with high subsidence rates (e.g. the Kings River fluvial fan) sequences are relatively thick and apices of subsequent sequences are vertically stacked. Areas with relatively low subsidence rates (e.g. the Tuolumne River fan) produced laterally stacked sequences. Rivers that experienced a significant increase in sediment supply to stream discharge ratios due to direct connection to outwash from glaciated portions of the Sierra Nevada developed high accommodation space and relatively thick sequences with deep incised valleys. Conversely, rivers that were not connected to glaciated regions (e.g. the Chowchilla River fan) and, thus, experienced a relatively minor change in sediment supply to discharge ratios during climate change events, produced thinner sequences that lack deep incised valleys. Local base-level connection to sea level, via the axial San Joaquin River, produced deeper incised valleys than those of internally drained rivers. Finally, narrow basin width allowed glacially connected fans to completely fill available accommodation space, thus producing smaller fans that lack preservation of distal, interglacial deposits. Evaluation of these controls allows prediction of sequence geometries and facies distributions for other San Joaquin Basin fans for input into future hydrogeological models.

Fluvial fans of the eastern San Joaquin Valley, California (Fig. 1), show evidence of depositional and erosional response to Quaternary climate change. We use the term 'fluvial fan', to distinguish fans that are dominated by perennial fluvial processes from alluvial fans that are dominated by debris-flow or sheet-flood processes. Stanistreet & McCarthy (1993) classified such fans as losimean or braided river fans. Episodes of basinwide aggradation occurred on the San Joaquin Basin fluvial fans in response to significant increases in sediment supply during Quaternary glacial episodes (Arkley 1962; Helley 1966; Janda 1966; Marchand 1977; Atwater 1980; Marchand & Allwardt 1981; Lettis 1982, 1988; Harden 1987; Lettis & Unruh 1991; Weissmann *et al.* 2002*a*). Incision occurred on many of the fans during interglacial periods in response to decreased sediment supply and stream discharge, leaving the upper fluvial fans exposed to widescale soil development.

Recently, Weissmann *et al.* (2002*a*) recognized that this cyclic pattern led to the development of stratigraphic sequences on the Kings River fluvial fan, where a sequence is defined as 'a relatively conformable succession of genetically related strata bounded at its top and base by unconformities, or their correlative conformities' (Mitchum 1977, p. 210). In this sequence stratigraphic model for fluvial fans,

accommodation space (e.g. the space made for sediment accumulation, after Jervey 1988) and, thus, stratigraphic evolution, is controlled by several factors, including: (1) change in the ratio of sediment supply to stream discharge; (2) basin subsidence rate; (3) local base-level change; and (4) basin geometry (e.g. width). On the Kings River fluvial fan, Weissmann *et al.* (2002*a*) concluded that sequence development was controlled primarily by changes in the ratio of sediment supply to stream discharge in response to the glacial outwash cycles in the Sierra Nevada, while basin subsidence was sufficient to allow burial and preservation of successive aggradational events. Sequence-bounding unconformities on the Kings River fluvial fan are marked by relatively well-developed palaeosols (Weissmann *et al.* 2002*a*).

The Kings River fluvial fan presents one possible stratigraphic outcome from the combination of controlling factors listed above. However, how will the stratigraphy differ if sediment loads or stream discharges are significantly different to those observed on the Kings River fan? Will fans display different stratigraphic character if basin subsidence is lower or if local base-level control is more significant? Although all of the eastern San Joaquin Valley fluvial fans show similar sequence development in response to Quaternary climate change (Marchand & Allwardt

From: HARVEY, A.M., MATHER, A.E. & STOKES, M. (eds) 2005. *Alluvial Fans: Geomorphology, Sedimentology, Dynamics.* Geological Society, London, Special Publications, **251**, 169–186. 0305-8719/05/$15 © The Geological Society of London 2005.

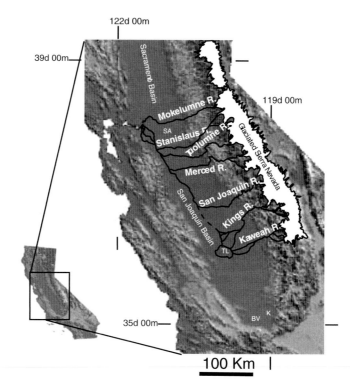

Fig. 1. Physiographical map of central California showing the approximate extent of Pleistocene glaciation, as reported by Wahrhaftig & Birman (1965), and positions of major rivers in the Central Valley. SA, Stockton Arch; TL, Tulare Lake sub-basin; BV, Buena Vista sub-basin; K, Kern Lake sub-basin.

1981), each fan's position in the San Joaquin Basin, along with the drainage basin connection to either glaciated or only non-glaciated regions in the Sierra Nevada, provides a different set of controls on stratigraphic evolution. Thus, we can observe differences in fan morphology and sequence geometry on other San Joaquin Basin fans in order to evaluate the influence of these four factors on accommodation space development.

In this paper we compare two fluvial fans in the eastern San Joaquin Basin – the Tuolumne River and Chowchilla River fluvial fans (Fig. 2) – to the Kings River fan in order to help answer these questions related to the influence of these various controls on sedimentation. Both the Kings River and Tuolumne River fans were influenced by direct connection to Sierra Nevada glacial outwash. However, the Kings River fan lies in the internally drained southern portion of the San Joaquin Basin, where subsidence rates are relatively high, the basin is wide and local base level is controlled by the elevation of Tulare Lake. In contrast, the Tuolumne River fan lies close to the northern outlet of the basin, where subsidence rates are lower, the basin is relatively narrow and local base level is controlled by the elevation at the conflu-

ence with the San Joaquin River, which, in turn, is controlled by sea level. Therefore, we can evaluate controls of subsidence rates, basin geometry and local base-level change on sequence form by comparing these fans. The Chowchilla River fan lies between the Kings River and Tuolumne River fans but was not directly connected to Sierra Nevada glaciations. Thus, this river experienced less sediment influx and lower discharge during glacial periods. Therefore, we can evaluate controls of sediment supply to stream power ratios on sequence geometries through comparison of the Chowchilla River fan to the Kings River and Tuolumne River fans.

Study area

The Great Valley of California marks an elongate asymmetric syncline that trends roughly NW–SE, originally formed as a Neogene forearc basin (Fig. 2) (Lettis & Unruh 1991). The Stockton Arch divides the Great Valley into two sub-basins – the San Joaquin Basin to the south and the Sacramento Basin to the north, with the inland delta of the Sacramento and San Joaquin rivers lying roughly above the

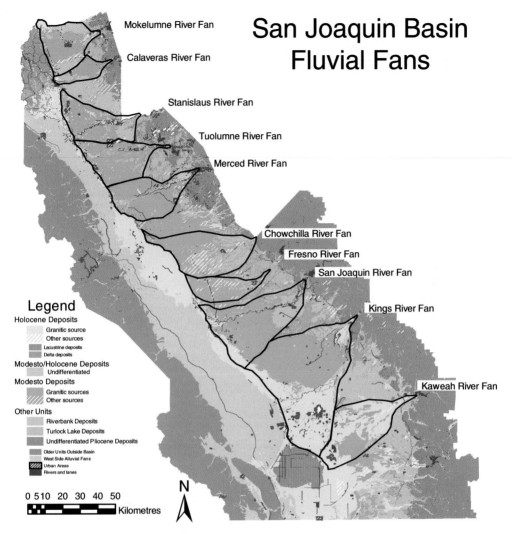

Fig. 2. Age relationships of exposed surfaces on fluvial fans in the eastern San Joaquin Valley based on soil surveys by Arkley *et al.* (1962), Ulrich & Stromberg (1962), Arkley (1964), Huntington (1971), Arroues & Anderson (1976), Stephens (1982), Nazar (1990), McElhiney (1992), Ferrari & McElhiney (2002), USDA/NRCS (2003) and Wasner & Arroues (2003) and mapping by Marchand & Allwardt (1981). Soil series correlation to Quaternary units were described by Helley (1966), Janda (1966), Marchand (1977), Marchand & Allwardt (1981), Lettis (1982, 1988), Harden (1987) and Huntington (1971, 1980). Digital data from the Natural Resources Conservation Service, US Department of Agriculture (www.nrcs.usda.gov).

Stockton Arch (Bartow 1991). The San Joaquin Valley is further divided into four sub-basins. The Tulare, Buena Vista and Kern Lake sub-basins are internally drained, and the northern San Joaquin sub-basin is drained by the northward-flowing San Joaquin River (Figs 1 & 2). The fluvial fans of this study lie in either the Tulare Lake sub-basin (Kings River fan) or the northern San Joaquin sub-basin (Tuolumne and Chowchilla River fans).

Several large fluvial fans formed along the eastern margin of the San Joaquin Basin where their rivers exit the Sierra Nevada (Fig. 2). Seven of these fans – the Mokelumne, Stanislaus, Tuolumne, Merced, San Joaquin, Kings and Kaweah River fans – have drainage basins that connect to glaciated portions of the Sierra Nevada (Fig. 1). The Calaveras, Chowchilla and Fresno River fans, along with minor fans produced by several smaller rivers, have drainage basins that tap only non-glaciated portions of the Sierra Nevada. No correlation exists between fluvial fan size and drainage basin area in the San Joaquin Valley (Fig. 3). However, fans tend to be smaller in the northern basin, reflecting the narrower basin width (Fig. 2).

The upper Cenozoic deposits of the eastern San Joaquin Valley fluvial fans have been divided into five allostratigraphic units – the Laguna of late Pliocene age, the North Merced Gravel of Pliocene or Pleistocene age, and the Turlock Lake, Riverbank and Modesto of Quaternary age (Marchand & Allwardt 1981; Lettis & Unruh 1991). Each unit consists of multiple successions of arkosic silt, sand and gravel related to aggradational response to glacial outwash, except for the North Merced Gravel which is a thin, widespread gravel interpreted to be a lag deposit on an early Pleistocene pediment surface (Marchand 1977; Marchand & Allwardt 1981; Lettis & Unruh 1991). Soils developed on surface exposures of each of these units display increasing maturity with age (Helley 1966; Janda 1966; Huntington 1980; Marchand & Allwardt 1981; Harden 1987), thus maps showing distributions of the units are readily produced from digital county soil surveys of the area (Fig. 2). Fans in the northern part of the study area display a westward-stepping shift of the fan apex with each successive aggradational event (Fig. 2). In contrast, the apex of the Kings River fan, located in the southern part of the study area, appears to have remained relatively constant. In the subsurface, unconformities identified by relatively mature palaeosols commonly mark the upper surface of each of these units (Davis & Hall 1959; Helley 1966; Janda 1966; Lettis 1982, 1988; Weissmann *et al.* 2002*a*).

Limited age dates from these deposits indicate that the Modesto unit was deposited between 10 and 40 ka, the Riverbank unit was deposited between 130 and 330 ka, and the Turlock Lake units prior to 600 ka (Marchand & Allwardt 1981; Harden 1987). These sparse dates indicate that aggradation occurred on the fluvial fans late in the glacial episodes, possibly during glacial maximum and recession (Marchand & Allwardt 1981; Harden 1987), with deposition in the basin centre initiating at the beginning of glaciation and continuing throughout the glacial episode (Atwater 1980; Atwater *et al.* 1986).

Fluvial fan sequence stratigraphy

Accommodation space on fluvial fans

Following definitions by Blum & Törnqvist (2000), Weissmann *et al.* (2002*a*) considered accommodation space to be a combination of *accumulation space* and *preservation space*. Accumulation space is defined as 'the volume of space that can be filled within present process regimes, and is fundamentally governed by the relationship between stream power and sediment load, and how this changes in response to geomorphic base level' (Blum & Törnqvist 2000, p. 20). The intersection point on the fan marks the dividing point between areas of positive and negative accumulation space on fluvial fans, where positive accumulation space (shown by deposition) occurs below the intersection point and negative accumulation space (shown by erosion) occurs above the intersection point (Weissmann *et al.* 2002*a*). As noted by Wright & Marriott (1993), maximum accumulation space is marked by the bankfull level of the river at grade as floodplain material can only aggrade to this level.

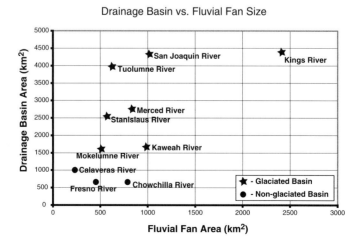

Fig. 3. Areas covered by fluvial fans (km²) and their adjacent Sierra Nevada drainage basins (km²), showing no correlation between these parameters in the San Joaquin Basin.

Preservation space is provided when 'subsidence lowers these deposits below possible depths of incision and removal' (Blum & Törnqvist 2000, p. 20). Subsidence in the San Joaquin Basin provided the preservation space for sequences of the study area fluvial fans. In addition, continued basin subsidence during interglacial periods produced space for deposition and preservation of subsequent glacial outwash events.

Our approach uses the geomorphic base level, or equilibrium profile, to define accumulation space, recognizing that prediction of past and new equilibrium profiles is difficult, if not impossible (Muto & Steel 2000; Viseras *et al.* 2003). However, we believe that the concept that changing equilibrium profile contributes to new accommodation space has significance. As will be shown for the San Joaquin Basin fans, the shift in equilibrium profile elevation due to sediment influx during glacial outwash periods (for upward shifts) or loss of sediment supply at the end of glacial outwash periods (for downward shifts) controls accumulation space cycles. Therefore, we believe the equilibrium profile concept to be important when considering sequence development on fluvial fans. Ultimately, however, we recognize that accommodation space is represented in the stratigraphic record as the thickness of fill for each sequence measured between the sequence boundaries, as suggested by Muto & Steel (2000).

This sequence stratigraphic approach for fluvial fans offers a means to evaluate stratigraphic evolution of a fluvial fan succession that is comparable to similar approaches used in the marine depositional environments (e.g. Posamentier *et al.* 1988; Posamentier & Vail 1988; Van Wagoner *et al.* 1990). An advantage of applying sequence stratigraphic concepts rather than focusing only on cycles of aggradation and degradation is that this conceptual approach incorporates basin subsidence and local base-level controls along with sediment supply and discharge variability in determining overall stratigraphic architecture.

Geomorphology and sequence development on the Kings River fluvial fan

The Kings River fluvial fan is the largest in the study area, with Modesto deposits covering approximately 2410 km^2 (Figs 3 & 4A). Basin subsidence rates west of the Kings River fan (beneath the Tulare Lake bed) may be as high as 1.1 m ka^{-1}, based on the depth to the Corcoran Clay, a wide-spread lacustrine unit deposited approximately 615 000 years ago in the basin (Lettis & Unruh 1991). The fan has a gentle convex-upwards cross-fan profile and a slightly concave-upwards longitudinal profile along stream channels (Fig. 5A). The Kings River currently lies

within a 10 m-deep incised valley, with the modern fan intersection point located approximately 34 km downgradient from the fan apex (Fig. 4A). Contrasting with other fans in the San Joaquin Basin, the upper part of this incised valley is broad (8 km wide) and filled with a small (85 km^2) and relatively high gradient (0.0032) fluvial fan (Fig. 4A & 5A). The gradient of this small 'internal' fluvial fan matches that of the lower portion of the river canyon; however, below this internal fan the Kings River has a relatively low gradient of approximately 0.0005 (Fig. 5A). In contrast, the upper-fan Modesto-age surface surrounding this incised valley has a steeper gradient of 0.0013 (Fig. 5A).

Above the fluvial fan, the Kings River drainage basin primarily passes through the Sierra Nevada batholith (granitic rocks). This drainage basin covers approximately 4385 km^2, with maximum elevations in the basin being around 4100 m and elevation at the fan apex around 140 m. The distance from the Sierra divide down to the fan apex is approximately 130 km, with relatively steep gradients (0.12) near the top of the basin, grading down to more gentle gradients (0.006) in the lower reaches of the mountains.

The Kings River currently empties into Tulare Lake, which covered a relatively large area until drained in the 1950s for agriculture. As the depositional lobe was prograding across the lakebed surface (prior to upstream dam construction on the Kings River), this local base level appears to have had minimal control on the Kings River profile. The present configuration of the Kings River fan represents typical interglacial fan morphology (Weissmann *et al.* 2002a).

In the subsurface, laterally extensive palaeosols that cap the Lower Turlock Lake, Upper Turlock Lake and Riverbank units can be correlated across most of the upper Kings River fluvial fan (Fig. 6A) (Weissmann *et al.* 2002a, 2004; Bennett 2003). These palaeosols dip gently basinward, with slightly higher dips on deeper units, indicating continued basinal subsidence (Weissmann *et al.* 2002a). The palaeosol surfaces converge at the apex of the fluvial fan, indicating that the intersection points for successive depositional events were located near the fan apex (Fig. 6A).

The exposed upper Kings River fan consists primarily of Modesto-age open-fan deposits with a large area of Riverbank deposits exposed along the northern side of the fan (Fig. 4A). Five Modesto-age palaeochannels radiate outward from an intersection point located near the fan apex (NE of Sanger) (Huntington 1980; Weissmann *et al.* 2002a). The gradients of these palaeochannels, approximately 0.0013, are twice as steep as the modern Kings River channel (Fig. 5A). Modesto open-fan deposits consist of sand and gravel representing channel deposits, with silty sand and sandy mud in overbank

A. Kings River Fluvial Fan

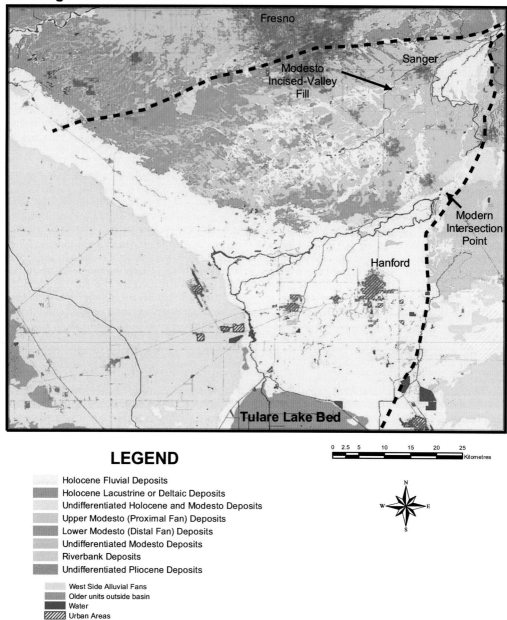

LEGEND

Holocene Fluvial Deposits
Holocene Lacustrine or Deltaic Deposits
Undifferentiated Holocene and Modesto Deposits
Upper Modesto (Proximal Fan) Deposits
Lower Modesto (Distal Fan) Deposits
Undifferentiated Modesto Deposits
Riverbank Deposits
Undifferentiated Pliocene Deposits

West Side Alluvial Fans
Older units outside basin
Water
Urban Areas

0   2.5   5   10   15   20   25
                                        Kilometres

Fig. 4. Interpreted soil surveys showing surface exposures of stratigraphic units on: (**A**) the Kings River fluvial fan (modified from Huntington 1971; Arroues & Anderson 1976, USDA/NRCS 2003);

deposits (Weissmann *et al.* 2002*a*). The silty deposits are typically laminated–massive and appear to consist of relatively unweathered silt that is similar in character to glacial flour, indicating that these deposits originated as glacial outwash material (Huntington 1980; Weissmann *et al.* 2002*a*).

Ground-penetrating radar (GPR) surveys collected over large portions of the Kings River fan (approximately 75 km of surveys), along with core and well logs, show that the Modesto deposits are relatively thin, displaying an average thickness of approximately 5–6 m (Burow *et al.* 1997; Bennett 2003).

B. Tuolumne River Fluvial Fan

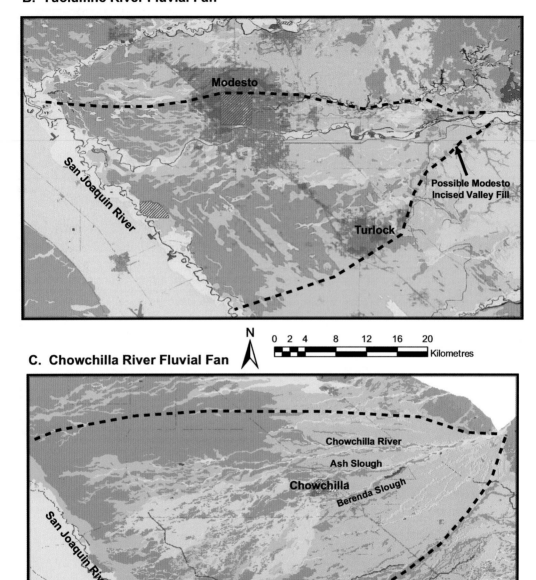

C. Chowchilla River Fluvial Fan

(**B**) the Tuolumne River fluvial fan (modified from Arkley 1964; McElhiney 1992; Ferrari & McElhiney 2002); and (**C**) the Chowchilla River fluvial fan (modified from Arkley *et al.* 1962; Ulrich & Stromberg 1962; Nazar 1990). Also shown is the presence of hardpan in the soil and the presence of 'unrelated subsoil', indicating the existence of the underlying sequence in the shallow subsurface.

A. Kings River Fluvial Fan

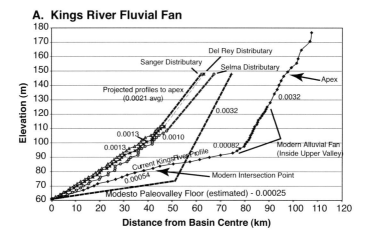

B. Tuolumne River Fluvial Fan

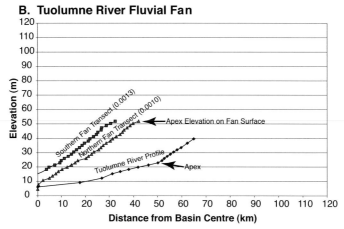

C. Chowchilla Fluvial Fan

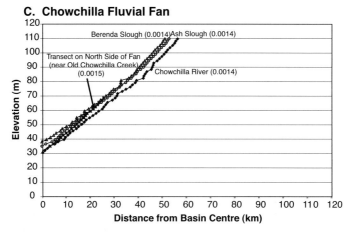

Fig. 5. Gradients on: (**A**) the Kings River fluvial fan; (**B**) the Tuolumne River fluvial fan; and (**C**) the Chowchilla River fluvial fan. These gradients were measured from the US Geological Survey 7.5-minute quadrangle maps. River gradients were measured along the channel, and other gradients were measured in along transects on the fan surface.

A) Kings River Fan

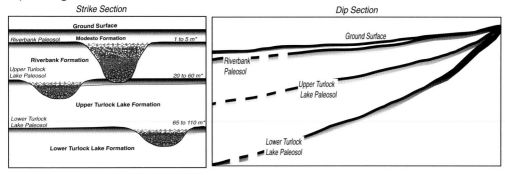

B) Tuolumne River Fan

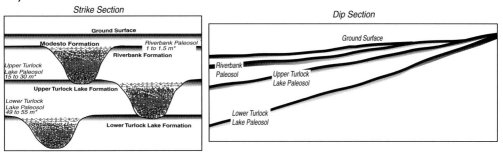

C) Chowchilla River Fan

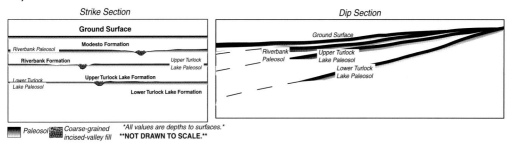

Fig. 6. Schematic cross-sections through: (**A**) the Kings River fluvial fan (based on Weissmann *et al*. 2002*a*, 2004); (**B**) the Tuolumne River fluvial fan (based on Burow *et al*. 2004); and (**C**) the Chowchilla River fluvial fan (based on Helley 1966).

The Modesto palaeochannels appear to have locally removed the Riverbank soil through erosion, forming breaks in this stratigraphic marker (Bennett 2003). However, the GPR survey indicates that this palaeosol is otherwise laterally continuous.

An exception to the 5–6 m Modesto thickness is found in an area where a palaeovalley existed during the interglacial period between the Riverbank and Modesto depositional events (Weissmann *et al*. 2002*a*, 2004). The geometry of this palaeovalley-fill was determined from soil surveys, core and well-log data. The 2 km palaeovalley-fill width is indicated on the soil survey, where older Riverbank deposits surround an elongate Modesto deposit (Fig. 4A) (Weissmann *et al*. 2002*a*, 2004). In addition, in the area that the soil survey indicates presence of this valley, palaeosols that cap the Riverbank and upper Turlock Lake deposits are absent in well logs and cores, indicating that these surfaces were truncated by erosion and incision during the valley formation. Cored wells from the valley-fill show that the bottom of the valley-fill is composed of a unit up to 8 m thick that consists of cobbles held within a very coarse sand matrix (Weissmann *et al*. 2002*a*, 2004). This cobble unit grades upward into a thick (*c*. 20 m) succession of very-coarse- to coarse-grained, cross-stratified–massive sand that is overlain by a mix of coarse-grained cross-stratified–massive sand interbedded with massive laminated silt. Sedimentary textures and structures in this incised-valley-fill are consistent with deposition in a large, fluvial system. The valley-fill unit is typically coarser grained at and near the surface than surrounding deposits. Although water-well logs are typically low quality and carry considerable uncertainty, drillers usually identify the thick basal cobble unit accurately as a significant change in drilling character occurs at this contact (e.g. drilling becomes rough and drilling fluid loss may occur due to the very high permeability of this bed), thus we can trace this valley-fill deposit downfan at least 20 km from the fan apex.

Kings River fluvial fan sequence stratigraphy. In the case of the Kings River fluvial fan, accumulation space developed on the fan during periods of glacial outwash, when increased sediment supply caused aggradation and steepening of the river gradient to accommodate the large sediment load (Weissmann *et al.* 2002a). Conversely, a significant decrease in sediment load and discharge at the end of glacial periods and the beginning of interglacial periods caused a loss of accumulation space (Huntington 1971, 1980; Weissmann *et al.* 2002a, 2004). This loss of accumulation space led to fan incision, a basinward shift of the fan intersection point, and a decrease in river gradient and river profile elevation (Fig. 5A). Throughout the interglacial period, deposition was confined to the distal portions of the fan while the upper fan was exposed to erosion, soil development or modification by eolian processes. The soils that formed during the interglacial periods mark the sequence bounding unconformities in the study area.

Rapid aggradation (e.g. accumulation space increase) occurred in response to higher sediment supply during the glacial events (Lettis 1982, 1988; Weissmann *et al.* 2002a). Initially, the incised valley filled with a fining-upwards succession of relatively coarse-grained channel and overbank deposits. Continued deposition due to high sediment supply to discharge ratios eventually filled the incised valley, after which the intersection point stabilized near the fan apex, as shown by the Modesto palaeochannels radiating outward from a proximally located intersection point (Fig. 4A). This led to unconfined, open-fan deposition across the entire fluvial-fan surface (Weissmann *et al.* 2002a, 2004).

The end of glaciation led to repetition of this stratigraphic cycle, with the subsequent decrease in the sediment supply to discharge ratio, fan incision, basinward shift in deposition and soil development on the upper alluvial fan.

The influence of controls on accommodation space

We use the Kings River fluvial fan stratigraphy as a basis for comparison to observe the influence of different controls on accumulation space and the resulting stratigraphic architecture of other San Joaquin Basin fluvial fans. The Kings River fan represents fan development in a portion of the basin that is relatively wide and has higher subsidence rates (and, thus, greater preservation space) than other portions of the basin. In addition, the river is directly connected to glaciations in the Sierra Nevada, therefore sediment supply and discharge in the Kings River were significantly increased during periods of glacial outwash. Finally, local base level does not appear to significantly influence the stratigraphic architecture of this fan (Weissmann *et al.* 2002a, 2004).

In the following section, we compare the morphology and stratigraphy of the Kings River fluvial fan to the Tuolumne River and Chowchilla River fans. The Tuolumne River fluvial fan represents a system affected by lower subsidence rates and stronger local base-level control. The Chowchilla River fluvial fan represents a system that was not directly connected to glacial outwash, therefore it did not experience the extreme sediment supply and stream discharge fluctuations observed on the Kings River or Tuolumne River fans.

Tuolumne River fluvial fan – influence of basin subsidence and local base level

The Tuolumne River fluvial fan surrounds the city of Modesto, California, and, with an area covering 630 km², is significantly smaller than the Kings River fluvial fan (Figs 2, 3 & 4B). The Kings River and Tuolumne River drainage basins experienced extensive Quaternary glaciation (Wahrhaftig & Birman 1965), and both drainage basin sizes, stream gradients and elevation differences are similar (Fig. 3). The Tuolumne River drainage basin, however, does cross a metamorphic terrain at lower elevations, and is surrounded by slightly lower peaks (approximately 3950 m in elevation) than are found in the Kings River drainage basin. As the two drainage basins are comparable, we assume that both fans experienced similar sediment-supply and stream-discharge variability. Basin subsidence rates near the Tuolumne River fan, however, are significantly lower than rates in the Kings River fan area (approximately $0.3 \, \text{m ka}^{-1}$ under the distal fan based on depth to the Corcoran Clay; Lettis & Unruh 1991). In addition, the San Joaquin Basin near the Tuolumne River is about half the width of the basin near the Kings River fan

(Fig. 2). Finally, the modern Tuolumne River is currently tied to the local base level of the San Joaquin River. Therefore, we expect observed differences between the Kings River and Tuolumne River fans to reflect impacts of the different basin subsidence rates, basin width and local base-level controls on the river systems.

Tuolumne River fan geomorphology and stratigraphy

The modern Tuolumne River lies in a relatively deep valley through the fluvial fan (c. 30 m near the apex), has a relatively low gradient and a profile that is slightly concave-upwards (Fig. 5B). In contrast with the Kings River, the toe of the Tuolumne River fan is truncated by the San Joaquin River, therefore no modern intersection point or depositional lobe exists on the fan (Fig. 4B). Thus, no portion of the exposed fan is affected by Tuolumne River flooding, and modern sediment transport is confined to the incised valley. The fact that the gradient of the Tuolumne River near the fan toe is similar to that of the San Joaquin River in the confluence reach and that the river is held in an incised valley at the fan toe indicates that the Tuolumne River profile is currently tied to the local base level of the San Joaquin River. We assume that the current fan morphology reflects a typical interglacial configuration of the Tuolumne River fan.

The soil surveys on the Tuolumne River fan show that the relatively young Modesto deposits lie above and west of Riverbank deposits, and Riverbank sediments lie above and west of Turlock Lake deposits (Fig. 4B) (Arkley 1964). In addition, similar palaeosol-bounded stratigraphic sequences to those observed on the Kings River fan appear to exist in the subsurface (Fig. 6B) (Davis & Hall 1959; Burow et al. 2004). As observed on the Kings River fan, the Modesto deposits of the Tuolumne fan consist of open-fan deposits with discrete channel sands surrounded by silt-dominated fine-grained floodplain sediments (Arkley 1964); however, large abandoned channels are not present on the exposed Modesto surface of the Tuolumne River fan. The presence of shallow 'unrelated subsoils' under Modesto sediments near the Riverbank exposures (Fig. 4B), described in soil surveys on the fan (Arkley 1964), indicate that the top of the Riverbank unit dips gently towards the basin centre. In addition, recent GPR surveys (Bennett 2003) and drilling in the area (Burow et al. 2004) indicate that the Modesto deposits form a thin (approximately 1–4 m) veneer over the gently basinward-dipping Riverbank palaeosol. Although well-log quality is relatively poor in this area and the sequence boundaries are difficult to trace in the subsurface, several recently drilled wells indicate that the Upper Turlock Lake sequence dips beneath the Riverbank deposits of the Tuolumne fluvial fan at a depth of approximately 15 m NE of the city of Modesto (Burow et al. 2004).

Gradients observed on the Modesto surface of the Tuolumne River fan are similar to those observed for the Modesto of the Kings River fan (approximately 0.0013) (Fig. 5B). The Modesto fan gradient is approximately the same as the gradient of the bedrock-controlled modern Tuolumne River above the fan apex. As observed on the Kings River fan, the Modesto fan gradient is significantly greater than the current river gradient of 0.0004, indicating a steepening of the stream profile in adjustment to the high-sediment supply during periods of glacial outwash.

The Modesto surface lies approximately 30 m above the modern Tuolumne River at the fan apex, and 4.5 m above the river at the fan toe. Assuming that similar incised valleys existed during previous interglacial periods, significant valley and canyon filling must have occurred during past glacial outwash periods. Two of these units have been tentatively identified in the subsurface – the Modesto and the Riverbank incised-valley-fill units (Lansdale et al. 2004). Soil surveys, topographic maps and sparse well-log data indicate the presence of the Modesto incised-valley-fill south of the current Tuolumne River position. As on the Kings River fan, soil surveys show a linear Modesto feature between Riverbank terraces (Fig. 4B). Some well logs within this area lack the Riverbank palaeosol in the shallow subsurface, indicating truncation of that surface, and show a presence of a thick gravel at approximately 30–33 m depth that is capped by sand-dominated deposits. Finally, at the location where this Modesto incised-valley-fill intersects the modern river valley, the modern valley narrows significantly downvalley, indicating more recent valley development below this point. Soil textures and GPR surveys do not show the presence of large Modesto-age channels across the Tuolumne River fan, as observed on the Kings River fan (Bennett 2003), thus the Modesto channel system on the Tuolumne River fan probably did not sweep across the entire fan surface.

A deeper incised-valley-fill unit is indicated by geophysical well logs beneath the city of Modesto, where a thick coarse-grained unit was identified between about 18 and 48 m depth (Burow et al. 2004; Lansdale et al. 2004). Well logs within this possible valley-fill indicate the presence of cobbles at the base of this unit. Additional work is needed, however, to delineate both of these incised-valley-fill units.

Tuolumne River fan stratigraphic evolution. Sequence development on the Tuolumne River fan appears to be similar to that of the Kings River fan. However, significant differences in stratigraphic character exist on the Tuolumne River fan, including: (1) thinner Modesto deposits are present on the fan (log data are insufficient to evaluate deeper units at this time); (2) successive aggradational events are laterally stacked, with apices being westward stepping (Fig. 6B); and (3) a deeper incised valley cuts the modern Tuolumne River fan, thus neither an intersection point nor a modern depositional lobe exist on this fan. The sequence geometry differences on the Tuolumne River fan reflect differences in accommodation space development, controlled by three possible factors: (1) relatively low subsidence rates coupled with the narrowness of this portion of the San Joaquin Basin; (2) the San Joaquin River local base level; and (3) slightly lower elevations (and thus less glacial erosion) in the Tuolumne River drainage basin.

First, because subsidence rates in this part of the basin are relatively low and the overall basin width at this location is relatively narrow, less preservation space was produced during intervening interglacial periods; therefore, the overall amount of accommodation space available during each successive glacial outwash event would be less than that observed on the Kings River fan. This lower subsidence rate is especially noteworthy near the fan apex on the Tuolumne River fan, where it resulted in the westward-stepping succession of fan apexes and sequences rather than vertical stacking as observed on the Kings River fan. In addition, the Tuolumne area subsidence rate was half that of the Kings River system and resulted in Modesto deposits that are approximately half the thickness of the same deposits on the Kings River fan. As the Modesto surface of the Tuolumne River fan has a similar gradient to that of the Kings River, and the Modesto surface fills the San Joaquin Basin, we assume that the available accommodation space in this part of the basin was filled by the Tuolumne River fan. In addition to differences in subsidence rates, the San Joaquin Basin is significantly narrower in the Tuolumne River fan area than in the Kings River fan area; thus, less overall accommodation space was available for Tuolumne River deposits. Therefore, lower subsidence rates appear to affect overall sequence thickness, while basin width controlled the ultimate fan sizes.

A second control on accommodation space availability may be the influence of the San Joaquin River local base level. The San Joaquin River lies at a lower elevation than the Tulare Lake bed of the Kings River, and its profile is controlled in part by sea level. Because of this connection, a deeper incised valley was formed through the Tuolumne River fan. Therefore, no active depositional lobe exists on the modern fan, indicating that no accumulation space currently exists on the fan. Assuming this fan configuration represents typical interglacial fan morphology, we expect that past interglacial sediments bypassed

the fan and are not preserved in the sedimentary record of the fan. Similar base-level control on fan incision was described by Muto (1987, 1988).

One might argue that confinement of the river to a deep incised valley would result in higher stream power and less fine-grained sediment accumulation, thus resulting in thinner Modesto deposition on the Tuolumne River fan. This argument would suggest that much of the fine-grained glacial outwash volume would bypass the fan until the valley filled to the point of overtopping. This could significantly influence the overall sediment volume deposited on the fan, thus potentially impacting the thickness of stratigraphic sequences. As the Modesto incised valley on the Kings River appears to have been of similar depth (possibly because the Modesto valley on the Kings River fan was connected to the San Joaquin River rather than the Tulare Lake Basin), the Kings River fan would have experienced the same constraints. Therefore, we do not believe local base level to be a significant influence on differences in accumulation space between the Kings River and Tuolumne River fans.

Finally, the amount of sediment load and discharge may have been different on the Tuolumne River system than that of the Kings River system, as the Tuolumne River drainage basin has lower elevations. This may have resulted in less glacial erosion and, therefore, less outwash material and a thinner Modesto unit. Thus, overall accumulation space may have been less on the Tuolumne River fan during outwash periods. The Modesto deposits, however, fill the San Joaquin Basin width, are truncated by the San Joaquin River and have the same gradient on the Tuolumne River fan as observed on the Kings River fan, indicating that the Modesto deposits filled the accommodation space that was available on the Tuolumne River fan. Accommodation space available on the Tuolumne River fan during Modesto deposition was less than that available on the Kings River fan due to the relatively narrow basin width in the Tuolumne River area. Therefore, we do not believe differences in sediment supply significantly influenced overall depositional patterns. Instead, basin size and subsidence rates appear to have been the primary controlling factors on sequence geometry of the Tuolumne River fan.

Chowchilla River fluvial fan – influence of sediment supply and stream discharge

The Chowchilla River fluvial fan surrounds Chowchilla, California, and, covering an area of approximately 788 km^2, is significantly smaller than the Kings River fluvial fan but larger than the Tuolumne River fan (Fig. 3). Even though the Chowchilla River drainage basin was not glaciated and is significantly smaller (658 km²), depositional sequences similar to those observed on the Tuolumne River and Kings River fluvial fans have been identified (Fig. 4C) (Helley 1966; Marchand & Allwardt 1981).

The Chowchilla River drainage basin drains similar granitic terrain as the Kings River fan; however, some metamorphic rocks are exposed in the lower portion of the drainage basin. Maximum elevations surrounding the Chowchilla River drainage basin are significantly lower (2100 m) and overall stream gradients are lower (0.09 in the upper reaches and 0.0012 in the lower reaches of the drainage basin) than the Tuolumne or Kings rivers.

The basin subsidence rate near the distal Chowchilla River fan is between that of the Tuolumne River fan and Kings River fan areas (approximately 0.35 m ka⁻¹: Lettis & Unruh 1991). West of the Chowchilla River fan, the San Joaquin River valley is broad and the San Joaquin River splits into several channels and sloughs, indicating that the Chowchilla River fan and opposing fans on the west side of the San Joaquin Valley do not completely fill the San Joaquin Basin. Therefore, accommodation space was still available at the end of the last aggradational cycle, indicating that accommodation space was not limited by basin subsidence rates or basin width as observed on the Tuolumne River fan. Instead, depositional cycle geometries appear to be primarily related to accumulation space variability (e.g. sediment supply and stream discharge history). As the Chowchilla River drainage basin was not glaciated, sediment-supply and stream-discharge volumes were not as great as that of rivers directly connected to glaciers (Helley 1966; Marchand & Allwardt 1981). Therefore, comparison of the Chowchilla River fan to the Kings River and Tuolumne River fans offers an opportunity to observe the influence of sediment supply and stream discharge on overall sequence geometry.

Chowchilla River fan geomorphology and stratigraphy. In contrast to the Kings River and Tuolumne River fans (and other glaciated river fans in the San Joaquin Basin), the current Chowchilla River is not held in a deep incised valley, but instead it appears to be a slightly underfit stream in the Modesto-age distributary channels (Fig. 4C) (Helley 1966). The Chowchilla River splits into three distributary channels, called the Chowchilla River, Ash Slough and Berenda Slough, approximately 6 km west of the Sierra Nevada mountain front (Fig. 4C). The modern river has approximately the same gradient as the surrounding Modesto fan surface (Fig. 5C).

Where the three distributary channels split, they are surrounded by Riverbank deposits and lie approximately 4 m below the Riverbank surface. Further downfan, the distributary channels lie

approximately 2 m below the Modesto terrace. Large floods on the Chowchilla River, however, historically overtopped these channels, spilling onto the distal Modesto open-fan deposits (Helley 1966). Near the toe of the fluvial fan, the Chowchilla River and distributary sloughs are held in natural levées and disappear into alluvium before reaching the San Joaquin River (Helley 1966), indicating that deposition occurs on the distal-fan surface during the current interglacial period.

As on the Tuolumne River fan, successive sequences are westward stepping, with Modesto deposits lying west of Riverbank deposits, and Riverbank deposits lying west of Turlock Lake and Pliocene deposits (Figs 4C & 6C) (Arkley *et al.* 1962; Ulrich & Stromberg 1962; Helley 1966; Marchand & Allwardt 1981). Helley (1966) reported that the Riverbank deposits east of Chowchilla form a relatively thin veneer over Turlock Lake deposits, filling older Turlock Lake channels and leaving narrow, low remnant Turlock Lake terraces. The Riverbank unit thickens to the west from 3–6 m thick near Chowchilla to approximately 30 m thick in the San Joaquin Basin. Similarly, the Modesto deposits overly the Riverbank deposits and are thin near the town of Chowchilla but thicken westward. Deposits in these units consist of a mixture of granitic- and metamorphic-derived sediments, reflecting the drainage basin source area, although granitic sources dominate the fan sediment (Helley 1966). Because the drainage basin is not connected to glaciated areas, floodplain silts of the Chowchilla River fan do not have the grey, fresh look of glacial flour (Helley 1966).

Chowchilla River fan sequence development. The presence of correlative Quaternary aggradational units to those on the Tuolumne River and Kings River fans indicate that the aggradational events had a similar timing to those observed on other fans in the basin (Helley 1966; Marchand & Allwardt 1981). In contrast to the sequences observed on the Tuolumne and Kings River fans, the Chowchilla's are relatively thin and lack the deep incised-valley-fill. Because accommodation space remained after each aggradational cycle, successive sequences prograded westward into the basin, thus each sequence is significantly thicker toward the basin centre.

Accumulation space cycles on the Chowchilla River fan could not have been controlled by the large sediment supply and discharge related to glacial outwash. Instead, Helley (1966) suggested that drainage basin erosion, and thus sediment supply to the Chowchilla River, was accelerated by higher rainfall, longer flow duration and greater surface runoff due to cooler temperatures. Marchand (1977) later suggested that Chowchilla River fan cycles were most probably controlled by sediment-supply variability related to hillslope stability changes in the drainage basin as vegetation type and density changed with Quaternary climate change. Similar climate-change-induced aggradational and degradational responses have been reported on fans in the SW United States by Bull (1991).

Following depositional patterns described by Bull (1991), we believe that accumulation space was created towards the end of glacial periods by increases in sediment supply released from drainage basin hillslopes during a change from higher vegetation density, that existed during wetter glacial periods, to thinner vegetation density of the warmer and dryer interglacial periods, Stream discharge was probably slightly higher during the glacial periods; however, Modesto channel size does not appear to be significantly greater than that of the modern channel system (Helley 1966).

The lack of a large interglacial incised valley on the Chowchilla River fan, along with the observation that the interglacial Chowchilla River has a similar gradient to that of the river during glacial aggradational periods, indicates that the amount of accumulation space change between glacial and interglacial periods is relatively small. Thus, it appears that a subtle change in sediment supply caused the accumulation space gain. In addition, accumulation space increase during glacial periods appears to be relatively small, as shown by thin Modesto deposits over Riverbank and thin Riverbank over Turlock Lake deposits on the upper-fan surface.

Discussion

San Joaquin Basin sequences

The Kings River, Tuolumne River and Chowchilla River fluvial fans experienced cyclic aggradational–degradational events controlled by Quaternary climate variability, with upper-fan exposure dominating interglacial periods and aggradation occurring during glacial periods. This cyclicity produced palaeosol-bounded sequences, where soils developed during interglacial low accumulation space periods. A predictable distribution of facies exists within different portions of these sequences. For example, in the upper fan we expect to see discrete channel deposits surrounded by overbank fines in the open-fan deposits, while in the palaeovalley fill deposits we expect to observe a general fining-upwards succession of relatively coarse-grained channel deposits. In addition, distal-fan deposits, emplaced below the interglacial intersection point, consist of discrete channel deposits surrounded by overbank fines, similar to open-fan units of the upper fan; however, rivers responsible for deposition are

smaller and sediment consists of material derived from weathered hillslopes (Weissmann *et al.* 2002*a*).

Although sequences on these fans are similar in character, subtle differences in fan character exist. These differences are related to variability in: (1) local basin subsidence rates; (2) basin size; (3) sediment supply to stream discharge ratio changes from climatic cyclicity; and (4) local base-level control. Table 1 summarizes the influences of subsidence rate and sediment supply/discharge variability on sequence geometry for fans of the eastern San Joaquin Basin. Local base-level control appears to influence sequence geometry by controlling the depth of incision for the interglacial interfan valley, where fan connection to sea level via the axial San Joaquin River produced deeper incised valleys than fans that are connected to internally drained portions of the basin.

Based on this analysis, we can predict sequence geometries on other fans of the San Joaquin Basin. For example, the Mokelumne, Stanislaus, Merced and San Joaquin River fluvial fans appear to have a similar character to the Tuolumne River fan (Fig. 2). These fans developed in areas with relatively low subsidence rates, have deep modern incised valleys, do not have an active depositional lobe and connect to the axial San Joaquin River. Therefore, we expect sequences that have relatively thick incised-valley-fill units, are westwards stepping and have minimal preservation of the interglacial fan deposits in the basin. In contrast, the Kaweah River fan drains into the Tulare Lake Basin, thus it experienced subsidence rates similar to the Kings River fan. Although the Kaweah River drainage basin is significantly smaller (Fig. 3), we expect to observe vertically stacked sequences, relatively thinner incised-valley-fill units and preservation of interglacial distal-fan deposits in the basin. Finally, fans associated with rivers that are not connected to glaciated portions of the Sierra Nevada (e.g. the Calaveras River, Duck Creek and Fresno River fans) display a similar geomorphic character to the Chowchilla River fan. Thus, we expect sequences to lack incised-valley-fill deposits and to be laterally stacked. In addition, interglacial deposits should be preserved near the basin centre on these fans, and an unconformity between the glacial and interglacial deposits is not expected in the distal portions of the fan.

The ability to predict sequence geometry and form should significantly aid future hydrogeological studies of the fans in the eastern San Joaquin Basin. As shown by Burow *et al.* (1999) and Weissmann *et al.* (2002*b*, 2004), facies distributions and sequence geometries significantly influence groundwater flow and contaminant transport on the Kings River fan. Additionally, the sequence stratigraphic approach provides an important framework for modelling facies distributions in fluvial fans (Weissmann & Fogg 1999). Future work will focus on testing the

Table 1. *Summary of effect of subsidence rate and sediment supply/stream discharge variability on sequence development and geometries.*

Subsidence rate	
Low	**High**
1. Laterally stacked sequences developed – apex of each successive aggradational event is located basinward of previous aggradational event	1. Vertically stacked sequences developed – fan apex location is relatively stable between successive aggradational events
2. Intersection point may move off fan during periods of low accumulation space, thus the fan toe may be truncated and the entire fan surface exposed to soil development	2. Intersection point shifts basinwards during periods of low accumulation space, although it remains on the fan. Thus, an active depositional lobe exists during low accumulation space periods
3. Thinner sequences deposited as less new accommodation space is added between successive aggradational events	3. Relatively thick sequences deposited as more new accommodation space is added between successive aggradational events
Sediment supply/stream discharge variability	
Non-glaciated basin response	**Glaciated basin response**
1. Less accumulation space developed during aggradation event	1. Relatively high accumulation space developed during aggradational event
2. Shallow channel incisions formed during periods of low accumulation space	2. Deep incised valleys formed during periods of low accumulation space
3. Upper fan may receive occasional large-event floods during periods of low accumulation space	3. Upper-fan surface inactive during periods of low accumulation space, thus exposed to soil development and erosion
4. Relatively thin sequences that are stacked laterally and fill the basin laterally	4. Relatively thick sequences that may be vertically stacked
5. Minimal stream gradient difference between periods of low accumulation space and periods of high accumulation space	5. Significant stream gradient difference between periods of low accumulation space and periods of high accumulation space

predictive ability of this sequence stratigraphic approach in the context of such groundwater studies.

Comparison to fan sequences in other areas

Cyclicity and sequence development on alluvial fans has been identified by many workers (e.g. Pierce & Scott 1982; Ritter *et al.* 1993; Reheis *et al.* 1996; Leeder *et al.* 1998; Zehfuss *et al.* 2001), yet these studies and the work presented here seem to conflict regarding timing of aggradation events on fans. Several examples exist that indicate aggradation timing similar to that of the San Joaquin Valley fans, where aggradation occurred during glacial periods on fans in southern Idaho (Pierce & Scott 1982), on the Cedar Creek fan in southern Montana (Ritter *et al.* 1993) and on fans in the Owens Valley, located on the east side of the Sierra Nevada (Zehfuss *et al.* 2001). In all of these examples, river systems

feeding the fans were directly tied to glaciated drainage basins, thus sediment supply increased significantly with sediment discharge from glaciers. As observed on the Kings and Tuolumne River fans, aggradation occurred due to this glacially supplied sediment increase.

Aggradational cycles on other fans appear to show aggradational cycles related to vegetation change in the drainage basin above the fan (e.g. Bull 1991; Reheis *et al.* 1996; Leeder *et al.* 1998). On these fans, climatic changes during the Quaternary affected vegetation density, and thus slope stability, in the drainage basin. Aggradation occurred on fans in the Great Basin of Nevada during the transition from cool, wet conditions to more arid at the end of glacial cycles because vegetation density decreased significantly, thus releasing sediment to the fans (Bull 1991; Reheis *et al.* 1996; Leeder *et al.* 1998). We believe that aggradational response on the Chowchilla River fluvial fan is similar to these fans, where vegetation change at the end of glaciation in the Sierra Nevada resulted in sediment release and aggradation on the fan; however, age dating and further studies on stratigraphic relationships between deposits on fans under glaciated basins against those under non-glaciated basins in the San Joaquin Valley are required to verify this hypothesis. Conversely, on fans surrounding the northern Mediterranean, Leeder *et al.* (1998) indicated that decreased vegetation on hillslopes during cool glacial episodes resulted in higher sediment load to rivers feeding fans, thus aggradation occurred during glacial periods.

Although exact timing of sequence development differs between these fan systems, all show that increased sediment supply related to climate change influence timing of aggradational events. Therefore, one must be aware of controls on sediment supply in the drainage basin in order to predict timing of aggradational cycles. Significantly, however, these studies show that allogenic change can cause incised-valley filling and aggradation (e.g. with sediment supply increase) or incised-valley development and degradation (e.g. with decreased sediment supply or increased discharge). In all these cases, accumulation space developed because sediment supply increased either due to a connection to glacial outwash or connection to a drainage basin that experienced loss of vegetation cover. Accumulation space was lost on these fans with either the end of glacial outwash or stabilization of the vegetation cover. Resulting stratigraphy, however, is similar between the fans, with sequence-bounding palaeosols and incised-channel or- valley bases marking periods of low accumulation space, and incised-valley-fill and open-fan deposits marking aggradation episodes during periods of increased accumulation space. Preservation of sequences and stacking relationships of the sequences are controlled by basin geometry and subsidence rates. Therefore, we believe that the geometry of stratigraphic successions and distribution of facies on fluvial fans can be assessed through a sequence stratigraphic approach.

Conclusions

The stratigraphic character of fluvial fans of the eastern San Joaquin Basin is dependent on the fan's position in the basin and its river's drainage basin characteristics. Four factors appear to control accommodation space development on these fans: (1) basin subsidence rates; (2) the degree of change in sediment supply to discharge ratios; (3) local base-level elevation changes; and (4) basin width.

Relatively high basin subsidence rates provided more accommodation space for successive sequences, thus allowing deposition of thicker sequences. In addition, higher subsidence rates led to vertical sequence stacking near the apex rather than lateral sequence progradation.

Fluvial fans that experienced a large change in the sediment supply to discharge ratio (e.g. those from glaciated river basins) developed sequences that consist of a relatively large incised-valley-fill and extensive open-fan deposits. On these fans the intersection point shifted to a position near the apex during periods of high sediment supply (e.g. high accumulation space periods during glacial outwash events), and the intersection point shifted basinwards during periods of low accumulation space (e.g. interglacial periods). Conversely, fans that did not experience large changes in the sediment supply to discharge ratio did not develop the deep incised valleys during low accumulation space periods. The intersection points on these fans, however, did appear to shift towards the apex during high accumulation space events. Additionally, these fans appear to have experienced relatively continuous deposition on the distal portion of the fan since an incised valley is not developed.

Local base level appears to control the ultimate depth of the interglacial incised valleys on glaciated fans. Deeper valleys developed along rivers that were connected to the San Joaquin River, which is ultimately tied to sea level. In contrast, relatively shallow valleys formed on rivers that are tied to internally drained (and higher elevation) portions of the basin.

Finally, the basin width affects the fan size in the San Joaquin Basin. Fans in the narrower, northern portion of the basin filled available accommodation space and, thus, are constrained by the basins width. These fans tend to be smaller than their southern counterparts, where the basin is significantly wider.

Acknowledgment is made to the Donors of the American chemical Society Petroleum Research Fund, for support of this research (PRF#37731-G8). This work was also funded in part by Michigan State University. We also benefited greatly from discussions with G. Fogg, J. Mount, K. Burow, D. Warke and W. Lettis. Comments by A. Hartley, C. Viseras and A.M. Harvey enhanced the manuscript significantly.

References

ARKLEY, R.J. 1962. The geology, geomorphology, and soils of the San Joaquin Valley in the vicinity of the Merced River, California. *California Division of Mines and Geology, Bulletin*, **182**, 25–31.

ARKLEY, R.J. 1964. *Soil Survey of the Eastern Stanislaus Area, California.* US Department of Agriculture, Soil Conservation Service, Series 1957, **20**.

ARKLEY, R.J., COLE, R.C., HUNTINGTON, G.L., CARLTON, A.B. & SMITH, G.K. 1962. *Soil survey of the Merced Area, California.* US Department of Agriculture, Soil Conservation Service, Series 1950, **7**.

ARROUES, K.D. & ANDERSON, C.H., JR. 1976. *Soil Survey of Kings County, California*. US Department of Agriculture. US Government Printing Office.

ATWATER, B.F. 1980. *Attempts to correlate late Quaternary climate records between San Francisco Bay, the Sacramento–San Joaquin Delta, and the Mokelumne River, California*. PhD Thesis, University of Delaware.

ATWATER, B.F., ADAM, D.P. *ET AL.* 1986. A fan dam for Tulare Lake, California, and implications for the Wisconsin glacial history of the Sierra Nevada. *Bulletin of the Geological Society of America*, **97**, 97–109.

BARTOW, J.A. 1991. *The Cenozoic Evolution of the San Joaquin Valley, California*. US Geological Survey, Professional Paper, **1501**.

BENNETT, G.L. 2003. *Estimating the nature and distribution of shallow subsurface paleosols on fluvial fans in the San Joaquin Valley, California*. MS Thesis, Michigan State University.

BLUM, M.D. & TÖRNQVIST, T.E. 2000. Fluvial responses to climate and sea-level change: a review and look forward. *Sedimentology*, **47**, 2–48.

BULL, W.B. 1991. *Geomorphic Responses to Climatic Change*. Oxford University Press, New York.

BUROW, K.R., PANSHIN, S.Y., DUBROVSKY, N.M., VAN BROCKLIN, D. & FOGG, G.E. 1999. *Evaluation of Processes Affecting 1,2-Dibromo-3-Chloropropane (DBCP) Concentrations in Ground Water in the Eastern San Joaquin Valley, California: Analysis of Chemical Data and Ground-water Flow and Transport Simulations*. US Geological Survey, Water-Resources Investigations Report, **99–4059**.

BUROW, K.R., SHELTON, J.L., HEVESI, J.A. & WEISSMANN, G.S. 2004. *Hydrogeologic Characterization of the Modesto Area, San Joaquin Valley, California*. US Geological Survey, Scientific Investigations Report, **2004–5232**.

BUROW, K.R., WEISSMANN, G.S., MILLER, R.D. & PLACZEK, G. 1997. *Hydrogeologic Facies Characterization of an Alluvial Fan Near Fresno, California, Using Geophysical Techniques*. US Geological Survey, Open-File Report, **97–46**.

DAVIS, S.N. & HALL, F.R. 1959. *Water Quality of Eastern Stanislaus and Northern Merced Counties, California*. Stanford University Publications, Geological Sciences, **6**.

FERRARI, C.A. & MCELHINEY, M.A. 2002. *Soil Survey of Stanislaus County, California, Western Part*. US Department of Agriculture, Natural Resources Conservation Service, Washington, DC.

HARDEN, J.W. 1987. Soils developed in granitic alluvium near Merced, California. *Bulletin of the US Geological Survey*, **1590-A**.

HELLEY, E.J. 1966. *Sediment transport in the Chowchilla River basin, Mariposa, Madera, and Merced Counties, California*. PhD Thesis, University of California, Berkeley.

HUNTINGTON, G.L. 1971. *Soil Survey, Eastern Fresno Area, California*. US Department of Agriculture, Soil Conservation Service. US Government Printing Office, Washington, DC.

HUNTINGTON, G.L. 1980. *Soil–land form relationships of portions of the San Joaquin River and Kings River alluvial depositional systems in the Great Valley of California*. PhD Thesis, University of California, Davis.

JANDA, R.J. 1966. *Pleistocene history and hydrology of the upper San Joaquin River, California*. PhD Thesis, University of California, Berkeley.

JERVEY, M.T. 1988. Quantitative geological modeling of siliciclastic rock sequences and their seismic expressions. *In*: Wilgus, C.K., Hastings, B.S. *ET AL.* (eds) *Sea-level Changes: An Integrated Approach*. SEPM (Society for Sedimentary Geology), Special Publications, **42**, 47–69.

LANSDALE, A.L., WEISSMANN, G.S. & BUROW, K.R. 2004. Influence of coarse-grained incised valley fill on ground-water flow in fluvial fan deposits, Stanislaus County, California, USA. *Eos, Transactions of the American Geophysical Union*, **85**, Fall Meeting Supplement, Abstract H21E-1057.

LEEDER, M.R., HARRIS, T. & KIRKBY, M.J. 1998. Sediment supply and climate change: implications for basin stratigraphy. *Basin Research*, **10**, doi: 10.1046/j.1365-2117.1998.00054.x.

LETTIS, W.R. 1982. *Late Cenozoic Stratigraphy and Structure of the Western Margin of the Central San Joaquin Valley, California*. US Geological Survey, Open-File Report, **82–526**.

LETTIS, W.R. 1988. Quaternary geology of the Northern San Joaquin Valley. *In*: GRAHAM, S.A. (ed.) *Studies of the Geology of the San Joaquin Basin*. SEPM Pacific Section, **60**, 333–351.

LETTIS, W.R. & UNRUH, J.R. 1991. Quaternary geology of the Great Valley, California. *In*: Morrison, R.B. (ed.) *Quaternary Nonglacial Geology: Conterminous U.S.* Geological Society of America, Geology of North America, **K-2**, 164–176.

MARCHAND, D.E. 1977. The Cenozoic history of the San Joaquin Valley and the adjacent Sierra Nevada as inferred from the geology and soils of the eastern San Joaquin Valley. *In*: SINGER, M.J. (ed.) *Soil Development, Geomorphology, and Cenozoic History of the Northeastern San Joaquin Valley and Adjacent Areas, California*. University of California Press, Guidebook for Joint Field Session, Soil Science Society of America and Geological Society of America, 39–50.

MARCHAND, D.E. & ALLWARDT, A. 1981. *Late Cenozoic Stratigraphic Units, Northeastern San Joaquin Valley, California*. Bulletin of the US Geological Survey, **1470**.

MCELHINEY, M.A. 1992. *Soil Survey of San Joaquin County, California*. US Department of Agriculture, Soil Conservation Service, Washington, DC.

MITCHUM, R.M., JR. 1977. Part eleven: glossary of terms used in seismic stratigraphy. *In*: PAYTON, C.E. (ed.) *Seismic Stratigraphy – Applications to Hydrocarbon Exploration*. AAPG, Memoirs, **26**, 205–212.

MUTO, T. 1987. Coastal fan processes controlled by sea level changes: a Quaternary example from the Tenryugawa fan system, Pacific coast of Central Japan. *Journal of Geology*, **95**, 716–724.

MUTO, T. 1988. Stratigraphical patterns of coastal-fan sedimentation adjacent to high-gradient submarine slopes affected by sea-level changes. *In*: NEMEC, W. & STEEL, R.J. (eds) *Fan Deltas: Sedimentology and Tectonic Settings*. Blackie, Glasgow, 84–90.

MUTO, T. & STEEL, R.J. 2000, The accommodation concept in sequence stratigraphy: some dimensional problems and possible redefinition. *Sedimentary Geology*, **130**, 1–10.

NAZAR, P.G. 1990. *Soil Survey of Merced County, California, Western Part*. US Department of Agriculture, Soil Conservation Service, Washington, DC.

PIERCE, K.L. & SCOTT, W.E., 1982. Pleistocene episodes of alluvia-gravel deposition, southeastern Idaho. *In*: BONNICHSEN, B. & BRECKENRIDGE, R.M. (eds) *Cenozoic Geology of Idaho. Idaho Bureau of Mines and Geology, Bulletin*, **26**, 685–702.

POSAMENTIER, H.W. & VAIL, P.R. 1988. Eustatic control on clastic deposition II – sequence and systems tract models. *In*: WILGUS, C.K., Hastings, B.S. *ET AL.* (eds) *Sea-level Changes: An Integrated Approach*. SEPM (Society for Sedimentary Geology), Special Publications, **42**, 125–154.

POSAMENTIER, H.W., JERVEY, M.T. & VAIL, P.R. 1988. Eustatic control on clastic deposition I—conceptual framework. *In*: WILGUS, C.K., HASTINGS, B.S. *ET AL.* (eds) *Sea-level Changes: An Integrated Approach*. SEPM (Society for Sedimentary Geology), Special Publications, **42**, 109–124.

REHEIS, M.C., SLATE, J.L., THROCKMORTON, C.K., MCGEEHIN, J.P., SARNA-WOJCICKI, A.M. & DENGLER, L. 1996. Late Quaternary sedimentation on the Leidy Creek fan, Nevada–California: geomorphic responses to climate change. *Basin Research*, **12**, 279–299.

RITTER, J.B., MILLER, J.R. *ET AL.* 1993. Quaternary evolution of Cedar Creek alluvial fan. Montana. *Geomorphology*, **8**, 287–304.

STANISTREET, I.G. & MCCARTHY, T.S. 1993. The Okavango Fan and the classification of subaerial fans. *Sedimentary Geology*, **85**, 114–133.

STEPHENS, F.G. 1982. *Soil Survey of Tulare County, California, Central Part*. US Department of Agriculture, Soil Conservation Service, Washington, DC.

ULRICH, R. & STROMBERG, L.K. 1962. *Soil Survey of the Madera Area, California*. US Department of Agriculture, Soil Survey Series 1951, **11**.

USDA/NRCS. 2003. *Soil Survey of Fresno County, California, Western Part*. (Unpublished survey.) US Department of Agriculture, Natural Resources Conservation Service, Washington, DC. SSURGO data set from www.nrcs.usda.gov

VAN WAGONER, J.C., MITCHUM, R.M., CAMPION, K.M. & RAHMANIAN, V.D. 1990. *Siliciclastic Sequence Stratigraphy in Well Logs, Cores, and Outcrops: Concepts for High-resolution Correlation of Time and Facies*. AAPG, Methods in Exploration Series, **7**.

VISERAS, C., CALVACHE, M.L., SORIA, J.M. & FERNÁNDEZ, J. 2003, Differential features of alluvial fans controlled by tectonic or eustatic accommodation space. Examples from the Betic Cordillera, Spain. *Geomorphology*, **50**, 181–202.

WAHRHAFTIG, C. & BIRMAN, J.H. 1965. The Quaternary of the Pacific Mountain System in California. *In*: WRIGHT, H.E., JR & FREY, D.G. (eds) *The Quaternary of the United States, A Review Volume for the VII Congress of the International Association for Quaternary Research*. Princeton University Press, Princeton, NJ, 299–340.

WASNER, A.R. & ARROUES, K.D. 2003. *Soil Survey of Tulare County, California, Western Part*. US Department of Agriculture, Natural Resources Conservation Service, Washington, DC.

WEISSMANN, G.S. & FOGG, G.E. 1999. Multi-scale alluvial fan heterogeneity modeled with transition probability geostatistics in a sequence stratigraphic framework. *Journal of Hydrology*, **226**, 48–65.

WEISSMANN, G.S., MOUNT, J.F. & FOGG, G.E. 2002*a*. Glacially driven cycles in accumulation space and sequence stratigraphy of a stream-dominated alluvial fan, San Joaquin Valley, California, U.S.A. *Journal of Sedimentary Research*, **72**, 270–281.

WEISSMANN, G.S., YONG, Z., LABOLLE, E.M. & FOGG, G.E. 2002*b*. Modeling environmental tracer-based groundwater ages in heterogeneous aquifers. *Water Resources Research*, **38**, 1198–1211.

WEISSMANN, G.S., ZHANG, Y., FOGG, G.E. & MOUNT, J.F. 2004. Influence of incised-valley-fill deposits on hydrogeology of a glacially-influenced, stream-dominated alluvial fan. *In*: BRIDGE, J. & HYNDMAN, D.W. (eds) *Aquifer Characterization*. SEPM (Society for Sedimentary Geology), Special Publications, **80**, 15–28.

WRIGHT, V.P. & MARRIOTT, S.B. 1993, The sequence stratigraphy of fluvial depositional systems: the role of floodplain sediment storage. *Sedimentary Geology*, **86**, 203–210.

ZEHFUSS, P.H., BIERMAN, P.R., GILLESPIE, A.R., BURKE, R.M. & CAFFEE, M.W. 2001. Slip rates on the Fish Springs fault, Owens Valley, California, deduced from cosmogenic ^{10}Be and ^{26}Al and soil development on fan surfaces. *Bulletin of the Geological Society of America*, **113**, 241–255.

Tertiary alluvial fans at the northern margin of the Ebro Basin: a review

GARY NICHOLS

*Department of Geology, Royal Holloway, University of London, Egham, Surrey TW20 0EX,
UK (e-mail: g.nichols@gl.rhul.ac.uk)
Also at the University Centre in Svalbard, P.O. Box 156, N-9171 Longyearbyen, Norway
(e-mail: gary.nichols@unis.no)*

Abstract: Tertiary alluvial fan deposits along the margin of the Ebro foreland basin north of
Huesca, Spain, are remarkable for the range of sedimentological and tectonic features preserved
within them. The fan deposits formed after the southern Pyrenean thrust front (the Guarga Thrust)
was established in this area, forming steep topography at the basin margin in the mid-Oligocene.
Some shortening continued during deposition of the earliest fans resulting in synsedimentary faults,
folds and unconformities. Clast compositions in the fan conglomerate beds record unroofing of the
thrust front and also reveal differences in bedrock provenance between adjacent, coeval fan
deposits. Bedrock provenance also influenced processes of deposition, with fans built up of detritus
derived from areas rich in gypsum and mudrock showing more evidence of debris-flow processes.
Deposition by debris flows also dominated the smallest fan body, but the majority of the fans were
the products of sedimentation from unconfined or poorly confined traction currents. These resulted
in sheets of conglomerate in the more proximal areas. Within individual fan bodies the proportion of
sandstone increases over a distance of up to 5 km to where the distal fan facies are seen as thin sandstone and mudstone beds. The fringes of the fan bodies interfinger with lacustrine and fluvial facies,
which indicate a temperate–semi-arid palaeoclimate. Vertical aggradation of the fan deposits due
to rising base level in the Ebro Basin in the Oligocene and early Miocene was followed by deep incision following a late Miocene base level-fall. This led to the partial erosion of the fan deposits and
their spectacular exposure in the modern landscape.

The northern margin of the Ebro Basin in northern
Spain (Fig. 1) has one of the most complete examples
of a fan-fringed basin margin of any pre-Quaternary
exposures. Along a 40 km-section of basin margin,
adjacent to the western External Sierras, conglomerate bodies are exposed and can be interpreted as the
remnants of late Oligocene–Early Miocene alluvial-
fan deposits (Fig. 2). Exposures of alluvial-fan
deposits at the margin of the Ebro Basin provide an
excellent opportunity to study the following aspects
of alluvial fan sedimentology:

- relationships between the structural and stratigraphic development of fans, and the structural
  evolution of a thrust-dominated basin margin;
- evidence for climatic and source-area bedrock
  controls on processes of deposition on alluvial
  fans;
- the three-dimensional facies relationships in the
  fan bodies, with clear proximal–distal variations
  in bed thickness and grain size;
- interfingering of fan deposits with fluvial and
  lacustrine sediments.

The quality of the exposure and the quantity of
geological information from a number of fan deposits
make this area an outstanding case study in pre-
Quaternary alluvial-fan sedimentation.

Tectonostratigraphic setting

Regional tectonics

The Pyrenees is an orogenic belt that developed in the
Cenozoic as a result of crustal shortening between the
Eurasian plate and the Iberian sub-plate (ECORS-
Pyrenees Team 1998; Choukroune & ECORS-
Pyrenees Team 1989; Muñoz 1992; Teixell 1996).
Along the southern side of the axial zone of the oro-
genic belt a series of thrusts developed, carrying
allochthonous units southwards. The decollement at
the base of these thrust units lies in Triassic beds, and
in the central part of the southern Pyrenees, north of
Huesca (Fig. 1), the Guarga thrust sheet formed
a frontal ramp, bringing Triassic, Cretaceous and
Palaeogene strata to the surface. The present-day
expression of this frontal ramp is a range of hills, the
External Sierras (also referred to as the 'Sierras
Exteriores' and 'Sierras Marginales') that became an
emergent structure in the middle–late Oligocene
(Puigdefábregas & Soler 1973; Nichols 1987a;
Hogan & Burbank 1996; Teixell 1996). These hills
form the northern margin of a foreland basin, the Ebro
Basin, which extends south to the Corillera Iberica
and east to the Catalan mountain range (Fig.1).

The Ebro foreland basin formed by the loading of
thrust sheets in the southern Pyrenean zone on the

From: HARVEY, A.M., MATHER, A.E. & STOKES, M. (eds) 2005. *Alluvial Fans: Geomorphology, Sedimentology, Dynamics.*
Geological Society, London, Special Publications, **251**, 187–206. 0305-8719/05/$15 © The Geological Society of
London 2005.

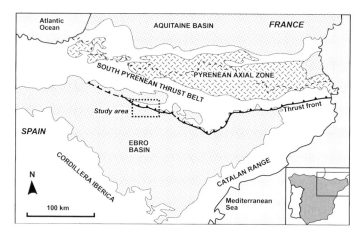

Fig. 1. Tectonic and stratigraphic setting of the south Pyrenenan thrust belt and the adjacent Ebro foreland basin (from Teixell 1996).

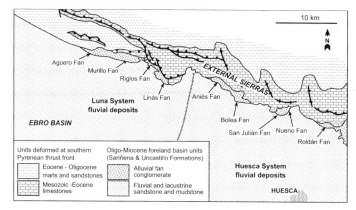

Fig. 2. The western External Sierras and adjacent parts of the Ebro foreland basin. The approximate extent of the nine alluvial fan bodies within the Uncastillo and Sariñena formations (Late Oligocene–Early Miocene) is shown (from Lloyd *et al.* 1998).

Iberian crust: flexural subsidence in the southern Pyrenees began in the early Eocene with the formation of the Tremp-Graus, Ainsa and Jaca basins (Puigdefàbregas 1975; Puigdefàbregas & Souquet 1984). The fill of these basins became allochthonous as the thrusting progressed southwards, and the depocentre shifted to the present area of the Ebro Basin in the middle Oligocene.

Stratigraphy

The stratigraphic framework for the area is summarized in Figure 3. Triassic through to early Oligocene units are all allochthonous in this part of the southern Pyrenees and are exposed as a deformed complex in the External Sierras. The Uncastillo and Sariñena formations are late Oligocene–early Miocene in age

(Quirantes 1978; Arenas *et al.* 2001), and are composed of continental facies of conglomerate, sandstone and mudstone. These deposits are the fill of the Ebro Basin in this area and lie unconformably on the older rocks. Deformation in these younger units is limited to a region close to the northern margin of the basin (Arenas *et al.* 2001).

Puigdefàbregas (1975) mapped out bodies of conglomerate and showed that they were a series of isolated alluvial-fan deposits; he also noted that the sandstone and mudstone beds were the products of deposition in fluvial channel and overbank environments in the northern part of the Ebro Basin. The relationship between the fan deposits and the other fluvial facies was demonstrated by Hirst & Nichols (1986) on the basis of clast compositions and sandstone petrography that showed that the small conglomerate bodies were derived directly from the

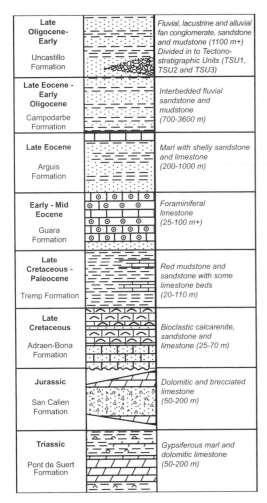

Late Oligocene-Early / Uncastillo Formation	Fluvial, lacustrine and alluvial fan conglomerate, sandstone and mudstone (1100 m+) Divided in to Tectono-stratigraphic Units (TSU1, TSU2 and TSU3)
Late Eocene - Early Oligocene / Campodarbe Formation	Interbedded fluvial sandstone and mudstone (700-3600 m)
Late Eocene / Arguis Formation	Marl with shelly sandstone and limestone (200-1000 m)
Early - Mid Eocene / Guara Formation	Foraminiferal limestone (25-100 m+)
Late Cretaceous - Paleocene / Tremp Formation	Red mudstone and sandstone with some limestone beds (20-110 m)
Late Cretaceous / Adraen-Bona Formation	Bioclastic calcarenite, sandstone and limestone (25-70 m)
Jurassic / San Calien Formation	Dolomitic and brecciated limestone (50-200 m)
Triassic / Pont de Suert Formation	Gypsiferous marl and dolomitic limestone (50-200 m)

Fig. 3. The stratigraphy of the western External Sierras and the Ebro Basin (from Puigdefàbregas 1975; Nichols 1987a; Arenas et al. 2001).

adjacent thrust front, whilst the fluvial deposits were derived from further to the north. More detailed stratigraphic studies by Hogan (1993), Arenas (1993) and Arenas et al. (2001) have demonstrated that parts of the alluvial fan successions are coeval with the fluvial deposits, which can in turn be partly correlated with lacustrine facies further to the south.

In this review, only the section of the Ebro Basin margin that lies west of the city of Huesca is covered (between 0°21′W, 42°16′N and 0°45′W, 42°22′N): this part of the Pyrenean thrust front is the western External Sierras. The bodies of Oligo-Miocene conglomerate and associated deposits exposed along the Ebro Basin margin will be referred to as 'fan bodies', each of which has been named after a nearby geographical feature (Fig. 2). The interpretation of these strata as the deposits of distinct alluvial fans has been

Table 1. *Summary of the alluvial fan bodies along the Ebro Basin margin, western External Sierras (from Lloyd et al. 1998; Nichols & Thompson 2005)*

Fan	Area (km²)	Main depositional processes
Agüero	6.0 km²	Clast-rich debris flow (Stage 1), waterlain (Stages 2 and 3)
Murillo	5.4 km²	Waterlain gravel and sand
Riglos	0.8 km²	Clast-rich debris flow
Linás	13.5 km²	Waterlain gravel and sand
Aniés	2.3 km²	Waterlain gravel and sand
Bolea	2.6 km²	Waterlain gravel and sand
San Julián	3.5 km²	Waterlain gravel and sand
Nueno	2.5 km²	Muddy debris flow
Roldán	8.5 km²	Waterlain gravel and sand

the subject of a number of papers (e.g. Hirst & Nichols 1986; Nichols 1987b; Lloyd et al. 1998; Nichols & Hirst 1998; Arenas et al. 2001). Of the nine alluvial fan bodies along this section of the Ebro Basin margin, detailed studies have only been published on four: Agüero (Nichols 1987b), Roldán (Nichols & Hirst 1998), Nueno and San Julián (Nichols & Thompson 2005). Descriptions of the other fans have been presented in unpublished PhD theses by Hirst (1983), Nichols (1984), Arenas (1993) and Lloyd (1994).

Description of the fan bodies

A brief description of each of the nine fan bodies is provided here as a context for the interpretation of them in following sections. The dimensions of most of these bodies have been documented in Lloyd et al. (1998), from which Table 1 is largely derived.

Agüero fan body

The most westerly of the fan bodies lies adjacent to the village of Agüero, from which it takes its name (Fig. 4). It is deeply dissected by a small stream, the Barranco de Pituelo, which forms a gorge with 400 m relief on the west side and 300 m on the east, where pinnacles of conglomerate, known locally as 'mallos', tower over 250 m above the village. This is the only fan body with the base of the succession exposed, and the oldest part of the succession is deformed in a series of synsedimentary folds and unconformities (see Nichols 1987b; Lloyd et al. 1998).

Murillo fan body

A prominent mass of conglomerate over 400 m thick forms a hill 1191 m high at Peña Ruaba, 2 km north of

Fig. 4. The village of Agüero and the 'mallos' of the adjacent fan body.

the village of Murillo de Gállego (Fig. 5). It is not dissected by any modern drainage, but the conglomerate body appears to be the core of an alluvial fan that was of similar dimensions to its neighbour at Agüero. A broad anticlinal structure affects the northern part of the body, and a tighter monoclinal fold deforms the southern part (see Nichols 1984 and below).

Riglos fan body

Los Mallos de Riglos is a well-known geomorphological feature of this area (Fig. 6). The two main pinnacles of coarse conglomerate tower over 300 m above the village of Riglos, and there are also views of the conglomerate body from the gorge of the Rió Gállego that display the relationship between the proximal-fan deposits and the contorted limestone beds of the Guara thrust front (Fig. 7). Despite their impressive appearance, these pinnacles are the remnants of the smallest of the alluvial fans in the western External Sierras (Lloyd *et al.* 1998).

Linás fan body

There is an almost continuous fringe of conglomerate along the basin margin (Fig. 6) between the main pinnacle at Riglos (Mallo Pisón) and the ridge of San Miguel on which lies the Ermita de San Miguel and the ruins of the Castillo de Marcuello. San Miguel is made up of sandstone and conglomerate beds, which form the Linás alluvial fan body (Fig. 8), named after the village of Linás de Marcuello, 1 km to the south of the Castillo. The bedding is near horizontal in all but the most proximal parts of the fan where beds dip south adjacent to the limestone bedrock of

the thrust front. Over 200 m of vertical section are well exposed, varying from dominantly conglomerate facies in the NW part of the body to pebbly sandstone beds beneath the Castillo.

Ebro Basin margin between Linás and Aniés

The Ebro Basin margin between the villages of Linás de Marcuello and Aniés, 9 km to the east, is apparently devoid of any Oligo-Miocene conglomerate units. Where the contact between the Uncastillo Formation and the deformed strata of the thrust front is exposed near Loarre there are small, localized patches of chaotic breccia made up of blocks up to 2 m across. These are interpreted as local talus deposits that were deposited in local depressions or gullies in the limestone bedrock of the basin margin (Nichols 1984). Between these breccias and extending up to 5 km out into the basin are coeval, gravelly sandstone and fine sandstone facies. Beds of the coarser facies have scoured bases and may be lens-shaped, whilst the fine sandstone shows ripple cross-lamination. These deposits may represent the remains of one or more fan bodies of mainly sandy sediment close to the basin margin.

Aniés fan body

The Aniés fan body is inconspicuous. It lies 2.5 km NE of the village, and horizontal beds of conglomerate form a unit approximately 200 m thick and over 1 km across. These beds lie against External Sierras bedrock with a steep unconformity. They appear to be similar in character to the beds exposed in the Bolea fan body to the east.

Fig. 5. The Murillo fan body viewed from the east. On the right-hand side an overturned succession of Triassic–middle Eocene strata dip to the north: on the left-hand side the trace of a monoclinal fold in distal-fan deposits can be traced.

Fig. 6. 'Los Mallos de Riglos' viewed from the south and adjacent parts of the basin margin showing a fringe of conglomerate across to the Linás fan body on the far right.

Bolea fan body

The Bolea fan body is exposed in a mass over 200 m thick and extends for 2.5 km E–W on the hillside above the Ermita de Trinidad, 3 km NE of the village of Bolea. The exposures are dominantly alternating beds of polymict conglomerate and sandstone, the former occurring in sharp-based, graded units about 1 m thick. Finer gained deposits, representing the more distal fan facies, are exposed only at the SE fringe of the outcrop area.

San Julián fan body

The distinction between this and the adjacent Nueno fan body has only recently been recognized

(Nichols & Thompson 2005). In earlier studies (Lloyd *et al.* 1998) the two were considered to the deposits of a single alluvial fan. The San Julián fan is best exposed in the sides of the Barranco de San Julián, which cuts a gorge about 200 m deep through the proximal fan facies. There are further exposures and a cross-section of more proximal, locally deformed facies to sandy, distal deposits in the Barranco de Fenés, 1 km west of the Ermita de San Julián (Fig. 9).

Nueno fan body

One kilometre north of the village of Nueno, at the mouth of the gorge of the Rió Isuela, distinctive red beds of sandstone and conglomerate are exposed on

Fig. 7. The Riglos fan body viewed from the gorge of the Río Gállego. Adjacent to the conglomerate mass are beds of the Guara Formation that are tightly folded and show imbricate, out-of-sequence thrusting (top left of view).

Fig. 8. Beds of undeformed conglomerate and sandstone making up the Linás fan body viewed from the Castillo de Marcuello.

the hillside (Fig. 10). These beds are the deposits of the Nueno fan and they can be traced westwards towards the San Julián fan to a region where the deposits of the two fans interfinger. The Nueno fan deposits contain a much lower proportion of limestone clasts than all the other fans (Nichols & Thompson 2005) and form a less resistant outcrop.

Roldán fan body

The two 'mallos' that form the Salto de Roldán either side of the Rió Flumen can be seen from the plain east of Huesca city, over 10 km away (Fig. 11). The 400 m-deep gorge and the near-vertical southern

faces of the pinnacles form spectacular scenery in the most completely exposed of the alluvial fan bodies. In addition to accessible outcrop of proximal facies in the pinnacles (Fig. 12), horizons can be traced laterally over a distance of over 2.5 km to the SW into the most distal deposits of this alluvial fan (Nichols & Hirst 1998).

Correlation of fan deposits

The correlation scheme used here is based on the tectono-stratigraphic units (TSUs) of Arenas *et al.* (2001) and the chronology of Hogan & Burbank (1996). The lowest, palaeovalley-confined part of

Fig. 9. Proximal facies on the western side of the San Julián body showing synsedimentary deformation in the form of a reverse fault and associated growth strata.

Fig. 10. Beds of the Nueno fan body on the western side of the valley of the Río Isuela.

the Agüero succession lies within TSU1, whilst the middle part, up to the base of the conglomerate pinnacles at this locality, are considered to be TSU2. Both TSU1 and TSU2 are late Oligocene in age (Arenas *et al.* 2001). TSU2 is also represented in the lower, deformed parts of the fan successions at Murillo and Riglos. The main pinnacles at Agüero and Riglos, and the upper half of the Murillo body, are all mapped by Arenas *et al.* (2001) as TSU3 deposits (early Miocene), along with all of the fan successions at Linás, Aniés and Bolea. Extending the Arenas *et al.* (2001) correlation scheme to the east indicates that the fan deposits at San Julián, Nueno and Roldán are all part of TSU3. This correlation is achieved by tracing correlative depositional units between the fans and the interfan deposits.

The period of deposition is difficult to constrain. A magnetic polarity stratigraphy has been established at Agüero (Hogan & Burbank 1996) covering the upper part of the succession exposed there. These authors suggest that deposition of the upper conglomerate beds (*c.* 250 m thickness) took place over a period of just over 1 Ma in the latest Oligocene. More recent work by Arenas *et al.* (2001) has cast doubt on the correlations used by Hogan & Burbank (1996) and point to lacunae within the conglomerate successions. The magnitude of these lacunae between the TSUs is not known, and as both the biostratigraphy and the magnetostratigraphy provided only patchy information (Arenas *et al.* 2001), the periods of fan deposition and, hence, the rates of sedimentation cannot be usefully constrained.

Fig. 11. The conglomerate pinnacles of the 'Salto de Roldán'. In the foreground the rocks are fluvial deposits of the Huesca System, and in the background are folded strata in the External Sierras.

Fig. 12. The unconformity contact (dotted line) between the most proximal deposits of the Roldán alluvial fan and folded and thrust-faulted Triassic–early Eocene rocks.

Formation of the External Sierras thrust front and the basin-margin topography

As the thrust front of the Guarga Trust sheet became emergent by ramping up through the stratigraphy, the newly created topography would have initially been made of sandstone and mudstone units of the Campodarbe Formation and the underlying Arguis Formation (Fig. 13). This material would have been relatively easy to erode and, although a range of hills may have existed, it is unlikely to have formed steep relief at the margin of the Ebro Basin. Significant relief would not have formed until the shortening was sufficient to bring the limestone of the Guara Formation to the surface: this unit offered more resistance to weathering and erosion than the overlying Arguis and Campodarbe formations, and the rate of

denudation became less than the rate of uplift. A second factor in the formation of the relief would also have been the emergence of the Guarga thrust plane: once the thrust broke the surface, the zone of internal décollement that formed a relatively low-friction slip plane would have passed onto the rougher land surface which would have offered more resistance to slip. The effect of this increase in friction can be seen in the Rió Gállego gorge section through the External Sierras (Figs 7 & 14) where imbricate thrusts formed out-of-sequence in the hanging wall of the Guarga Thrust (Nichols 1987a, 1989).

It is evident from the clasts in the fan deposits that the Guarga Thrust became emergent and started to form a basin margin with steep relief in the late Oligocene (Fig. 13). Clasts eroded from the Guara limestone are recognized in the oldest beds of the

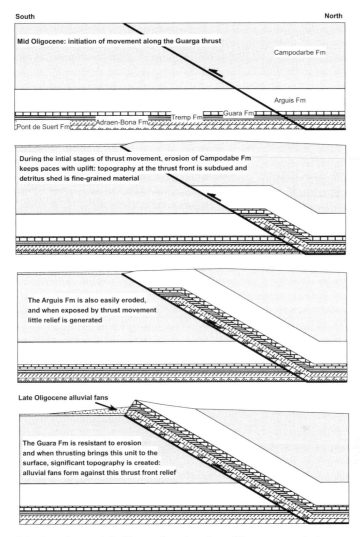

Fig. 13. Evolution of the thrust front and the history of erosion of pre-Oligocene strata in the western External Sierras.

Agüero fan (Nichols 1987*b*), the Riglos and Bolea fans, and in the basin-margin fan deposits at San Julián. Clast assemblages in all the fans are apparently devoid of material derived from the Arguis Formation. Sand eroded from the Campodarbe Formation is a probable source of some of the sandy facies in the fan deposits: other sources of arenaceous material in the External Sierras lithologies are relatively thin sandstone beds in the Guara Formation, the Palaeocene Tremp Formation and the Upper Cretaceous strata. Clasts from the limestone units in each of these formations are recognized in all of the fan deposits in varying proportions (Lloyd *et al.* 1998).

Basin-margin relief

The amount of relief that formed can be determined at the contact between the most proximal fan deposits and the bedrock in the External Sierras. The most impressive exposure of the unconformity is at Roldán, where 400 m of almost vertical relief can be seen (Fig. 12). The Roldán fan deposits have only undergone slight deformation (see below), so the unconformity surface represents a basin margin relief of at least 400 m. At other locations the unconformity is less completely exposed (all the fans from Nueno to Linás), or has been deformed (the fans between Riglos and Agüero).

Fig. 14. Proximal facies of angular, poorly sorted talus deposits of the Riglos fan body in unconformable contact with limestone beds of the Guara Formation. The unconformity has been folded, overturning the talus deposits in the centre of the picture and the limestone strata are repeated by imbricate thrusts (top of picture).

Where the unconformity between the fan deposits and the rocks of the External Sierras has not been deformed the structure of the thrust front in the early Miocene must have been more or less the same as it is today. The highest structural unit in the thrust front complex of the External Sierras is a right-way-up allochthonous succession (Nichols 1987a, 1989; Pocoví et al. 1990; Millán et al. 1995, 2000). Beds of the Guara Formation in this structural unit form much of the highest relief in the External Sierras today: they are also the youngest resistant strata (Fig. 13), and provided abundant clasts to the fans, so this same relief must have existed in Early Miocene times. It is therefore possible to reconstruct the palaeorelief at the basin margin, extending from the base of the undeformed fan deposits to the tops of the hills where the Guara Formation is exposed. At Roldán this relief was approximately 900 m, the lower 400 m of which was almost vertical.

The preservation of this very steep, localized relief as an unconformity surface indicates that there was little or no mass wastage of the slope by landsliding during the time it took for the fan to mantle the topography. This implies that the rock strength was great enough to maintain the very high slope angle (Burbank et al. 1996) and also that landsliding was not a dominant process under the relatively dry climatic regime, in contrast to the situation in a very humid climate (Hovius et al. 1997). Vertical accretion of fan deposits at the basin margin reduced this relief through time and the highest undeformed fan deposits are approximately 500 m below the highest parts of the External Sierras in this area. The tops of

the 'mallos' at Roldán are at 1124 m above sea level and 5 km to the north the summit of Pico de Águila, consisting of the Guara Formation, is 1629 m.

Fan aggradation and base level

During the Oligocene and Miocene, the Ebro Basin was endorheic, enclosed by the Iberian and Catalan ranges to the SW and SE, as well as the Pyrenees to the north (Vergés et al. 1995). Up to 3000 m of aggradation occurred in the southern Pyrenean foreland (Coney et al. 1996) as sediment filled the basin and raised the base level (Nichols 2004). Accommodation space was provided by the enclosed nature of the basin, in addition to any loading due to thrust-sheet stacking in the Pyrenean thrust belt (e.g. Teixell 1996) or sediment loading in the basin. The structural and stratigraphic relationships between the thrust front strata and all of the fans east of Riglos show that there was no differential subsidence between the thrust front and the basin because the contact is undeformed. Accumulation of deposits at the basin margin, therefore, occurred by passive filling forming a fan succession at least 400 m thick at Roldán, and at least 200 m on most of the other fans. Vertical profiles through the fan deposits show some subtle coarsening and fining-up trends (e.g. at Roldán; Hirst 1983), but in general the fan successions are characterized by aggradation (e.g. the upper part of Agüero; Nichols 1987b), and this is evident from the rather uniform appearance of the exposed, proximal-fan deposits (Figs 4–6, 8, 11 & 18). These aggradational patterns indicate that equilibrium was achieved between the rising base level and the processes of fan deposition (Nichols 2004).

Spacing of the fans and the structure of the basin margin

The average spacing of the apices of the fans over the 36 km between Agüero and Roldán is 4.5 km (see Fig. 2; Table 1), but the actual spacing varies from as little as 2.5 (between Agüero and Murillo, and between San Julian and Nueno) to 9 km between Linás and Aniés. Lloyd *et al.* (1998) considered that a structural control on the location of the apex of the fan could be argued for Agüero and Riglos, where there are transverse structures in the thrust belt (Nichols 1987*a*). The apex of the Roldán fan lies at the core of a N–S syncline that crosses the External Sierras (Puigdefàbregas 1975) and a structural control on the location of the drainage could be argued also in this case. A parallel structure north of Nueno and San Julián may also have played a role in controlling the positions of these two fans. The locations of the Bolea, Aniés, Linás and Murillo fans are not obviously controlled by any structural feature in the thrust front.

The regular spacing in drainage in orogenic belts recognized by Hovius (1996) is not evident in the External Sierras, although it must be emphasized that Hovius (1996) was considering drainage on the scale of the whole orogen, whereas the External Sierras are a small part of the Southern Pyrenean thrust belt. It is therefore difficult to draw any conclusions about the controls on the positions of these nine fans, except to say that some may be structurally controlled and some not, and that there is no obvious pattern to the spacing.

Depositional facies

The facies recognized in the alluvial fan deposits are summarized in Table 2, compiled from original observations and material present in previous works (Nichols 1987*b*; Nichols & Hirst 1998; Arenas *et al.* 2001; Nichols & Thompson 2005).

Rock falls

Cones of angular, poorly sorted talus derived from adjacent bedrock cliffs form a component of many alluvial fan deposits (Blair & McPherson 1994), particularly in the most proximal parts. The contact between folded units of Lower Eocene limestone (the Guara Formation) and proximal alluvial-fan deposits is exposed in the steep walls of the Rió Gállego gorge cut through the edge of the Riglos

fan body (Figs 7 & 14) and in the area between the two main 'mallos'. The conglomerate exposed contains boulders and angular blocks at least 2 m across, and there is no sign of bedding. It appears to be monomict, composed entirely of Guara Formation limestone clasts. These are interpreted as rock-fall deposits and are best exposed at the back of the pinnacles at Riglos. Similar deposits are also seen at the contact between the bedrock of early Eocene limestone and fan deposits exposed on the western margin of the San Julián fan. Here the palaeo-surface of Guara Formation is brecciated and reddened, with a localized talus of angular fragments that pass laterally over a few metres into waterlain facies of the San Julián fan.

Debris-flow deposits

The construction of an alluvial-fan body by debris-flow deposition has been recognized in many modern and ancient examples. In particular, some of the early, detailed studies of fan processes in the SW USA emphasized the importance of coherent flows in alluvial-fan sedimentology (e.g. Hooke 1967; Bull 1972, and references therein). In a more recent synthesis of alluvial-fan sedimentary processes, Blair & McPherson (1994) also consider debris flows to be one of the main mechanisms of fan deposition.

In the western External Sierras fans, deposits interpreted as the products of debris flows occur in a number of places, but overall form a relatively small percentage of the facies. At Riglos the exposed fan facies are poorly sorted, coarse conglomerates forming thick beds that are amalgamated in places (Fig. 15). The beds are ungraded, clast- or matrix-supported and the clasts are randomly oriented, with some elongate cobbles oriented with their long axes at a high angle to the bedding surface (Fig. 15). These characteristics suggest deposition by mud-poor debris flows. The Stage 1 deposits at Agüero were interpreted by Nichols (1987*b*) as sheetflood facies on the basis that they have a clast-supported fabric. However, re-examination of these valley-fill sediments has indicated that they are very similar to the facies exposed at Riglos and are also the products of more cohesive, hyperconcentrated flows.

Matrix-supported conglomerate textures are only seen in the Nueno fan body. At this locality the proximal facies are dominated by muddy conglomerate beds with a high proportion of clasts of detrital gypsum (Fig. 16). The control on the processes of transport and deposition on this fan is considered by Nichols & Thompson (2005) to be the high mudrock content of the source-area bedrock that resulted in more cohesive flows at Nueno compared to the adjacent and coeval San Julián fan.

Table 2. *Summary of the depositional facies in the alluvial fan bodies (modified from Nichols & Hirst 1998; Nichols & Thompson 2004; Arenas et al. 2001)*

Facies	Characteristics	Processes	Occurrence
Conglomerate with angular clasts	Poorly sorted, angular clasts, structureless, no bedding, localized lenses adjacent to bedrock	Rock fall	Proximal facies at Riglos, San Julián and along basin margin between Linás and Aniés
Conglomerate with randomly-oriented clasts	Clast-supported, subangular–subrounded randomly oriented clasts, thick-bedded (>2 m), sheet form	Clast-rich debris flow	Agüero (Stage 1), Riglos
Matrix, supported conglomerate	Matrix-supported, gypsum and limestone clasts in a sandy mudstone matrix, sheet form	Mud-rich debris flow	Nueno
Clast-supported, thickly bedded conglomerate	Clast-supported, thick beds (>2 m), horizontal or crude cross-stratification, sheet form, erosive basal contact	Poorly confined gravelly channel deposits: amalgamated bar units	Occur in varying proportions on all fans except Agüero (Stage 1), Riglos and Nueno
Clast-supported, thinly bedded conglomerate	Clast-supported, thin beds (<2 m), horizontally stratified, normally graded, sheet form	Poorly confined or unconfined flows on fan surface	As above
Lenses of conglomerate	Lenses of clast-supported conglomerate, erosive basal contact	Gravelly channel, bar deposits	As above
Pebbly sandstone	Pebbly sandstone, beds tens of cm–m thick, sheet form, graded, horizontally stratified or rarely cross-bedded	Sandy–pebbly poorly confined–unconfined flow deposits	As above
Sandstone	Massive, horizontally stratified or ripple-laminated sandstone, sheet beds, cm-bedded	Sandy poorly confined–unconfined flow deposits	As above
Interbedded sandstone and mudstone	Heterolithic sandstone and mudstone, cm-bedded, horizontal and ripple laminated	Unconfined flow	Margins of all fans

Waterlain deposits

Published studies of the western External Sierras fans have considered that sheetflood processes are the dominant style of deposition, for example at Agüero (Nichols 1987b) and at Roldán (Nichols & Hirst 1998). In the latter paper, the conglomerate beds are described as clast-supported with a matrix of very-coarse–medium sand: the clasts are moderately well rounded, and imbrication of oblate pebbles and cobbles common. The geometry of the beds is recorded as mainly laterally extensive sheets that can be traced over a distance of several hundred metres, and lensoid bodies interpreted as fills of scours. However, inspection of the internal geometries of the sheet bodies at Roldán has revealed inclined surfaces within the conglomerate that define lenses 1–2 m thick and 5–10 m wide (Fig. 17). Individual sheets (Fig. 18) are made up of laterally amalgamated lens-shaped bodies of conglomerate and sandstone. At Agüero conglomerate beds show horizontal stratification, imbrication and rare trough cross-bedding: the bases of many beds are strongly scoured with conglomerate filling minor channels (Nichols 1984). Similar features are seen in proximal conglomerates at San Julián (Nichols & Thompson 2005), Murillo, Linás (Nichols 1984) and Bolea. Re-inspection has also shown that medial facies at Bolea are pebbly sandstone, with pebbles occurring as layers or as discrete lenses, and at Linás the medial beds are m-thick beds of sandstone and pebbly sandstone, partly cross-bedded with sharp, scoured bases.

Comparison of these deposits with the detailed descriptions of sheetflood facies in Blair (1999a, b) suggests that the previous interpretations of the fan-depositional processes in this area may need to be revised. According to Blair (1999a) sheetflood deposits show two main facies: (a) couplets of coarse gravel in planar beds, 5–25 cm thick, rhythmically interstratified with gravelly sand of similar thickness; and (b) texturally similar gravel and sand that displays up-fan dipping, low-angle, cross-beds. Blair (1999a) considers these facies to be the products of deposition from antidune bedforms in high-discharge sheetfloods. Neither the thin couplets nor the antidune cross-bedding have been documented from the deposits in the western External Sierras fans. The presence of strongly scoured bases, lenses of conglomerate and some cross-bedding are more consistent with deposition from poorly confined, unstable, braided river channels on the fan surface, similar to the outwash fans of Boothroyd & Nummedal (1978) and the fans described by Nemec & Postma (1993). This suggests that the fans were built up by fluvial processes, with streams discharging from the External Sierras relief depositing gravel on bars within 1 or 2 km of the basin margin, and sand in shallow, poorly confined streams up to about 5 km from the margin (Fig. 19). Sheet bodies of gravel and sand were constructed by the lateral migration of the flow over the fan surface during a discharge event, or by the amalgamation of several events. Stream-channel processes do not fall within the spectrum of processes seen on the alluvial fans as defined by Blair & McPherson (1994), although the bodies of sediment built up at the Ebro Basin margin in the Oligo-Miocene are of similar size and shape to the fans of Blair & McPherson (1994).

Controls on processes of deposition

Rock-fall deposits are localized adjacent to steep topography at the basin margin where beds of Triassic–Early Eocene limestone formed cliffs in the late Oligocene and early Miocene. Within the fan deposits, debris-flow facies occur in two settings. At Riglos and in Stage 1 at Agüero, the drainage basin area would have been small and with steep relief, conditions that favour debris-flow processes (Blair & McPherson 1994). The change to waterlain deposits

Fig. 15. Poorly sorted, ungraded beds of conglomerate near the base of the main pinnacle at Riglos. These are interpreted as matrix-poor debris-flow deposits.

in the second depositional phase at Agüero may be interpreted as a consequence of the expansion of the drainage basin area that resulted in an increase in runoff in the enlarged basin: this is recorded in the fan succession as a change from monomict to

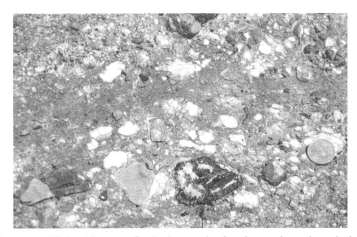

Fig. 16. Clasts of gypsum (white) and limestone floating in a matrix of sandy, gravely, mudstone in the Nueno fan. These are interpreted as debris-flow deposits derived from a drainage area rich in Triassic gypsiferous mudstone.

Fig. 17. Proximal facies of the Roldán fan body. The conglomerate beds have sharp, erosive bases and show a distinct lensoid geometry (picked out by dashed lines). They are interpreted as the deposits of poorly confined flow forming bars of gravel on the fan surface.

Fig. 18. Sheets of sandstone and conglomerate from medial parts of the Roldán fan body.

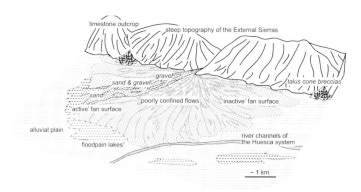

Fig. 19. Depositional model for the Roldán fan and other fans in this area that were predominantly formed of waterlain sediment (upper parts of Agüero, Murillo, Linás, Aniés, Bolea and San Julián).

polymict clast compositions (Nichols 1987*b*; Lloyd *et al.* 1998). The persistence of disorganized conglomerate facies throughout the succession at Riglos may indicate that the drainage basin remained relatively small and steep: the fan itself is the smallest along this section of the basin margin. The Nueno fan had a larger drainage basin, but the bedrock was dominated by mudrock and gypsum, resulting in flows with a higher proportion of fines and debris-flow characteristics (Nichols & Thompson 2005).

The predominance of waterlain facies in the alluvial fans of the Ebro Basin margin indicates that precipitation in the External Sierras in Oligo-Miocene time was high enough for non-viscous flows to occur from all but the smallest or most mudrock-dominated catchments. The flows were capable of scouring into the fan surface, but there is no evidence of deep incision forming an entrenched channel in any of the fans. This indicates that the fans did not build up above the level of the basin floor at the fan toes, and that through time aggradation on the fan and on the basin floor were at approximately the same rate. Further evidence for a low gradient across the fans is provided by the extensive sheets of gravel and sand produced by laterally migrating, poorly confined braided river channels. Low-surface gradient has been identified as a cause of fan-shaped bodies of sediment in other alluvial systems, albeit on a larger scale in examples such as the Kosi fan (Wells & Dorr 1987; Gohain & Parkash 1990). The downstream changes from gravel- to sand-dominated deposition over distances of only 1–2 km suggest rapid loss of stream power downflow: loss of water by infiltration into the fan surface and, to a probable lesser extent, evaporation is likely to have played an important role in this process.

Relationship to other facies in the Ebro Basin

Alluvial-fan deposits form only a small proportion of the Sariñena and Uncastillo formations, which consist mainly of fluvial channel and overbank deposits plus minor lacustrine facies (Hirst & Nichols 1986; Arenas *et al.* 2001). Interfingering of the fan facies with these other deposits can be seen in all except the Aniés fan, where exposure is poor. In the west, the Uncastillo Formation was mainly deposited by the Luna fluvial distributary system (Hirst & Nichols 1986; Nichols 1987*c*), channels of which show a palaeoflow direction towards the east and SE in the vicinity of the Agüero and Murillo fans. River-channel facies between each of the stages of fan deposits at Agüero (Nichols 1987*b*) indicate that the fluvial system deposited close to the basin margin between periods of fan progradation. The southern, distal parts of the Roldán fan interfinger with deposits of the Huesca fluvial distributary

system that locally shows a palaeoflow towards the west and WNW (Hirst & Nichols 1986; Nichols & Hirst 1998). A similar relationship is also exposed at the SW fringe of the San Julián fan (Nichols & Thompson 2005).

Between the Roldán and Nueno fans, the coeval facies are lacustrine (Nichols & Hirst 1998; Nichols & Thompson 2005), consisting mainly of thin, wave-rippled sandstone beds interbedded with calcareous mudstone plus some fine-grained limestone. These facies extend up to the basin margin between the two fan bodies, suggesting that there was a ponding of water in the interfan area. In the region between the Linás and Aniés fans a basin-margin facies of pebbly sandstone interfingers with thin beds of sandstone and mudstone of the Huesca and Luna fluvial distributary systems (Nichols 1984).

Palaeoclimate

The alluvial-fan deposits provide little direct evidence of the palaeoclimate. Waterlain depositional processes may be common on fans formed in very arid climatic settings, such as Death Valley, California (Blair 1999*a*, *b*), so the predominance of this style of sedimentation in the Ebro Basin fans may not be a significant palaeoclimatic indicator. The best indications of palaeoclimate are to be found in the coeval floodplain facies. Mature calcrete profiles are absent from the floodplain sediments in the Uncastillo and Sariñena formations, despite the abundant carbonate within the succession (Nichols & Hirst 1998). The pedogenic profiles present (Nichols 1984) are poorly developed vertisols, similar to those described by Bown & Kraus (1981) from the early Tertiary of Wyoming, and a similar, relatively humid climate is suggested. Organic material is very rarely preserved within the floodplain deposits and this suggests that floodplain conditions were oxidizing. Further indicators of palaeoclimate are the presence of gypsum in places, but these deposits appear to have been localized around the toes of some of the fans (e.g. Murillo, Nueno and Roldan) where waters rich in calcium sulphate derived from the Triassic Pont de Suert Formation ponded and evaporated. Desiccation cracks from within channel-fill deposits of the Luna System have been noted by Nichols (1987*c*) and are evidence of ephemeral flow in the channels, but the channel sandstone bodies generally have quite well-developed bar and dune structures indicating that the flow was generally continuous.

The late Oligocene–early Miocene palaeoclimate was therefore probably semi-arid–temperate, with net evaporation in the basin greater than the input of water from direct precipitation and flow from the Pyrenees. It was therefore somewhat similar to the present-day climate, with a summer monthly mean of up to 30 °C and seasonal rainfall.

Deformation features

Synsedimentary deformation

The thrust-generated topography at the basin margin had formed by the mid-Oligocene. However, thrust movement continued at some locations resulting in synsedimentary deformation of alluvial-fan deposits. The largest amount of deformation is in TSU1 deposits at Agüero (Stage 1) and in TSU2 strata in Agüero (Stages 2 and 3), Murillo and Riglos. There are some synsedimentary deformation features in TSU3 deposits at San Julián and Roldán.

These features are most completely displayed at Agüero (Nichols 1987b; Lloyd et al. 1998), and are also seen within the fan successions of Riglos, San Julián and Roldán. At Agüero the lowest strata, which lie within TSU1 and are the oldest alluvial fan deposits in the western External Sierras area, show a progressive unconformity (Fig. 20) (Riba 1976; Anadon et al. 1986). This structure is interpreted as the tilting of the depositional surface southwards during deposition (Nichols 1987b) that can be related to the development of the thrust structures in the adjacent part of the External Sierras (Pocoví et al. 1990; Lloyd et al. 1998) during the late Oligocene. A syndepositional fold in the middle and upper parts of the succession here (Fig. 21) are evidence of deformation during TSU2 and the early part of TSU3, and can also be related to movements in the adjacent thrust front (Nichols 1987b; Lloyd et al. 1998).

Conglomeratic strata in the Murillo fan body that span the same age range (TSU2 and TSU3) also

Fig. 20. A progressive unconformity at the base of the Agüero fan body. The oldest beds, on the right, lie unconformably on Eocene sandstone beds and are overturned at an angle of 60°. The bedding dip gradually changes to vertical through 100 m of sedimentary section.

show a broad syncline structure, but the evidence of growth during sedimentation is not so clear, with only a suggestion of thickening into the syncline and wedging of the beds northwards (Fig. 5).

Fig. 21. Growth folds in the middle and upper parts of the Agüero fan body (TSU2 and TSU3).

However, there has been a considerable amount of post-depositional deformation (see below). On the opposite side of the valley of the Rió Gállego the contact between the Riglos conglomerate body and the folded and thrust-faulted limestone beds at the front of the Guarga Thrust is spectacularly exposed (Figs 7 & 14).

With the exception of some very locally tilted beds in the most proximal parts of the Linás fan, there is no evidence of synsedimentary deformation in any of the fans between Riglos and San Julián. According to the correlation scheme of Arenas et al. (2001) the fan deposits exposed in this part of the basin margin are younger (all within TSU3) than those to the west. In San Julián fan deposits either side of the Barranco Fenes (Fig. 11), S-dipping beds of conglomerate occur in the lower part of the proximal fan succession. These beds are part of a monoclinal fold that formed by thrust movement in the limestone beds which form the basement to the fan. The thinning of the beds over the upper axis of the monocline indicates that the fold grew gradually during fan deposition, progressively rotating part of the fan depositional surface. The growth fold structures can only be identified on the west flank of the San Julián fan: equivalent strata on the eastern side are not exposed. Further to the east, the Nueno fan succession also shows no evidence of syndepositional growth structures at equivalent levels. The structure is, therefore, very localized and the amount of displacement on a fault to create 60 m of palaeo-relief would have been between 120 and 230 m, assuming a low-angle thrust fault dipping between 15° and 30° to the north.

A growth structure of similar dimensions and geometry can be seen in the lower part of the Roldán fan where there is a small fault-related growth fold within the most proximal part of the fan succession at the base of the eastern pinnacle (visible from the western side of the gorge). In this case a reverse fault cuts up through about 50 m of conglomerate, terminating in a monoclinal fold, above which there is about 40 m of growth strata to the level where the fold dies out (Friend et al. 1989).

Post-depositional deformation

West of Linás the Fuencalderas anticline is a broad open fold about 4 km south of the basin margin within the Uncastillo Formation that affects beds in TSU3 (Arenas et al. 2001). This fold provides evidence of shortening in the area continuing into the early Miocene. Closer to the thrust front, alluvial-fan deposits within TSU3 are deformed at Agüero and Murillo. In the valley of the Barranco de Pituelo west of Agüero village a steep monocline structure deforms distal-fan facies, and the Luna System

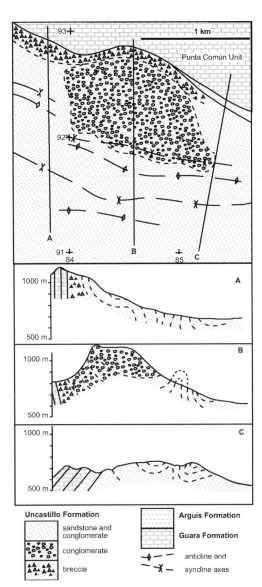

Fig. 22. Map and cross-sections through the Murillo fan body and adjacent units of the thrust front in the External Sierras.

sandstone and mudstone beds with which they interfinger (Nichols 1987b). This structure can be traced 4 km west where it affects the distal facies of the Murillo fan. Here the shortening is greater, with the central limb of the monocline overturned, dipping 70°–80° N (Figs 5 and 22).

The Murillo fan body displays an unusual example of the interaction between alluvial-fan formation and the deformation of the adjacent basin margin. The currently exposed mass of conglomerate covers an area of approximately 1 km² and is interpreted to be the core of an alluvial fan body with an area of 5.4 km^2 (Nichols 1984; Lloyd 1994; Lloyd *et al.* 1998). The conglomerate beds are folded into an open syncline, with the lowest beds exposed possibly deformed during deposition (cf. the coeval Stages 2 and 3 deposits of the adjacent Agüero fan). The youngest beds of the exposed succession can be traced from an open syncline in proximal fan conglomerates to the tight monocline in the distal fan deposits (Fig. 22). In three parallel cross-sections through this structure (Fig. 22) a lateral variation in fold geometry and the orientation of the adjacent strata in the thrust front can be seen. Adjacent to the apex of the fan, mapped as the centre of the conglomerate body, the lower Eocene, Palaeocene and upper Cretaceous strata dip at 80°–90°S. These beds are part of an allochthonous unit, the Punta Común Unit (Nichols 1987*a*; Arenas *et al.* 2001), that strikes 110°. To the west of the fan apex the beds are overturned, dipping 70°N, and to the east, in the Río Gíllego gorge, they are also overturned, dipping 60°–65°N (Fig. 22). At the conglomeratic apex of the fan the Punt Común Unit beds have undergone less folding than where the beds are adjacent to the more sand-rich lateral margins of the fan. The conglomeratic core of the alluvial fan has acted as a relatively rigid block, resulting in other strata, including those of the thrust front, wrapping around it. This relationship also suggests that the conglomerate beds were lithified prior to deformation.

To the east, the Linás fan deposits and all of the margin deposits as far as San Julián appear to be largely unaffected by post-depositional deformation, although exposures of the distal-fan deposits at Aniés and Bolea are poor. The strata at Roldán and Nueno show a dip of up to 9° to the north close to the basin margin, shallowing southward to horizontal. The origin of this deformation is not clear, but may simply be the expression of a limb of a broad anticline similar to the Fuencalderas anticline to the west.

Deformation timing and controls

The syn- and post-depositional shortening at the Ebro Basin margin in the late Oligocene was relatively minor: Lloyd (1994) calculated that the shortening across the Agüero fan, which shows the greatest degree of deformation, was approximately 2000 m. Several of the fans (Linás, Aniés, Bolea, Nueno) and the youngest parts of most of the others show no evidence of tectonic deformation. This means that there could only have been a small amount of thrust-related uplift in the External Sierras during fan deposition, and the relationship between the deformed pre-Oligocene strata and the marginal basin deposits has not changed significantly since the Oligocene. Movement along this part of the south Pyrenean thrust front had ceased by the end of the Oligocene.

Exhumation of the Tertiary alluvial fans

The pinnacles of conglomerate along the margin of the Ebro Basin are a striking feature of the geomorphology of the area (Figs 4, 6, 7 & 11) and their presence owes as much to late Neogene erosion as the processes of their formation discussed above. During the period of alluvial-fan deposition the Ebro Basin was endorheic, and lacustrine sedimentation in the basin centre continued until the middle Miocene (Calvo 1990; Garcia-Castellanos 2003) to fill the basin to at least 1000 m above present sea level. However, in the late Miocene (Tortonian) sediment started to accumulate at the site of the modern Ebro Delta (Evans & Arche 2002), indicating that the Ebro Basin changed to being externally drained at this time. There are over 2000 m of late Miocene and younger strata in the Ebro Delta succession, punctuated by an unconformity that marks the Messinian base-level fall in the Mediterranean (Evans & Arche 2002). As headward erosion by the Rio Ebro resulted in river capture throughout large parts of the southern Pyrenees the regional base-level fall was up to 1000 m (Garcia-Castellanos 2003; Nichols 2004). The rates of incision varied during the late Neogene, and there were periods of aggradation to the north of the Ebro Basin (Jones *et al.* 1999), but the overall effect has been one of deep incision into the basin-margin succession and exposure of the Oligo-Miocene alluvial fans.

Conclusions

- The emergence of the relief at the basin margin in the late Oligocene occurred once relatively easily eroded Middle–Late Eocene strata had been removed to expose the more resistant Early Eocene limestone beds (the Guara Formation).
- Thrust-related shortening in the External Sierras thrust front continued during the late Oligocene resulting in syndepositional deformation of the earliest fan deposits exposed in the western part of the area.
- Deposition on most of the fans was by poorly confined flows producing sheet bodies of sand and gravel by lateral amalgamation of channel and bar deposits. Hyperconcentrated and debris-flow deposits are recognized in instances where the catchment area is interpreted to have been relatively small (Riglos and Stage 1 of the Agüero fan) or where the hinterland bedrock

was dominated by mudrock and evaporites (Nueno).

- The palaeoclimate is inferred to have been semi-arid–temperate. Rainfall in the External Sierras was sufficient to provide enough water for transport and deposition from traction currents on the fan surfaces. However, a degree of aridity is indicated by the presence of gypsum as clasts, the formation of gypsum nodules in alluvial-plain palaeosols and the evidence for desiccation of coeval river channels in the Luna System. Lacustrine limestone beds interfingering with the distal-fan facies at Roldán indicate that the alluvial plain could not have been arid.
- The thick successions of alluvial-fan deposits formed as a consequence of sedimentation at the margins of a basin that experienced a rising base level throughout the period of fan deposition. The absence of deep incision in the fan successions indicates that aggradation on the fans and on the alluvial plains kept pace with each other.
- Exhumation of the fan deposits is a consequence of a late Miocene base-level fall that came about when the Ebro Basin switched from being endorheic to a basin with external drainage in the Tortonian.

Many colleagues have contributed ideas and comments about these spectacular deposits, but thanks are particularly due to P. Friend, who introduced me to this area, P. Hirst, who carried out the original study of the Roldán fan, and Royal Holloway MSci students B. Thompson and C. Chadwick, who carried out project work on selected fans. The author would also like to thank T. Elliott for his helpful comments on an earlier version of this manuscript.

References

ANADON, P., CABRERA, L., COLOMBO, M., MARZO, M. & RIBA, O. 1986, Syntectonic intraformational unconformities in alluvial fan deposits, eastern Ebro basin margin (N.E. Spain). *In*: ALLEN, P.A. & HOMEWOOD, P. (eds) *Foreland Basins*. International Association of Sedimentologists, Special Publications, **8**, 259–271.

ARENAS, C. 1993. *Sedimentología y paleogeografía del Tercario del margen pirenaico y sector central de la Cuenca del Ebro (zona aragonesa occidental)*. PhD Thesis, University of Zaragoza.

ARENAS, C., MILLÁN, H., PARDO, G. & POCOVÍ, A. 2001. Ebro Basin continental sedimentation associated with late compressional Pyrenean tectonics (north-eastern Iberia): controls on basin margin fans and fluvial systems. *Basin Research*, **13**, 65–89.

BLAIR, T.C. 1999*a*. Sedimentology and progressive tectonic unconformities of the sheetflood-dominated Hell's Gate alluvial fan, Death Valley, California. *Sedimentary Geology*, **132**, 233–262.

BLAIR, T.C. 1999*b*. Sedimentary processes and facies of the waterlaid Anvil Spring Canyon alluvial fan, Death Valley, California. *Sedimentology*, **46**, 913–940.

BLAIR, T.C. & MCPHERSON, J.G. 1994. Alluvial fans and their natural distinction from rivers based on morphology, hydraulic processes, sedimentary processes and facies assemblages. *Journal of Sedimentary Research*, **A64**, 450–589.

BOOTHROYD, J.C. & NUMMEDAL, D. 1978. Proglacial braided outwash: A model for humid alluvial-fan deposits. *In*: MIALL, A.D. (ed.) *Fluvial Sedimentology*. Canadian Society of Petroleum Geologists, Memoirs, **5**, 641–668.

BOWN, T.M. & KRAUS, M.J. 1981. Lower Eocene palaeosols (Willwood Formation, Northwest Wyoming, USA) and their significance for palaeoecology, palaeoclimatology and basin analysis. *Palaeography, Palaeoclimatology, Palaeoecology*, **34**, 1–30.

BULL, W.B. 1972. Recognition of alluvial fan deposits in the stratigraphic record. *In*: RIGBY, J.K. & HAMBLIN, W.K. (eds) *Recognition of Ancient Sedimentary Environments*. Society of Economic Paleontologists and Mineralogists, Special Publications, **16**, 63–83.

BURBANK, D.W., LELAND, J., FIELDING, E., ANDERSON, R.S., BROZOVIC, N., REID, M.R. & DUNCAN, C. 1996. Bedrock incisions, rock upOlift and threshold hillslopes in the northwestern Himalaya. *Nature*, **379**, 505–510.

CALVO, J.P. 1990. Up-to-date Spanish continental Neogene synthesis and palaeoclimate interpretation. *Revista de la Sociedad Geológica de España*, **6**, 29–40.

CHOUKROUNE, P. & ECORS-PYRENEES TEAM. 1989. The ECORS Pyrenean deep seismic profile reflection data and the overall structure of an orogenic belt. *Tectonics*, **8**, 23–39.

CONEY, P.J., MUÑOZ, J.A., MCCLAY, K.R. & EVENCHICK, C.A. 1996. Syntectonic burial and post-tectonic exhumation of the southern Pyrenees foreland fold–thrust belt. *Journal of the Geological Society, London*, **153**, 9–16.

ECORS-PYRENEES TEAM. 1998. The ECORS deep reflection survey across the Pyrenees. *Nature*, **331**, 508–511.

EVANS, G. & ARCHE, A. 2002. The flux of siliciclastic sediment from the Iberian Peninsula with particular reference to the Ebro. *In*: Jones, S.J. & Frostick, L.E. (eds) *Sediment Flux to Basins: Causes, Controls and Consequences*. Geological Society, London, Special Publications, **191**, 199–208.

FRIEND, P.F., HIRST, J.P.P., HOGAN, P.J., JOLLEY, E.J., MCELROY, R., NICHOLS, G.J. & RODRIGUEZ VIDAL, J. 1989. *Pyrenean Tectonic Control of Oligo-Miocene River Systems, Huesca, Aragon, Spain*. 4th International Conference on Fluvial Sedimentology. Excursion Guidebook 4. Publicacions del Servei Geològic de Catalunya.

GARCIA-CASTELLANOS, D., VERGÉS, J., GASPAR-ESCRIBANO, J. & CLOETINGH, S. 2003. Interplay between tectonics, climate, and fluvial transport during the Cenozoic evolution of the Ebro Basin (NE Iberia). *Journal of Geophysical Research*, **108**, B7, 2347, doi:10.1029/2002JB002073,2003.

GOHAIN, K. & PARKASH, B. 1990. Morphology of the Kosi Megafan. *In*: RACHOKI, A.H. & CHURCH, M. (eds) *Alluvial Fans: A Field Approach*. Wiley, Chichester, 151–178.

HIRST, J.P.P. 1983. *Oligo-Miocene alluvial systems in the*

northern Ebro Basin, Huesca Province, Spain. PhD Thesis, University of Cambridge.

HIRST, J.P.P. & NICHOLS, G.J. 1986. Thrust tectonic controls on alluvial sedimentation patterns, southern Pyrenees. *In*: ALLEN, P.A. & HOMEWOOD, P. (eds) *Foreland Basins*. International Association of Sedimentologists, Special Publications, **8**, 153–164.

HOGAN, P.J. 1993. *Geochronologic, tectonic and stratigraphic evolution of the southwestern Pyrenean foreland basin, northern Spain*. PhD Thesis, University of Southern California, Los Angeles.

HOGAN, P.J. & BURBANK, D.W. 1996. Evolution of the Jaca piggyback basin and emergence of the External Sierra, southern Pyrenees. *In*: FRIEND, P.F. & DABRIO, C.J. (eds) *Tertiary Basins of Spain: The Stratigraphic Record of Crustal Kinematics*, Cambridge University Press, Cambridge, 153–160.

HOOKE, R.LeB. 1967. Processes on arid-region alluvial fans. *Journal of Geology*, **75**, 438–460.

HOVIUS, N. 1996. Regular spacing of drainage outlets from linear mountain belts. *Basin Research*, **8**, 29–41.

HOVIUS, N., STARK, C.P. & ALLEN, P.A. 1997. Sediment flux from a mountain belt derived by landslide mapping. *Geology*, **25**, 231–234.

JONES, S.J., FROSTICK, L.E. & ASTIN, T.R. 1999. Climatic and tectonic controls on fluvial incision and aggradation in the Spanish Pyrenees. *Journal of the Geological Society, London*, **156**, 761–769.

LLOYD, M.J. 1994. *Sediment provenance studies in the Pyrenean foreland basin, Aragon, Spain*. PhD Thesis, University of Cambridge.

LLOYD, M.J., NICHOLS, G.J & FRIEND, P.F. 1998. Alluvial fan evolution at thrust fronts: the Oligo-Miocene of the southern Pyrenees. *Journal of Sedimentary Research*, **68**, 869–878.

MILLÁN, H., POCOVÍ, A. & CASAS, A. (1995) El frente de cabalgamiento surpirenaico en el extremo occidental de las Sierras Exteriores: sistemas imbricados y pliegues de despegue. *Revista de la Sociedad Geológica de España*, **8**, 73–90.

MILLÁN, H., PUEYO, E.L., AURELL, M., LUZÓN, A., OLIVA, B., MARTÍNEZ, M.B. & POCOVÍ, A. 2000. Actividad tectónica registrada en los depósitos terciarios del frente meridional del Pirineo central. *Revista de la Sociedad Geológica de España*, **13**, 279–300.

MUÑOZ, J.A. 1992. Evolution of a continental collision belt: ECORS-Pyrenees Crustal Balanced Cross-Section. *In*: MCCLAY, K.R. (ed.) Thrust Tectonics. Chapman & Hall, London, 235–246.

NEMEC, W. & POSTMA, G. 1993. Quaternary alluvial fans in southwestern Crete: sedimentation processes and geomorphic evolution. *In*: MARZO, M. & PUIGDEFÁBREGAS, C. (eds) *Alluvial Sedimentation*. International Association of Sedimentologists, Special Publications, **17**, 235–276.

NICHOLS, G.J. 1984. *Thrust tectonics and alluvial sedimentation, Aragón, Spain*. PhD Thesis, University of Cambridge.

NICHOLS, G.J. 1987a. The structure and stratigraphy of the Western External Sierras of the southern Pyrenees. *Geological Journal*, **22**, 245–259.

NICHOLS, G.J. 1987b. Syntectonic alluvial fan sedimentation, southern Pyrenees. *Geological Magazine*, **124**, 121–133.

NICHOLS, G.J. 1987c. Structural controls on fluvial distributary systems – the Luna System, Northern Spain. *In*: ETHRIDGE, F.G., FLOREZ, R.M. & HARVEY, M.D. (eds) *Recent Developments in Fluvial Sedimentology*. SEPM, Special Publications, **39**, 269–277.

NICHOLS, G.J. 1989. Structural and sedimentological evolution of part of the west central Spanish Pyrenees in the Late Tertiary. *Journal of the Geological Society, London*, **146**, 851–857.

NICHOLS, G.J. 2004. Sedimentation and base level controls in an endorheic basin: the Tertiary of the Ebro Basin, Spain. *In*: *Tertiary Geology: Recent Advances. Boletín Geológico y Minero, España*, **115**, 427–438.

NICHOLS, G.J. & HIRST, J.P.P. 1998. Alluvial fans and fluvial distributary systems, Oligo-Miocene, northern Spain: contrasting processes and products. *Journal of Sedimentary Research*, **68**, 879–889.

NICHOLS, G.J. & THOMPSON, B. 2005. Bedrock lithology control on contemporaneous alluvial fan facies, Oligo-Miocene, southern Pyrenees, Spain, *Sedimentology*, **52**, 571–585.

POCOVÍ, A., MILLÁN, N., NAVARRO, J.J. & MARTINEZ, M.B., 1990, Rasgos estructurales de la Sierra de Salinas y zona de los Mallos (Sierras exteriores, Prepirineo, provincias de Huesca y Zaragoza). *Geogaceta*, **8**, 36–39.

PUIGDEFÁBREGAS, C. 1975. La sedimentación molásica en la Cuenca de Jaca. *Pirineos*, **104**, 1–188.

PUIGDEFÁBREGAS, C. & SOLER, M. 1973. Estructura de las Sierra Exteriores pirenaicas en el corte del río Gállego (provincia de Huesca). *Pirineos*, **109**, 5–15.

PUIGEDEFÁBREGAS, C. & SOUQUET, P. 1984. Tecto-sedimentary cycles and depositional sequences of the Mesozoic and Tertiary from the Pyrenees (France, Spain). *Tectonophysics*, **129**, 173–203

QUIRANTES, J. 1978. *Estudio sedimentologico y estratigrafico del Terciario continetal de los Monegros*. Doctoral Thesis, Institución Fernando el Católico (CSIC), Zaragoza.

RIBA, O. 1976. Syntectonic unconformities of the Alto Cardener, Spanish Pyrenees: a genetic interpretation. *Sedimentary Geology*, **15**, 213–233.

TEIXELL, A. 1996. The Ansó transect of the southern Pyrenees: basement and cover geometries. *Journal of the Geological Society, London*, **153**, 301–310.

VERGÉS, J., MILLÁN, H. *ET AL*. 1995. Eastern Pyrenees and related foreland basins: pre-, syn- and post-collisional crustal-scale cross-sections. *Marine and Petroleum Geology*, **12**, 903–915.

WELLS, N.A & DORR, J.A. JR. 1987. A reconnaissance of sedimentation on the Kosi alluvial fan of India. *In*: ETHRIDGE, F.G., FLOREZ, R.M. & HARVEY, M.D. (eds) *Recent Developments in Fluvial Sedimentology*. SEPM, Special Publications, **39**, 51–61.

Source area and tectonic control on alluvial-fan development in the Miocene Fohnsdorf intramontane basin, Austria

MICHAEL WAGREICH[1] & PHILIPP E. STRAUSS[2]

[1]*Department of Geological Sciences, University of Vienna, Althanstrasse 14, A-1090 Vienna, Austria (e-mail: michael.wagreich@univie.ac.at)*
[2]*OMV AG, Gerasdorfer Strasse 151, A-1211 Vienna, Austria*

Abstract: Middle Miocene alluvial fans in the intramontane Fohnsdorf Basin of the Eastern Alps originated along normal faults and linked strike-slip faults in a continental half-graben setting. The fans display considerable facies differences. Debris flows of the Rachau fan are characterized by a sandy matrix and large boulders, whereas debris flows of the Apfelberg fan are characterized by higher silt and clay content and smaller clasts. Key control of debris-flow facies is the lithology contrast in the fan source areas. Sand, pebbles and large outsized boulders originated predominantly from the resistant augengneiss- and amphibolite-dominated hinterland of the Rachau fan, whereas a significant higher proportion of mud and silt and smaller boulders have been derived from the Apfelberg fan catchment, which was dominated by mica schists and marble.

Alluvial fans are recognized as sensitive recorders of the evolution of piedmont basins and their margins (e.g. Heward 1978; Nemec & Postma 1993; Blair & McPherson 1994; Lloyd *et al.* 1998). Alluvial fans constitute a widespread facies in intramontane basins due to strong local uplift and subsidence along faults within actively deforming orogens. Fans in intramontane settings show a high degree of diversity because of complex basin geometries, contrasting source areas over short distances, and different tectonic movements that influence fan geomorphology and facies.

This paper describes Neogene alluvial-fan deposits within a large intramontane basin of the Eastern Alps, the Fohnsdorf Basin of Styria (Sachsenhofer *et al.* 2000; Strauss *et al.* 2001). The purpose of this paper is to document the origin of significant facies differences between adjacent fans. Fan formation and stratigraphic evolution is attributed to faulting along basin margins, whereas fan sedimentology seems to be strongly controlled by different source-area lithologies.

Geological and palaeogeographical overview

The Miocene Fohnsdorf Basin is one of several intramontane, fault-bounded basins within the Eastern Alps of Austria (Sachsenhofer *et al.* 2000, 2003; Strauss *et al.* 2001, 2003). It is situated on metamorphic complexes of the Austroalpine tectonic unit that underwent a complex evolution of thrusting and metamorphism during Cretaceous–Palaeogene times. Intramontane basins developed during the Oligocene–Miocene along strike-slip faults as a response to lateral eastwards extrusion of central

parts of the Eastern Alps (e.g. Ratschbacher *et al.* 1991). The Fohnsdorf Basin formed along one of these major strike-slip fault systems, the Mur-Mürz Fault (Decker & Peresson 1996), that bordered one of the extruding blocks. The fault linked these en echelon basins to the contemporaneous large pull-apart structure of the Vienna Basin (Ratschbacher *et al.* 1991; Decker & Peresson 1996).

The Fohnsdorf Basin subsided during the Early–Middle Miocene as a pull-apart along overstepping, E-W-trending strike-slip fault of the Mur–Mürz fault system (Fig. 1). Coarse and fine siliciclastics, coal seams and rare layers of limestone were deposited during the first basin stage (Strauss *et al.* 2001). Subsequently, tectonic stresses changed, and the strike-slip basin evolved into a half-graben with major subsidence concentrated in the southern part of the basin. During this time (MN 6, Middle–Late Badenian, Strauss *et al.* 2003) the Apfelberg Formation was deposited in the southern part of the basin (Figs 1 & 2). This more than 1000 m-thick formation is mainly composed of weakly consolidated conglomerate beds ('Blockschotter' of Polesny 1970).

Stratigraphic–structural setting of the Apfelberg Formation

The distribution of the Apfelberg Formation is limited to the SE part of the Fohnsdorf Basin. Good exposures can be found in the proximal, hilly part at the basin margins along road cuts and incised creeks, whereas the more distal parts of the Apfelberg Formation in the basin centre are largely covered by Quaternary–recent sediments, and outcrops are rare and patchy. Strata of the Apfelberg Formation have

From: Harvey, A.M., Mather, A.E. & Stokes, M. (eds) 2005. *Alluvial Fans: Geomorphology, Sedimentology, Dynamics.* Geological Society, London, Special Publications, **251**, 207–216. 0305-8719/05/$15 © The Geological Society of London 2005.

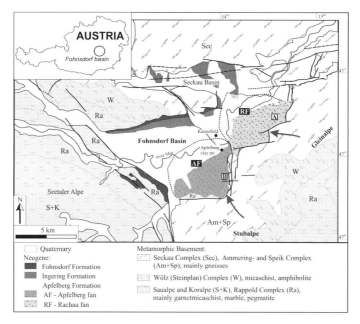

Fig. 1. Geological overview of the Fohnsdorf Basin and the basement complexes in the Eastern Alps of Austria. A is the position of Rachau fan logs of Figure 3; B is the position of proximal Apfelberg fan log of Figure 6.

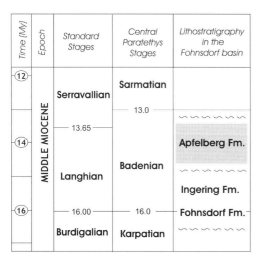

Fig. 2. Simplified stratigraphic chart of the Fohnsdorf Basin, based on Strauss *et al.* (2001, 2003), indicating the position of the Apfelberg Formation.

a tectonic dip of 10°–40° to the north; steeper inclinations are predominantly concentrated at the basin margins in the south. The sediments are largely undeformed with the exception of strata deposited directly adjacent to the faults at the basin margins (Strauss *et al.* 2001). Contacts with the underlying metamorphic basement are often faulted. Where

original sedimentary contacts are preserved, the underlying gneisses and mica schists record the effects of extensive palaeoweathering under subtropical climatic conditions, for example a strong loosening and significant reddening of the bedrock.

The Apfelberg Formation at the SE margin of the Fohnsdorf Basin consists of a wedge-shaped conglomerate complex that tapers towards the NW. Mapping indicates a maximum thickness of more than 1000 m (Strauss *et al.* 2001). Based on conspicuous material and facies differences, a NE fan, the Rachau fan, has been distinguished from a SW fan complex, the Apfelberg fan (Strauss *et al.* 2003). Both fans display an elongate fan-like aerial extent, although later erosion has modified this, especially along northern fan margins due to southward tilting of the basin. Reconstructed fan areas based on today's outcrops, facies mapping and the distribution of clast lithologies (Polesny 1970) are of the order of 20–35 km². The Rachau fan is mainly characterized by clasts of orthogneisses and amphibolites from the Gleinalpe area to the east and SE (see Fig. 1), whereas clasts of the Apfelberg fan are composed predominantly of lithologies that can be matched with the Stubalpe units to the south. The boundary between the two fans is loosely defined due to poor outcrop conditions in this area. However, based on clast lithologies and heavy mineral data (Polesny 1970), an interfingering of conglomerate beds of the two fans can be mapped, defining a transitional zone between the fans. In a distal position the alluvial fans grade into a lacustrine delta complex, e.g. in the Apfelberg clay pit (Strauss *et al.* 2003).

Faults along the southern and SE basin margin record several deformational phases. On the basis of cross-cutting relationships an older phase of sinistral strike-slip movement could be separated from a younger phase of normal faulting, followed by compression (Strauss *et al.* 2001). Normal faulting along E–W-trending, steeply N-dipping fault planes could be attributed to NNW–SSE extension, with linked N–S-trending strike-slip faults (Fig. 1).

Sedimentology

The two fans of the Apfelberg Formation are distinguished by significant lithology and facies contrasts. In general, both fans can be classified as debris-flow-dominated.

Rachau fan

Outcrops of sediments of the Rachau fan are mainly located in the proximal part of the fan, whereas the distal parts are largely covered. In proximal areas, the deposits of the Rachau fan comprise mainly stacked sheets of thick-bedded, unsorted–very poorly sorted, matrix- to clast-supported conglomerates with maximum clast sizes ranging from cobbles to large boulders (Fig. 3). The debris is subangular–subrounded, comprising fragments of local bedrock, mainly augengneisses, orthogneisses, amphibolites and minor paragneisses and quartzites (Fig. 4a, b). The matrix is an unsorted mixture of angular finer grained gravel and sand. The clay content of the matrix is below 5%. Pebbles with modal diameters of 10–30 cm are dispersed within this matrix. Outsized clasts are common and reach diameters of more than 3 m. These extremely outsized boulders are concentrated in a few distinct beds. Between the conglomerate beds thin lenticular sand–granule layers are present.

Conglomerate bed boundaries are often indistinct and amalgamation is a common feature. Thicknesses of individual conglomerate beds range from 40 cm to more than 3 m. Where outcrop conditions permit, observations of individual beds over tens of metres reveal a sheet-like–broadly lenticular geometry. Most beds are internally massive and appear structureless. Crude horizontal stratification defined by subhorizontally oriented clasts is very rare. The bases of thick boulder beds are planar and show no obvious erosion. Some of the thinner beds with lenticular geometry display erosional, convex-downward bases and locally show low-relief scouring of a few tens of centimetres. Inversely graded

basal parts, where the largest boulders in the bed are excluded from the lower third part, have been observed on several beds, and most outsized clasts are concentrated in the upper half of the beds. The fabric is mainly disorganized with no preferred clast orientation, and large boulders in nearly vertical positions are present. The tops of beds are often crudely normal graded and show transitions to interbedded sandstones. Large cobbles and boulders project above the surfaces of many beds.

The boulder conglomerate beds are sometimes separated by roughly horizontally stratified thinner conglomerate beds up to 80 cm thick, with organized fabric and rare lenses of fine–coarse sandstone. Weakly lithified sandstones display planar lamination locally marked by pebble stringers. These facies makes up less than 10% of the measured sections (Fig. 3). No distinct coarsening- or fining-upwards trends in maximum particle size could be observed.

Interpretation. The dominant boulder-bearing conglomerate facies of the Rachau fan is interpreted to represent deposits of a wide variety of debris-flow types from mudflows to largely cohesionless debris flows and transitions to sheetfloods (e.g. Postma 1986; Blair & McPherson 1994). The disorganized fabric, the presence of large outsized clasts, matrix-supported bed intervals and clasts in vertical position indicate deposition on a debris-flow-dominated alluvial fan (cf. Hubert & Filipov 1989; Blair & McPherson 1994). Although the number of measured beds is rather low and the variance is high, a crude correlation of maximum particle sizes (MPS, mean of 10 largest clasts of bed) and bed thickness (BTh, Nemec & Steel 1984) could be observed (Fig. 5), pointing also to debris-flow depositional processes. The conglomerates of the Apfelberg Formation represent mainly deposits of low cohesive debris flows, as determined by the low clay content of the beds and the predominance of a sandy matrix in most of the beds; mudflows with a strong clay-matrix support are not present due to the generally low clay content. The competence of these low cohesive flows was considerable, as large boulders were transported by flows with a dominantly sandy matrix and clay contents of around 5% (cf. Rodine & Johnson 1976). Inversely graded bed bases and the concentration of outsized clasts in the upper portion of the beds are attributed to the upwards movement out of basal shear horizons (Hubert & Filipov 1989). Above the shearing layer the coarse debris moved probably as a semi-rigid, high-strength plug. Crude normal grading and pebbly sandstones may indicate deposition by surging debris flows according to Nemec & Steel (1984). Normally graded bed tops with transitions to sandy layers indicate transitions from debris-flow transport to turbulent water flow in the late stages of flood events, a common feature of debris flows

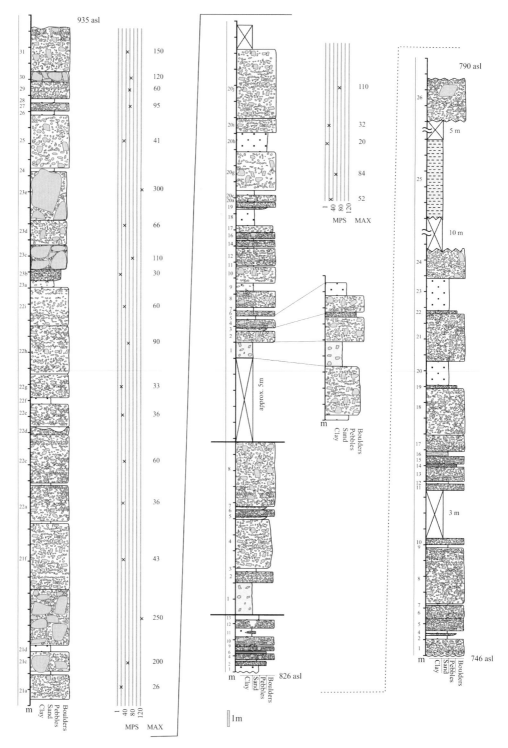

Fig. 3. Sedimentary logs through the Rachau fan deposits. MPS, mean particle size of 10 largest clasts of bed; MAX, diameter of largest component; asl, above sea level.

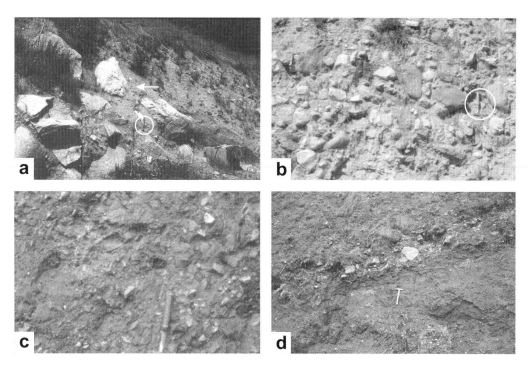

Fig. 4. Outcrop photographs of fan deposits of the Apfelberg Formation. (**a**) 3 m-thick debris-flow boulder bed (bed 23e of Fig. 3). Note the predominance of light-coloured augengneiss blocks. Top of the boulder bed with the sandy matrix interpreted as reworking and later infill of sand from the top of the layer. (**b**) Typical debris-flow conglomerate of the Rachau fan with predominantly clast-support, no preferred fabric and light-coloured sandy matrix (bed 21f of Fig. 3). (**c**) Typical debris-flow deposit of the proximal Apfelberg fan (bed 2 of Fig. 6) with maximum particle size up to 30 cm, and a dark, mica-rich and more clay-rich matrix compared to the debris flows of the Rachau fan. (**d**) Debris flows of the distal Apfelberg fan displaying generally smaller maximum clast sizes and predominant matrix-support in the lower bed.

(e.g. Blair & McPherson 1992). Some beds also show signs of later reworking by water flow, such as winnowing of the uppermost part of the beds and infiltration of a better sorted sandy matrix from the top.

Minor water-laid conglomerate and sandy layers are interbedded between the thick debris flows, and record either water flow in the late stages of flood events or reworking by intermittent stream flow and sheetfloods. These conglomerates are distinguished by their smaller clast sizes and bed thicknesses, crude horizontal stratification and local imbrication of clasts, and by a more lenticular geometry with some indications of erosion and channelling at their bases. Observed channels are a maximum of a few metres in diameter. Transport mechanism may include transitional types between dilute debris flows and streamflow, including sheetfloods (e.g. Wells & Harvey 1987; Blair 1999).

Apfelberg fan

Proximal sections (Fig. 6) of the Apfelberg fan are dominated by conglomerates with bed thicknesses generally below 2 m. Clasts are mainly composed of garnet mica schists and mica-rich paragneisses, and varying amounts of marbles, amphibolites, quartzites and pegmatites. Clast sizes are generally below 20 cm; outsized clasts reach diameters of 45 cm and are extremely rare (Fig. 4c, d). Matrix-supported fabrics dominate in the proximal parts of the fan; in distal settings both matrix- and clast-supported fabrics are present. Detrital muscovite, chlorite and biotite flakes derived from mica schists and paragneisses are a conspicuous constituent of the matrix, and account for the dark grey–greenish colour. The fabric is generally disorganized with no preferred clast orientation. Crude horizontal stratification was observed rarely. Bed bases are generally planar and non-erosive, but scoured bases are also present. Both inverse and normal grading is extremely rare; outsized clasts occur mainly in the upper portions of beds.

In more distal positions of the fan, coarse conglomerates make up about 10–20% of the sections. Lenticular channel fills composed of conglomerates and sandstones are intercalated within fine-grained massive conglomerates with maximum particle sizes

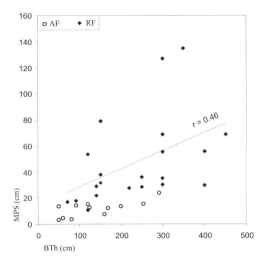

Fig. 5. MPS/BTh diagrams depicting bed thickness v. mean maximum particle size (arithmetic mean of 10 largest clasts) after Nemec & Steel (1984). Diamonds indicate measurements from the Rachau fan (RF) debris flows; circles indicate measurements from the Apfelberg fan (AF). Note that most of the debris flows of the Apfelberg fan fall into the lower left-hand corner.

of 5–10 cm. Grain-size curves (Fig. 6) document the poorly sorted nature of these conglomerates, interpreted as debris flows below, compared to the channelized deposits. The debris-flow matrix consists of an unsorted mixture of 10–25% of granules–pebbles, 30–50% of sand, 15–35% of silt and 5–15% of clay (Fig. 6). These distal fan deposits interfinger with planar stratified sheet sandstones, coal and tuff layers, including thin non-marine mollusc shell beds with vertebrate remains (Strauss *et al.* 2003).

Interpretation. The majority of the beds in the proximal portion of the Apfelberg fan consist of debris flows. Distinctive features for debris flows are matrix-supported conglomerates, outsized clasts in upright position, and high silt and clay contents in the matrix. Grain-size curves are similar to recent debris-flow deposits in terms of their largely unsorted texture and the significant clay contents (e.g. Hubert & Filipov 1989; Blair & McPherson 1998; Blair 1999). However, the debris flows show significant differences in grain size and matrix content to those from the Rachau fan. Debris flows of the Apfelberg fan display higher overall matrix contents, higher silt and clay proportions of the matrix (Fig. 6), lower maximum particle sizes and lower bed thickness values (Fig. 5). Matrix-supported beds are more common than clast-supported ones compared to the Rachau fan, which points to transportation by more cohesive debris flows.

In distal-fan areas, fluvial channels with conglomerate and sandstone fills and finer-grained braid plain deposits, including coal layers, predominate over debris-flow deposits. Transitions to delta-plain deposits of a lacustrine fan-delta environment are present (Strauss *et al.* 2001, 2003).

Discussion

Alluvial fans generally form in settings where a hinterland with a steep relief lies adjacent to a lower gradient basin, separated by a strong change in the slope gradient, for example by a synsedimentary active fault (Heward 1978). Higher-gradient mountainous streams in the uplifted hinterland can transport coarse detritus to the faulted basin margin of a low-relief plain, where alluvial-fan deposition occurs (e.g. Lloyd *et al.* 1998). The occurrence of outsized clasts of a few metres in diameter throughout the whole succession of the Rachau fan calls for a high-gradient, constantly exposed hinterland providing a mountainous source area. Synsedimentary tectonic movements, as demonstrated by Strauss *et al.* (2001), along the prominent southern basin margin faults of the Fohnsdorf Basin provided the relief necessary for continuous fan development. Normal faulting along older sinistral E–W-trending strike-slip faults due to NNW–SSE extension resulted in an asymmetric, southwards-deepening half-graben, which was filled by the wedge-shaped clastic fans of the Apfelberg Formation (Fig. 7), a typical situation for a continental half-graben (Leeder & Gawthorpe 1987). A backstepping of normal faults and synsedimentary cracked pebbles could be verified for this stage (Strauss *et al.* 2001, 2003). Adjacent to the faults, the fan sediments were tectonically tilted to about 20°–40°s, thus the primary slope gradients cannot be reconstructed.

Although several factors can influence alluvial-fan sedimentation and facies, such as climate, catchment type, relief and tectonic setting, the fan facies of the Fohnsdorf Basin seem to be dependent mainly on lithology variations in the hinterland of the fans. The fans have developed contemporaneously under essentially identical climatic and tectonic conditions. A similar catchment type with a generally similar relief can be inferred for both fans based on their adjacent position and general relief reconstructions for the Miocene (e.g. Frisch *et al.* 2001). The Rachau and Apfelberg fans have broadly similar areas and originated contemporaneously along normal or oblique basin-margin faults.

The major difference between the Rachau and the Apfelberg fans, which is regarded as the critical factor causing the differences in debris flows and thus fan facies, was apparently the lithology of the bedrock in

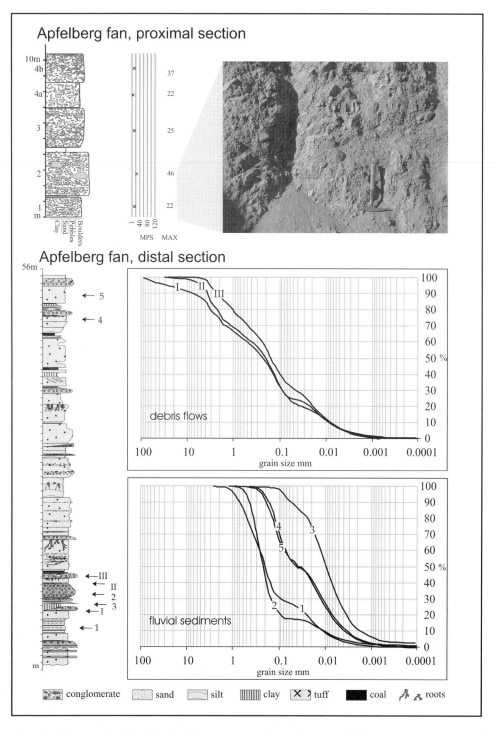

Fig. 6. Proximal and distal profiles of the Apfelberg fan with typical grain-size cumulative curves to distinguish debris flows and fluvial-channel conglomerates in the Apfelberg clay pit (standard sieve methods, sediment balance for sand fraction, and sedigraph analyser for the silt and clay fraction).

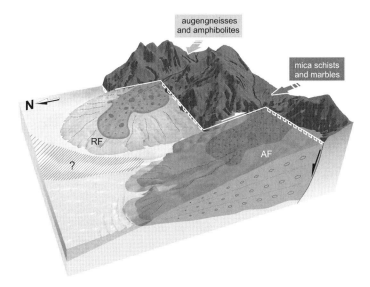

Fig. 7. Conceptual depositional model for the fans of the Apfelberg Formation in the intramontane Fohnsdorf Basin, indicating positions of basin-margin faults and evolution of fans due to faulting and different source areas. The distal portion of the Apfelberg fan grades into a lacustrine fan-delta plain (scale about 1:100 000).

the source areas of the fans. Striking differences exist in the clast lithologies (Table 1), the matrix composition and the heavy mineral compositions (Table 2) of the fan deposits. Clasts of the Rachau fan consist mainly of durable augengneisses, amphibolites and orthogneisses of the Gleinalpe area (mainly Amering and Speik metamorphic complexes; see Fig. 1), whereas the Apfelberg fan comprises material from the Stubalpe units (including the Rappold Complex and the Steinplan Complex), mainly soft mica schists, mica-rich paragneisses and quartzites, marbles, pegmatites and minor amphibolites. These differences were also noted by Polesny (1970) based on heavy mineral samples that display a significantly higher amount of green hornblende in sands from the Rachau fan, whereas the Apfelberg fan displays extremely garnet-rich assemblages. The change in the source-area type is due to a major fault–thrust contact between the Gleinalpe and the Stubalpe units, and the considerably higher Miocene erosional surface. Striking differences in source lithologies also control the type and grain-size distribution of the debris-flow matrix. The matrix of the Apfelberg fan debris-flows is more clay- and silt-rich, reaching up to 40% of the total size distribution. This is interpreted as a consequence of the breakdown and strong weathering of prevailing foliated mica-rich lithologies in the source area; whereas the more sand-dominated matrix of the Rachau fan deposits is the product of the weathering of augengneisses and orthogneisses into single sand-sized quartz and feldspar grains.

The restriction of large outsized boulders, up to 3 m in diameter, to the Rachau fan is a rather

unexpected phenomenon because this suggests that the largely non-cohesive debris flows of the Rachau fan had an apparently higher flow competence than the more clay-rich cohesive debris flows of the Apfelberg fan. This may be explained by two factors – either by a steeper relief and, thus, a higher fan gradient due to stronger tectonic movements along this segment of the basin-margin fault array, or again as a source-area lithology effect due to the higher resistivity of the Rachau fan lithologies against weathering. Given the tectonic reconstructions (Strauss *et al.* 2001, 2003) and the indications for strong normal faulting along the Apfelberg fan fault segment, which suggests no significant differences in Miocene fault movements in the hinterland of the two fans, we conclude that source lithologies played the main role. Strong weathering within the humid, subtropical climate of the Middle Miocene in this area (e.g. Steininger *et al.* 1989) may have reduced considerably the clast sizes of the easily weathered mica schists of the Apfelberg fan source area compared to the resistant gneiss and amphibolite lithologies in the Rachau fan hinterland. The gneisses reacted to tectonic stresses by rather widely spaced jointing and to weathering by slow grain-to-grain disintegration along joints into sand. This spheroidal weathering produced resistant, and thus significantly larger, blocks than the weathering of the strongly foliated, soft mica schists and paragneisses of the hinterland of the Apfelberg fan. The conspicuous rounding of these resistant gneiss boulders took place within the catchments, where these blocks are inferred to have been exposed for long

Table 1. *Clast composition data, Rachau fan (R-samples) and Apfelberg fan (A samples). 100 pebble counts, size fraction 2–20 cm. Data partly from Polesny (1970) and Worsch (1972) (Others include quartz, pegmatites, serpentinite)*

	Augengneiss	Amphibolite	Other gneisses	Quartzites	Mica schists	Marbles	Others
R1	52	17	21	6	3	0	0
R2	60	29	7	5	1	0	0
R3	52	26	14	7	1	0	0
A0	11	4	70	3	0	1	10
A1	0	0	0	3	78	19	0
A2	0	0	4	2	58	28	6
A3	0	0	4	2	55	36	1
A4	0	2	25	2	2	57	12
A5	8	7	45	20	2	7	11
A6	5	11	58	12	1	5	8
A7	0	3	0	1	47	43	6
A8	0	0	12	0	49	26	12
A9	0	6	4	1	33	50	5

Table 2. *Mean heavy mineral composition of sands from the Rachau fan (R, mean of three samples) and the Apfelberg fan (A, mean of six samples). Data from Polesny (1970)*

	Garnet	Hornblende	Zoisite	Epidote	Disthene	Titanite	Rutile	Tourmaline	Apatite	Amphibole
R	14.3	66.6	3.9	11.2	1.5	2.1	0.6	0.6	1.2	0.5
A	74.8	11.8	2.7	9.2	0.3	1.1	0.6	1.5	0.8	0.1

periods in stream beds, as normal floods would not have been able to transport them.

According to Wells & Harvey (1987) debris-flow-dominated successions are typical for alluvial fans with relatively small catchment areas and high slope gradients. Maximum slope gradients may be estimated as 2°–8° based on comparable debris-flow-dominated recent fans (e.g. Hubert & Filipov 1989; Blair 1999). Debris flows of the Apfelberg Formation fans probably originated at the transition from mountainous streams into the unconfined basin by failure of unsorted gravelly sediment of the hinterland as a result of the rapid addition of water, for example during strong rains. The hinterland of the Apfelberg fan can be expected to contain a higher proportion of muddy soils and a flatter morphology as a result of weathering of softer lithologies in comparison to the Rachau fan.

Bedrock lithology differences are, thus, regarded as the key factor causing the contrasting debris-flow facies of the two fans. In this respect fan development in the Fohnsdorf Basin displays a similar hinterland control to that of recent fans in Death Valley (Blair 1999) and the Rocky Mountains (Blair 1987).

This work is based on the diploma thesis of P. Strauss in the Fohnsdorf Basin and work within the Austrian Science Fund project FWF-P14370. The University of Vienna, the Styrian Government and the Stratigraphic Commission of the Academy of Sciences are thanked for financing this work. Thanks are due to G. Nichols and A.M. Harvey for constructive reviews.

References

BLAIR, T.C. 1987. Tectonic and hydrologic controls on cyclic alluvial fan, fluvial, and lacustrine rift-basin sedimentation, Jurassic–lowermost Cretaceous Todos Santos Formation, Chiapas, Mexico. *Journal of Sedimentary Petrology*, **57**, 845–862.

BLAIR, T.C. 1999. Cause of dominance by sheetflood vs. debris-flow processes on two adjoining alluvial fans, Death Valley, California. *Sedimentology*, **46**, 1015–1028.

BLAIR, T.C. & MCPHERSON, J.G. 1992. The Trollheim alluvial fan and facies model revisited. *Bulletin of the Geological Society of America*, **104**, 762–769.

BLAIR, T.C. & MCPHERSON, J.G. 1994. Alluvial fans and their natural distinction from rivers based on morphology, hydraulic processes, sedimentary processes, and facies assemblages. *Journal of Sedimentary Research*, **A64**, 450–489.

BLAIR, T.C. & MCPHERSON, J.G. 1998. Recent debris-flow processes and resultant form and facies of the Dolomite alluvial fan, Owens Valley, California. *Journal of Sedimentary Research*, **68**, 800–818.

DECKER, K. & PERESSON, H. 1996. Tertiary kinematics in the Alpine–Carpathian–Pannonoian System: links between thrusting, transform faulting and crustal extension. *In*: WESSELY, G. & LIEBL, W. (eds) *Oil and Gas in the Alpidic Thrustbelts and Basins of Central and Eastern Europe*. European Association of Geoscientists and Engineers, Special Publications, **5**, 69–77.

FRISCH, W., KUHLEMANN, J., DUNKL, I. & SZÉKELY, B. 2001. The Dachstein paleosurface and the Augenstein Formation in the Northern Calcareous Alps – a mosaic stone in the geomorphological evolution of the Eastern Alps. *International Journal of Earth Sciences*, **90**, 500–518.

HEWARD, A.P. 1978. Alluvial fan sequence and mega sequence models: with examples from Westphalian D–Stephanian B coalfields, Northern Spain. *In*: MIALL, A.D. (ed.) *Fluvial Sedimentology*. Canadian Society of Petroleum Geologists, Memoirs, **5**, 669–702.

HUBERT, J.F. & FILIPOV, A.J. 1989. Debris-flow deposits in alluvial fans on the west flank of the White Mountains, Owens Valley, California, U.S.A. *Sedimentary Geology*, **61**, 177–205.

LEEDER, M.R. & GAWTHORPE, R.L. 1987. *Sedimentary Models for Extensional Tilt-block/Half-graben Basins*. Geological Society, London, Special Publications, **28**, 139–152.

LLOYD, M.J., NICHOLS, G.J. & FRIEND, P.F. 1998. Oligo-Miocene alluvial fan evolution at the southern Pyrenean thrust front, Spain. *Journal of Sedimentary Research*, **68**, 869–878.

NEMEC, W. & POSTMA, G. 1993. Quaternary alluvial fans in southwestern Crete: sedimentation processes and geomorphic evolution. *In*: MARZO, M. & PUIGDEFÀBREGAS, C. (eds) *Alluvial Sedimentation*. International Association of Sedimentologists, Special Publications, **17**, 235–276.

NEMEC, W. & STEEL, R.J. 1984. Alluvial and coastal conglomerates: their significant features and some comments on gravelly mass-flow deposits. *In*: KOSTER, E.H. & STEEL, R.J. (eds) *Sedimentology of Gravels and Conglomerates*. Canadian Society of Petroleum Geologists, Memoirs, **10**, 1–31.

POLESNY, H. 1970. *Beitrag zur Geologie des Fohnsdorf-Knittelfelder und Seckauer Beckens*. PhD thesis, University of Vienna.

POSTMA, G. 1986. Classification for sediment gravity-flow deposits based on flow conditions during sedimentation. *Geology*, **14**, 291–294.

RATSCHBACHER, L., FRISCH, W. & LINZER, H. 1991. Lateral extrusion in the Eastern Alps. Part 2: structural analysis. *Tectonics*, **10**, 257–271.

RODINE, J.D. & JOHNSON, A.M. 1976. The ability of debris, heavily freighted with coarse clastic materials, to flow on gentle slopes. *Sedimentology*, **23**, 213–234.

SACHSENHOFER, R.F., BECHTEL, A., REISCHENBACHER, D. & WEISS, A. 2003. Evolution of lacustrine systems along the Miocene Mur–Mürz fault system (Eastern Alps, Austria) and implications on source rocks in pull-apart basins. *Marine and Petroleum Geology*, **20**, 83–110.

SACHSENHOFER, R.F., KOGLER, A., POLESNY, H., STRAUSS, P. & WAGREICH, M. 2000. The Neogene Fohnsdorf Basin: basin formation and basin inversion during lateral extrusion in the Eastern Alps. *International Journal of Earth Sciences*, **89**, 415–430.

STEININGER, F.F., RÖGL, F., HOCHULI, P. & MÜLLER, C. 1989. Lignite deposition and marine cycles. The Austrian Tertiary lignite deposits – a case history. *Sitzungsberichte der Österreichischen Akademie der Wissenschaften, mathematisch-naturwissenschaftliche Klasse, Abteilung I*, **197**, 309–332.

STRAUSS, P., WAGREICH, M., DECKER, K. & SACHSENHOFER, R.F. 2001. Tectonics and sedimentation in the Fohnsdorf-Seckau Basin (Miocene, Austria): From a pull-apart basin to a half-graben. *International Journal of Earth Sciences*, **90**, 549–559.

STRAUSS, P.E., DAXNER-HÖCK, G. & WAGREICH, M. 2003. Lithostratigraphie, Biostratigraphie und Sedimentologie des Miozäns im Fohnsdorfer Becken (Österreich). *In*: PILLER, W.E. (ed.) *Stratigraphia Austriaca*. Österreichische Akademie der Wissenschaften, Schriftenreihe der Erdwissenschaftlichen Kommissionen, **16**, 111–140.

WELLS, S.G. & HARVEY, A.M. 1987. Sedimentologic and geomorphic variations in storm-generated alluvial fans, Howgill Fells, northwest England. *Bulletin of the Geological Society of America*, **98**, 182–198.

WORSCH, E. 1972. Geologie und Hydrologie des Murbodens. *Mitteilungen der Abteilung für Geologie, Paläontologie und Bergbau am Landesmuseum Joanneum Graz*, 32, 1–111.

Upper Cretaceous–Palaeocene basin-margin alluvial fans documenting interaction between tectonic and environmental processes (Provence, SE France)

S. LELEU, J.-F. GHIENNE & G. MANATSCHAL

CGS-EOST, UMR 7517 CNRS-ULP, 1 rue Blessig, F-67084 Strasbourg, France
(e-mail: leleu@illite.u-strasbg.fr; ghienne@illite.u-strasbg.fr; manatschal@illite.u-strasbg.fr)

Abstract: Upper Cretaceous–Palaeocene alluvial-fan conglomerates exposed along the northern margin of the Arc Basin (Provence, SE France) preserve a continuum between undeformed basinal deposits and syntectonic alluvial-fan deposits. Based on the distribution of facies associations and growth structures in the alluvial-fan deposits, and using marker levels and erosional surfaces, the tectono-sedimentary evolution of the basin margin is discussed. On a long timescale, the stratigraphic pattern in the alluvial-fan deposits mainly records the tectonic activity in the catchment, and subordinate out-of-syncline thrusts in the basin margin. On an intermediate timescale, evolution in the drainage area controls the spatial evolution of the alluvial fans and some minor changes in depositional facies. High-frequency cycles record aggradation–stabilization sequences, resulting in vertically superimposed alluvial-fan bodies more probably tectonically controlled, whereas alternation between conglomerates–siltstones at the scale of interbedding most probably reflects climatic cycles.

Alluvial fans commonly occur in regions of active deformation and are therefore often used as a proxy for tectonic activity. Whereas on a Quaternary timescale alluvial-fan evolution is strongly controlled by climate changes (Wells *et al.* 1987; Harvey 1990) and, in some cases, by base-level changes (Harvey 2002), on a longer timescale alluvial-fan depositional successions are controlled by tectonically and climatically driven environmental changes, as well as processes acting in the catchment area (e.g. DeCelles *et al.* 1991). Therefore, studies dealing with the first-order stratigraphic architecture of syntectonic alluvial-fan successions have to take into account the temporal and spatial interaction of these different forcing factors acting in both catchment area and adjacent sedimentary basin.

This contribution presents a case study based on Upper Cretaceous–Palaeocene alluvial-fan conglomerates that have been deposited along a tectonically active basin margin in the Montagne Sainte Victoire area (northern Arc Basin, Provence, SE France). Excellent exposure, preserving continuum between undeformed basinal and alluvial-fan deposits, enables correlation between uplift and erosional processes occurring in the catchment area with sedimentation in the basin. In this paper the depositional architecture and temporal evolution of the syntectonic alluvial fans exposed in the Montagne Sainte Victoire area are described. Based on these observations, the relationships between the observed stacking pattern of the alluvial-fan deposits and the forcing factors are interpreted.

Geological overview

Tectonic setting

The tectonic structure of Provence has classically been described as a N-vergent fold-and-thrust belt that consists of a succession of wide (>15 km), E–W-trending synclinal troughs bounded by narrower (<10 km), strongly deformed anticlinal ranges (Lutaud 1935; Aubouin & Mennessier 1963) (Fig. 1). These structures have been referred to the Pyreneo-Provencal phase, including all the Late Cretaceous–middle Eocene (pre-Nummulitic) deformation events. This deformation phase has been commonly interpreted to be related to northward migration of the Iberia and Corso-Sardic blocks (Mattauer & Seguret 1971; Choukroune & Mattauer 1978; Debroas 1990; Olivet 1996). Lutaud (1935) proposed an initiation of the anticlinal ranges during the Late Cretaceous, followed by a major compressional event overprinting the former structures during the Eocene (Bartonian event), an idea that has been confirmed by later studies (e.g. Tempier 1987).

The Arc Basin is one of these E–W-trending synclinal troughs. It is filled with thick (>1000 m) Campanian–Eocene continental sediments overlying Jurassic–Upper Cretaceous (Coniacian–Santonian) shallow-marine carbonates (Durand & Guieu, 1983; Babinot & Durand 1984). The location of its western and eastern margins is mainly controlled by Neogene erosion, whereas the southern and northern margins are defined by Late Cretaceous–middle Eocene

From: HARVEY, A.M., MATHER, A.E. & STOKES, M. (eds) 2005. *Alluvial Fans: Geomorphology, Sedimentology, Dynamics.* Geological Society, London, Special Publications, **251**, 217–239. 0305–8719/05/$15 © The Geological Society of London 2005.

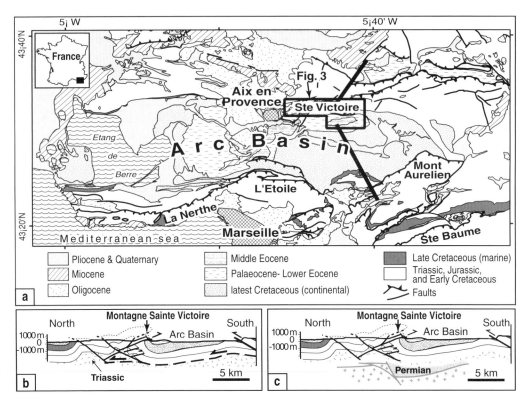

Fig. 1. (**a**) Geological map of SW Provence showing the location of the Arc Basin (modified after Rouire 1979), and two possible interpretations of the tectonic structure of the northern margin of the Arc basin as either related to (**b**) thrust tectonics with a décollement in the Triassic evaporates, or (**c**) reactivation of a previous basin (e.g. a Permian basin) underlying the Montagne Sante Victoire area.

compressional structures (the Etoile and Sainte Baume Ranges, and the Montagne Sainte Victoire Range, respectively) (Fig. 1a). Both the southern and northern margins preserve Upper Cretaceous–Palaeocene conglomerate successions (Fig. 2) representing ancient transverse alluvial fans draining into a basin controlled by a longitudinal drainage network.

The interpretation of the Late Cretaceous–Eocene compressional structures in Provence is still a matter of debate. Tempier (1987) proposed a fold-and-thrust belt in which the Mesozoic–Cenozoic sedimentary cover was decoupled from its basement along Triassic evaporates, with an estimated total shortening of the order of 25 km. In this interpretation the narrow anticlinal ranges formed over ramps during N-vergent thrusting (Fig. 1b). In alternative models, the anticlinal ranges were interpreted as large-scale pop-up structures resulting from the reactivation and inversion of inherited basement faults of either Jurassic (Chorowicz *et al.* 1989) or Permian age (Roure & Coletta 1996) in the foreland of an active fold-and-thrust belt (Fig. 1c). Biberon

(1988) proposed a S-vergent back-thrust for the Montagne Sainte Victoire, associated with an overall N-vergent thrust fault, an interpretation which is not supported by the observations presented in this paper.

The post-Eocene tectonic evolution in the studied area is dominated first by Oligocene extension, leading to the reactivation of NNE–SSW-trending faults such as the Aix Fault. Activity along this fault forming the eastern border of the Oligocene Aix Basin was coeval with the deposition of thick fluvial conglomerates. In the western zone of the study area all tectonic structures are sealed by a prominent planar wave-ravinement surface at the base of Miocene calcarenites (Fig. 2). This area has not been affected by the late Miocene Alpine deformation observed further to the north and to the east (Champion *et al.* 2000). Instead, the present-day landscape results from Miocene to present-day uplift related to opening of the Ligurian Sea and prominent erosional surfaces associated with the Messinian sea-level fall (Clauzon 1984).

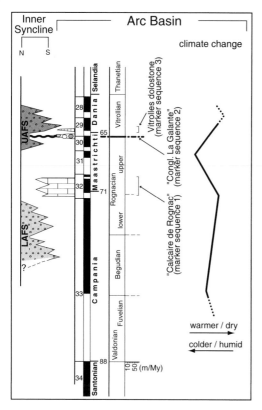

Fig. 2. Time section showing the ages and stratigraphic position of the Lower Alluvial Fan System (LAFS) and the Upper Alluvial Fan System (UAFS) relative to the three marker levels defined in this work, which are the Calcaire de Rognac Formation, the Conglomérat de la Galante and the Vitrolles dolostone. These units are shown within the regional stratigraphic framework as well as within the international time chart (for correlation see text). For a discussion of the climate change deduced from the sediments in the Arc Basin see the References and discussion in the text.

Stratigraphic record in the Arc Basin

In the Arc Basin, stratigraphic subdivisions correspond to local stages defined on the distribution of continental fossils (Medus 1972; Babinot & Durand 1980a, b) (Fig. 2). Integrated studies combining biostratigraphic and magnetostratigraphic investigations by Westphal & Durand (1990) and Cojan et al. (2003) enabled proposals of correlations of the regional continental stages with the international stratigraphic chart to be made. From bottom to top, the stratigraphic succession comprises: lacustrine and palustrine carbonates of the early Campanian (Valdonian and Fuvelian stages); lacustrine and fluvial sediments of the middle Campanian (Begudian stage); red fluvial siltstones and sandstones, overlain by lacustrine carbonates of the late Campanian (lower Rognacian stage); red fluvial siltstones of the Maastrichtian (upper Rognacian stage) and fluvial conglomerates, Conglomérat de la Galante Formation, of the latest Maastrichtian (lowermost Vitrollian stage); and red siltstones and dolocrete horizons of the early Danian (Vitrollian stage); and red siltstones and lacustrine limestone of the late Danian and Thanetian. Thick lacustrine limestones with intervening subordinate fluvial or lacustrine shales are classically attributed to the Eocene (Durand & Guieu 1983; Babinot & Durand 1984; Cojan 1993). Widespread and repetitive lake deposits essentially located in the vicinity of the Aix Fault (Cojan 1993) suggest that the Arc Basin may have behaved, at least temporarily, as an endoreic system (Colson & Cojan 1996).

According to the definition of the Campanian–Maastrichtian boundary by Gradstein et al. (1995) and to a Cretaceous–Tertiary boundary located a few metres above the basal Conglomérat de la Galante (Cojan et al. 2000), the Fuvelian, Begudian, Rognacian and lowermost Vitrollian stages correspond to the Campanian and Maastrichtian (Fig. 2). Most of the Vitrollian corresponds to the Danian, and the Thanetian local stage corresponds to the Selandian and Thanetian (Fig. 2). Sedimentation rates have been estimated at between 40 and 60 m/Ma^{-1} during the Campanian, and then decreased during Maastrichtian (Westphal & Durand 1990). Sedimentation rates have been estimated at around 20 m/Ma^{-1} for Danian times (Cojan pers. comm. 2003).

In the NE Arc Basin, three Marker Sequences can be traced from the basin towards the deformed basin margin. In this study, these Marker Sequences are considered as 'time markers', allowing a correlation between the basin and the basin-margin deposits. The lowermost marker sequence, Marker Sequence 1 (Fig. 2), is represented by the base of the Calcaire de Rognac Formation. Cojan (1989) found a prominent palaeosol horizon as a marker level in this limestone. The Calcaire de Rognac is isochronous in the study area representing a latest Campanian–earliest Maastrichtian age (Cojan et al. 2003). Marker Sequence 2 corresponds with the Conglomérat de la Galante Formation and its lower pseudo-gley palaeosols sequence (Cojan 1989), and has been dated as latest Maastrichtian (Cojan et al. 2000). Marker Sequence 3, which is found only in the most basinward sections, corresponds with an early Palaeocene palustrine dolostone horizon referred to as the Vitrolles dolostone (Colson & Cojan 1996) (Fig. 2).

Palaeoclimatic reconstructions based on palynological studies (Medus 1972; Ashraf & Erben 1986) and clay minerals (Sittler & Millot 1964) suggest a subtropical climate punctuated by semi-arid episodes prevailing during the latest Cretaceous and

Palaeocene in the Arc Basin (Colson & Cojan 1996) (Fig. 2). Isotopic studies on palaeosoils and dinosaurian eggshells (Cojan *et al.* 2003) provide evidence for a warming trend during the Late Campanian (Begudian and early Rognacian) culminating before the deposition of the Calcaire de Rognac Formation. A cooling trend is then recorded during the Maastrichtian (upper Rognacian), before a renewed warming trend that began in the latest Maastrichtian (uppermost Rognacian) followed by drier conditions as shown by numerous calcrete horizons in early Danian (Vitrollian) sediments (Fig. 2).

Architecture of the northern basin margin

Along the NE margin of the Arc Basin (Montagne Sainte Victoire area) Upper Cretaceous–Palaeocene alluvial fan systems and their transition to time-equivalent basinal successions are preserved (Fig. 3). These alluvial systems were fed from an emerging E–W-trending anticlinal range forming the northern border of the basin. Although the northern margin of the Arc Basin has been overprinted by the major Eocene compressional event related to the formation of a fold-and-thrust belt in the southern Provence, in some places primary Upper Cretaceous–Palaeocene depositional geometries are still preserved (e.g. Inner Syncline, Figs 4 & 5).

Previous studies by Corroy (1957), Billerey *et al.* (1959), Durand & Tempier (1962) and Chorowicz & Ruiz (1979) distinguished two superimposed alluvial-fan conglomerate successions, known as the 'Begudian–Rognacian breccia' and the 'Danian–Montian breccia'. These conglomerate units are separated by a major angular and erosional surface. Locally Maastrichtian siltstones are preserved beneath the erosional surface. These units show a rapid thinning basinwards, and the transition to time-equivalent basinal successions is preserved by the interfingering of alluvial and basinal facies. In this study, we define two superimposed alluvial-fan systems referred to as the Lower Alluvial Fan System (LAFS) and the Upper Alluvial Fan Systems (UAFS) (Figs 2 & 5). The two alluvial-fan systems form depositional sequences including the alluvial-fan deposits and their time-equivalent proximal basinal successions (Fig. 5). The base of the LAFS is not well constrained in age and the top of the LAFS corresponds to the base of the Calcaire de Rognac Formation, indicating a probable middle–late Campanian age for the LAFS (Fig. 2). The base of the UAFS is diachronous from north to south. The northern deposits overlay an erosional unconformity, truncating the LAFS and locally the Maastrichtian floodplain deposits. Based on the recognition of a palaeosol sequence (base of Marker Sequence 2), they are interpreted here to correlate basinward with the basal, essentially non-erosive surface underlying the Conglomérat de la Galante Formation (Marker Sequence 2) (Fig. 5). The base of the upper UAFS is post-Vitrolles dolostone (Marker Sequence 3) and these deposits are overlain by an early Thanetian limestone, indicating a Palaeocene age for the UAFS. The two units can be confidently described as chronostratigraphic units. This definition contrasts to that of the previously defined 'Begudian–Rognacian breccias' and 'Danian–Montian breccias', which were defined as lithostratigraphic units.

Based on changes in the Upper Cretaceous–Palaeocene depositional architecture and on the later Eocene tectonic overprint, the northern margin can be subdivided into three zones (Fig. 3): a western zone (le Tholonet–Roques Hautes) (Fig. 4); a central zone (Montagne Sainte Victoire); and an eastern zone. This paper focuses mainly on the alluvial-fan architecture preserved in the western zone (Figs 4 & 5).

In the western zone, the conglomerates of the LAFS and the UAFS are best exposed in the Inner Syncline (Fig. 3). Here conglomerate units can be traced over about 1 km perpendicular to, and over more than 5 km parallel to, the basin margin (Fig. 6b). Nevertheless, the architecture of the alluvial-fan deposits changes from east to west in this area. Conglomerate strata in the northern limb of this syncline display dips ranging from 10° to 70° and abut against subvertical–overturned Jurassic limestones (Figs. 5 & 7a). In the core and southern limb of the Inner Syncline, relatively thick floodplain deposits are locally present as lateral deposits of the alluvial fans. The Calcaire de Rognac Formation (Marker Sequence 1) and uppermost Maastrichtian siltstones are locally present between the LAFS and the UAFS (Figs 5 & 7b). The UAFS lies unconformably on either LAFS deposits (Fig. 7c) or uppermost Maastrichtian siltstones (Fig. 7b). At the southern border of the Inner Syncline, the LAFS is thrust (T1) onto the Calcaire de Rognac Formation, resulting in an imbricate fan structure, which has been reactivated and folded during later Eocene compression (Fig. 5). In the area of le Tholonet, the S-vergent thrust T1 is sealed by the conglomerates of the UAFS, the latter resting with a sharp erosional contact either onto the conglomerates of the LAFS or onto the Calcaire de Rognac Formation (Figs 4 & 7c).

In the central zone (Fig. 3), the Inner Syncline and the T1 thrust fault are not observed. Instead, a monoclinal structure is developed forming the continuation of the southern limb of the anticlinal structure described further to the west. The southern face of the Montagne Sainte Victoire comprises an overturned succession that consists of folded Jurassic–Lower Cretaceous carbonates unconformably overlain by conglomerates of the LAFS (Tempier & Durand 1981). Locally, the conglomerates form the infill of

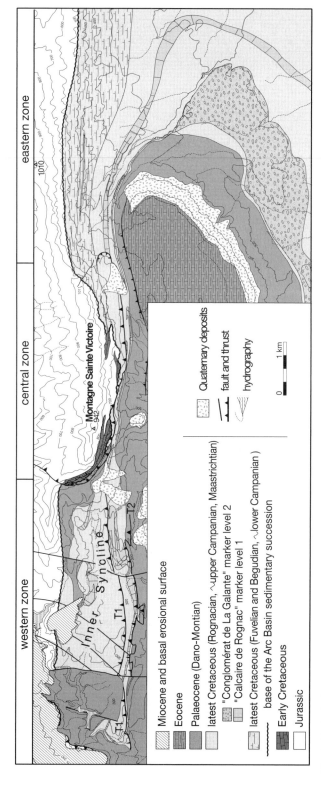

Fig. 3. Geological map of the NE margin of the Arc Basin.

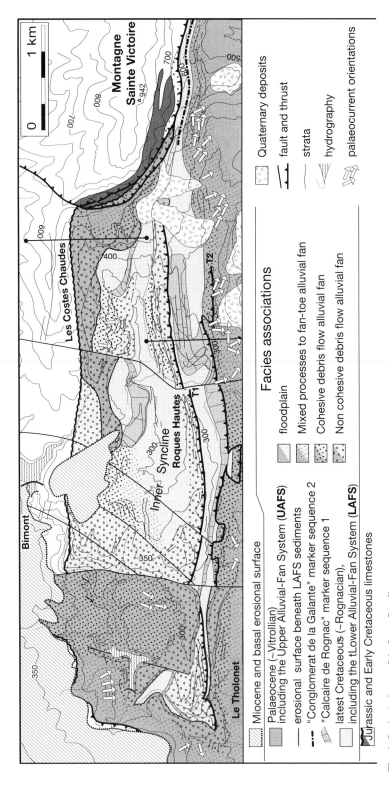

Fig. 4. Geological map of the Inner Syncline.

Montagne Sainte Victoire

▲ 942

Les Costes Chaudes

Bimont

Inner Syncline

Roques Hautes

Le Tholonet

0   1 km

T1

T2

Facies associations

Quaternary deposits

fault and thrust

strata

hydrography

palaeocurrent orientations

Miocene and basal erosional surface

Palaeocene (~Vitrollian)
including the Upper Alluvial-Fan System (**UAFS**)

erosional surface beneath UAFS sediments

"Conglomerat de la Galante" marker sequence 2

"Calcaire de Rognac" marker sequence 1

latest Cretaceous (~Rognacian),
including the tLower Alluvial-Fan System (**LAFS**)

Jurassic and Early Cretaceous limestones

floodplain

Mixed processes to fan-toe alluvial fan

Cohesive debris flow alluvial fan

Non cohesive debris flow alluvial fan

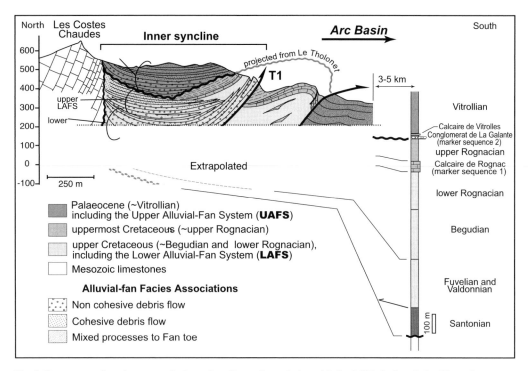

North Les Costes
Chaudes

Inner syncline

Arc Basin   South

projected from Le Tholonet

T1

3-5 km

Vitrollian

Calcaire de Vitrolles
Conglomerat de La Galante
(marker sequence 2)
upper Rognacian

Calcaire de Rognac
(marker sequence 1)

lower Rognacian

Extrapolated

250 m

Palaeocene (~Vitrollian)
including the Upper Alluvial-Fan System (**UAFS**)

uppermost Cretaceous (~upper Rognacian)

upper Cretaceous (~Begudian and lower Rognacian),
including the Lower Alluvial-Fan System (**LAFS**)

Mesozoic limestones

Alluvial-fan Facies Associations

Non cohesive debris flow

Cohesive debris flow

Mixed processes to Fan toe

upper
LAFS

lower

Begudian

Fuvelian and
Valdonnian

Santonian

100 m

Fig. 5. Reconstructed section across the Inner Syncline and correlation with the drill hole data derived from the Arc Basin.

1 km-wide palaeovalley. Previous workers (e.g. Corroy *et al.* 1964*a, b*; Tempier 1987) suggested that the Inner Syncline continued eastwards below an Eocene thrust fault related to the emplacement of the present-day Montagne Sainte Victoire (Fig. 3). The Jurassic–Lower Cretaceous limestones of the Montagne Sainte Victoire is now considered as para-autochtonous (displacement <400m) relative to the basin-margin succession (Gonzalez pers. comm. 2004). Following this new interpretation, the Montagne Sainte Victoire represents an anticlinal structure formed during Late Cretaceous–Palaeocene time at the northern margin of the Arc Basin. This interpretation is in agreement with the condensed upper Cretaceous succession observed in the mono-clinal structure south of the Montagne Sainte Victoire.

Combined with drill-hole data from the basin, the eastern zone permits better documentation of the stratigraphic evolution of the basin-margin succes-sions and their transition into their time-equivalent basinal successions (Fig. 6a). Here, the basin margin is exposed within a monoclinal structure that pro-gressively opens eastwards and terminates along a km-scale N–S-trending structure along which basal parts of the Arc Basin are exhumed (Figs 4 & 6a). Within the NE Arc Basin, Jurassic limestones are conformably overlain directly by early Campanian sediments, and Lower Cretaceous strata are not preserved.

Facies distribution and growth structures in alluvial-fan successions

Lithofacies associations

At the northern margin of the Arc Basin, the sedi-mentary succession comprises alluvial-fan deposits that interdigitate laterally and distally with basinal facies. The latter consist mainly of red siltstones rep-resenting floodplain deposits containing channel sandstones and several lacustrine limestone layers (Cojan 1993). Eight lithofacies are recognized (Table 1), grouped into five facies associations, each of which is representative of a distinct depositional environment. The variability between these facies associations reflects the range of processes at work in different alluvial fans.

- *Scree facies association*: this association is pre-served at the basin edge and consists of amal-gamated breccia beds several metres thick (lithofacies 1 in Table 1; Fig. 8a).

(a)

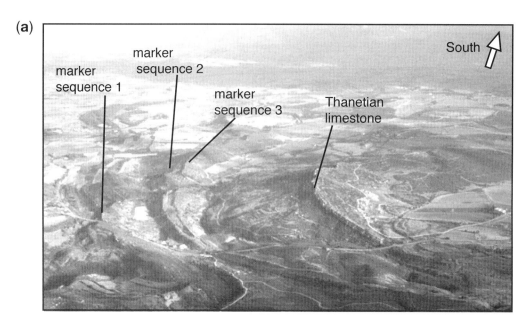

(b)

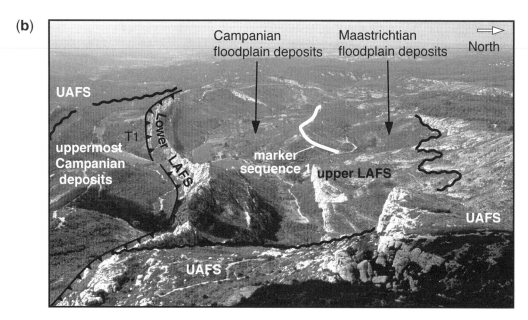

Fig. 6. (**a**) Photograph of the NE Arc Basin (view from the Montagne Sainte Victoire looking to the south) showing the major marker levels within the basin (for comparison see also Fig. 3). (**b**) Photograph of the Inner Syncline (view from the Montagne Sainte Victoire to the west), showing the major marker levels and the structures shown on the map in Figure 4 and in the section in Figure 5.

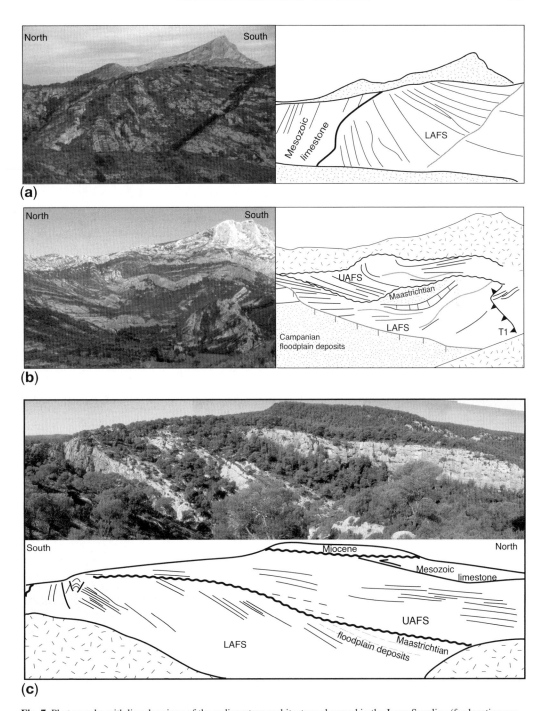

Fig. 7. Photographs with line drawings of the sedimentary architecture observed in the Inner Syncline (for location see Fig. 4). (**a**) View of the Bimont wedge showing alluvial fan conglomerates wedging towards and onlapping onto an erosional surface separating the alluvial fans from the Mesozoic limestones. (**b**) Sedimentary architecture in the Roques Hautes area showing the relation between the LAFS and the UAFS (for explanations see the text). (**c**) Erosional surface at the top of the LAFS sealed by the UAFS in the area of Tholonet (for details see the text).

Table 1. *Facies description and interpretation*

Lithofacies	Sedimentary features	Pedogenesis	Interpretation
1 – Breccias	Crudely stratified, mainly openwork breccias of angular very angular, moderately sorted gravels – cobbles. Each bed has a granulometric sorting		Scree deposits
2 – Amalgamated conglomerate beds (0.3–2 m thick)	Crudely stratified conglomerates of subrounded – angular, poorly sorted pebbles – cobbles, typically 5–10 cm, in diameter, rarely >50 cm. The framework is clast- to matrix-supported, depending from both depositional and compaction processes. The matrix is a sandy muddy siltstone. Locally, sparse blocks, aligned blocks, clast imbrications, basal grooves, small shallow channels, gravel mantles, openwork horizons	*Microcodium*, carbonated glaebules and mottling structures are occasional sedimentary features. Some carbonatization of beds occurred Calcimorph palaeosols are locally present between conglomerate beds underlying an unconformity	Each conglomerate bed represents an episodic non-cohesive debris flow. Aligned blocks form levée deposits. Limited post-floodwater winnowing results in the washing out of fine material, the formation of erosion surfaces and gravel mantles. Conglomerate bodies reflect aggradation on the medial part of individual alluvial-fan lobes Forms the dominant facies of the facies association of non-cohesive debris-flow-dominated alluvial fans
3 – Individual conglomerate beds (0.1–1.5 m thick)	Conglomerate lenses or beds of subrounded to angular, poorly sorted pebbles – cobbles, typically 5–10 cm in diameter, rarely >50 cm. Lenses are generally erosionally based (0.5–2 m in depth), with an upper flat or convex-upwards profile. The framework is clast- to matrix-supported, depending on both depositional and compaction processes. Relatively thin (<0.3 m) matrix-supported conglomerate form locally the basal part of some beds. The matrix is a sandy muddy siltstone or a silty sandstone. Clast imbrications and alignments of coarse clasts flanking finer grained conglomerates occur in places. Some of the coarser clasts protrude above the upper surface of the conglomerate beds. Some coarsening-upwards deposits occurred in place with an openwork fabric on the top	*Microcodoium* and carbonated glaebules are occasional sedimentary features	Individual conglomerate beds represent non-cohesive debris-flow deposits in the distal part of an alluvial fan. Basal matrix-supported horizon may represent initial deposition by more cohesive and sheared flows. Active post-depositional runoff cut gullies or channels, which are used as a guide by the next, essentially non-erosional debris-flow event. Runoff also reworks the matrix of the upper parts of the deposits. Pedogenesis is pervasive. Conglomerate sequence reflect aggradational event of the distal part of individual alluvial-fan lobe Forms one of the dominant facies of the facies association of mixed processes alluvial fans

Facies	Description	Pedogenic features	Interpretation
4 – Conglomerates with planar stratification	Crudely stratified, clast-supported conglomerates of subrounded – subangular, well-sorted gravels–pebbles, typically 2–5 cm in diameter. Gravels couplets are present, showing the alternation of clast-supported gravels and clast-supported pebbles with a matrix compose of silty sand		Bedload-dominated deposits resulting from hypercentrated flows or from gravel bars in wide distalfan channels. Forms one of the subordinated facies of the facies association of mixed processes alluvial fans
5 – Pebbly muddy siltstones	Massive, clast-rich, red siltstones including variable proportions of granules, gravels and pebbles (from 30 to 50%)	*Microcodium*, mottling structures and sparse carbonated glaebules are present	Cohesive debris-flow deposits originated either from the distant catchment area or from post-flood reworking of nearby clast- to matrix-supported conglomerates. Incipient pedogenesis. Dominant in the cohesive debris-flow facies association
6 – Granule-rich muddy siltstone	Massive red siltstones with dispersed granules – small gravels, including a number of horizons characterized by isolated–coalescent calc-areous nodules. *Microcodium* is locally pervasive. Occasional pebbles – cobbles, either isolated or in discontinuous lenses	*Microcodium* and mottling structures are present. Sparse – coalescent carbonated glaebules are omnipresent. Locally thick recarbonatization occurred	Mudflow deposits or amalgamation of discrete siltstone horizons of reworked, mainly fine-grained material from debris-flow- or bedload-dominated deposits. Thick recarbonatization beds indicate a groundwater high level in the fan-toe area. Forms one of the dominant facies in the mixed processes facies association
7 – Laminated sandstone	Horizontally laminated, sorted, medium-grained sandstones with occasional ripples	None	Waterlain sands resulting from the reworking of debris-flow or bedload-dominated deposits. Subordinated facies in the mixed processes facies association
8 – Red muddy siltstone	Massive red siltstones including a number of horizons characterized by isolated–coalescent calcareous nodules, or by prismatic limestone beds. Limy root cast or siltstone-filled, 0.5–1 cm in diameter, branching pipes are abundant in places. Occasional discontinuous horizons with pebbles. Locally, massive grey, organic-rich siltstone with mottling	Either mottling structures and sparse–coalescent carbonated glaebules, or multicoloured mottling structures are present	Floodplain deposits related to the longitudinal drainage system with pervasive pedogenesis either of calcimorph type or of pseudo-gley type

(a)   **(b)**

(c)   **(d)**

(e)   **(f)**

Fig. 8. (**a**) Crudely stratified breccia (lithofacies 1) of the scree facies association; (**b**) clast- to matrix-supported conglomerates (lithofacies 2) representing debris-flow deposits in the proximal alluvial-fan facies association; (**c**) lenses of clast- to matrix-supported conglomerates (lithofacies 3) interbedded with a pebbly muddy siltstone horizon (lithofacies 5) in the medial alluvial-fan facies association; (**d**) laminated sandstones (lithofacies 7) onlapping onto an isolated clast at the top of a conglomerate bed in the distal alluvial-fan facies association; (**e**) erosional gullie cut into granule-rich muddy siltstones (lithofacies 6) and filled by clast- to matrix-supported conglomerates (lithofacies 3) in the distal alluvial-fan facies association; (**f**) convex-up lens of clast- to matrix-supported conglomerates (lithofacies 3) resting with an erosional contact onto red muddy siltstone (lithofacies 8) with pedogenic prismatic structures in the fantoe alluvial-fan facies association.

- *Non-cohesive debris-flow facies association*: this facies association is characterized by amalgamated conglomerate beds (lithofacies 2 in Table 1; Fig. 8b). In places, granule-rich muddy siltstones (lithofacies 6) are interbedded with amalgamated conglomerates.
- *Cohesive debris-flow facies association*: this facies association is characterized by pebbly muddy siltstones (lithofacies 5 in Table 1; Fig. 8c) associated with amalgamated conglomerate beds (lithofacies 2) (<2 m) and locally with granule-rich muddy siltstones (lithofacies 6).
- *Mixed processes facies association*: this facies association is mainly made up of granule-rich muddy siltstones (lithofacies 6 in Table 1), including beds or lenses of individualized conglomerates (lithofacies 3 in Table 1) characterized by well-defined erosional gullies (Fig. 8e).

Laminated sandstones (lithofacies 7 in Table 1; Fig. 8d) and conglomerates with planar stratification (lithofacies 4 in Table 1) occasionally occur. Interbedded floodplain siltstones (lithofacies 8) indicate recurrent interfingering with basinal facies.

- *Fan-toe facies association*: this facies association mainly comprises red muddy siltstones (lithofacies 8 in Table 1) with subordinate granule-rich muddy siltstones (lithofacies 6) that present locally thick carbonatisation (Fig. 8f).

Alluvial-fan deposits are organized in repetitive facies successions developed on two orders of magnitude. Such succession is best developed within the upper UAFS where individual conglomerate lenses and sheets (lithofacies 2) are intercalated with siltstones (lithofacies 6 and 8). The logged section (Fig. 9) presents a global *mixed processes facies association*.

Small-scale facies successions attain 25–40 m in thickness (Fig. 9). Within these sediments, siltstone beds thicken up at the expense of conglomerates, which are arranged into a fining-upwards package of strata. These are capped by mature palaeosols. Larger scale facies successions comprise one or more small-scale facies succession and attain 50–60 m in thickness (Fig. 9). They are limited by mature palaeosols associated with the thick accumulation of siltstones at the base and at the top of the succession.

Distribution of alluvial-fan deposits

The Lower Alluvial Fan System (LAFS). The LAFS, up to 250 m thick in the core of the Inner Syncline, rests unconformably on previously folded and eroded Jurassic limestones, with local palaeovalleys (Fig. 7a). Palaeocurrent data obtained from pebble imbrication (lithofacies 2 in Table 1) indicate a southward dispersal pattern from the catchment area in the north towards the floodplain in the south (Fig. 4).

The LAFS conglomerates can be subdivided into a lower and an upper subsystem (Fig. 5). In the northern limb of the Inner Syncline these subsystems are vertically stacked and poorly differentiated. A thick, northward-wedging succession of amalgamated individual conglomerate bodies (5–15 m thick) comprises the *non-cohesive debris-flow facies association*. Further to the south, in the core and the southern limb of the Inner Syncline, the two subsystems are separated by floodplain sediments. Here the lower part of the LAFS comprises a 75–100 m-thick conglomerate sheet formed by vertically superimposed and laterally juxtaposed individual conglomerate bodies (Fig. 6b) that can be traced over about 5 km parallel to the basin margin. They essentially comprise the *non-cohesive debris-flow facies associ-*ation grading upwards into a succession comprising *cohesive debris flows* and *fan-toe facies associations*.

In the upper part of the LAFS, which is well differentiated in the core of the Inner Syncline, the spatial distribution of alluvial-fan deposits is markedly different. Here, isolated, non-coalescent conglomerate depocentres are observed instead of a widespread conglomerate sheet (Fig. 6b). Vertically stacked conglomerate bodies are bounded by erosional, locally angular disconformities and/or by several dm-thick calcimorph palaesols. They comprise the *non-cohesive debris-flow facies association*, grading laterally and over short distances into *fan-toe facies associations* and grading upward into a *cohesive debris-flow facies association*. Floodplain sediments constitute the sedimentary fill between conglomeratic depocentres. The conglomerates of the LAFS are onlapped by limestone beds of the Calcaire de Rognac Formation and subsequently overlain by Upper Rognacian floodplain sediments (Fig. 7b).

The Upper Alluvial Fan System (UAFS). Conglomerates of the UAFS are exposed in the Inner Syncline, and can be traced basinwards across the anticlinal structure (e.g. le Tholonet or east of Roques Hautes) (Fig. 4). The UAFS can also be divided into a lower and an upper subsystem.

In the northern limb of the Inner Syncline, conglomerate beds belonging to the UAFS are poorly differentiated. At the basin edge, they rest unconformably on Jurassic limestones locally overlain by breccias of the *scree facies association*. Immediately southward, a northward wedging thick succession comprises the *non-cohesive debris-flows facies association*, forming the lower UAFS. Locally, this succession fills deep incisions within uppermost Maastrichtian floodplain siltstones or LAFS deposits (Fig. 7b). The lower UAFS preserves structures of levées and channels of proximal parts of the alluvial fans.

In the southern limb of the Inner Syncline, the lower UAFS forms a thick (up to 150 m) succession comprising the *non-cohesive debris flows facies association*. The succession thins and onlaps southwards onto previously tilted and eroded uppermost Maastrichtian floodplain sediments (Figs 5 & 7b, c). These depositional geometries, with no evidence of synsedimentary deformation in the Roques Hautes zone, indicate the existence of a palaeorelief before the onset of deposition of the UAFS. The palaeorelief and related thrust faults were sealed during deposition of the UAFS, as indicated by the aggradation of subhorizontal conglomerate beds (*mixed processes facies association*) forming the upper UAFS (Figs 5 & 7).

In the Inner Syncline, a well-developed mature calcimorph palaeosol accumulation is recognized in between lower and upper UAFS, with a local

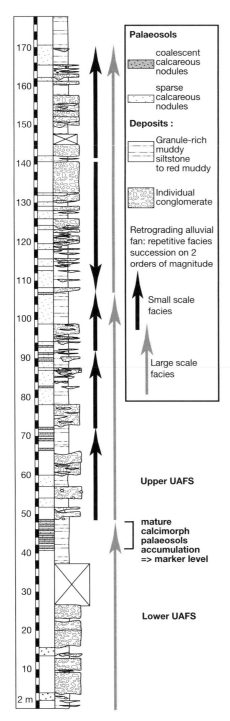

Fig. 9. Logged section of the uppermost UAFS.

unconformity in Le Tholonet area. This mature palaesol accumulation is a marker level that can be traced towards the south, and is found in the basal deposits of the southern monoclinal structure (Fig. 9).

The upper UAFS deposits grade vertically and southward from *non-cohesive debris-flows facies association* to *mixed processes facies association*. They form widespread alluvial fans. In the monoclinal structure the transition to the undeformed basin deposits is formed by high-frequency successions of *mixed processes* and *fan-toe facies associations*, which records groundwater carbonatization in the uppermost horizons. Palaeocurrent data obtained from pebble imbrication (lithofacies 2) and gully orientations (lithofacies 3) indicate an overall southward dispersal. However, palaeocurrent measurements in the outer syncline, east of Roques Hautes, show a persistent SE dispersal (Fig. 4).

Evidence for growth structures

Excellent outcrop conditions and a weak overprint by the Eocene compressional event ensure that the depositional architecture of the Late Cretaceous–Palaeocene alluvial fans is well preserved and exposed in the Inner Syncline between Le Tholonet and the Montagne Sainte Victoire (Figs 4, 6b & 10). The presence of growth structures in the LAFS and UAFS supports the idea that the alluvial fans formed during active deformation in the northern margin of the Arc Basin.

Growth structures in the (LAFS). In the northern limb of the Inner Syncline individual conglomerate bodies thin northward, resulting in a noticeable stratal wedging (Figs 5, 7a & 10). Along the basal unconformity separating the Mesozoic limestones from the conglomerates, the angle between limestone and conglomerate beds increases from 10° to 70°. The conglomerates display a progressive decrease in stratal dips upwards and southwards in the section, which is consistent with syndepositional gradual rotation of the northern fold limb (Hardy & Poblet 1995) during deposition of the LAFS.

In the southern limb of the Inner Syncline, the basal conglomerates of the LAFS overlie a thrust fault (T1) with a concordant contact. In places, higher units in the LAFS thin and onlap southward onto progressively steeper and older strata (Figs 5, 6b & 10). Even if thinning may be in part related to the alluvial-fan-toe geometry, the angular disconformities indicate that the southern limb of the Inner Syncline was gradually tilted during deposition of the upper subsystem of the LAFS. These geometrical relationships suggest that the T1 thrust fault became

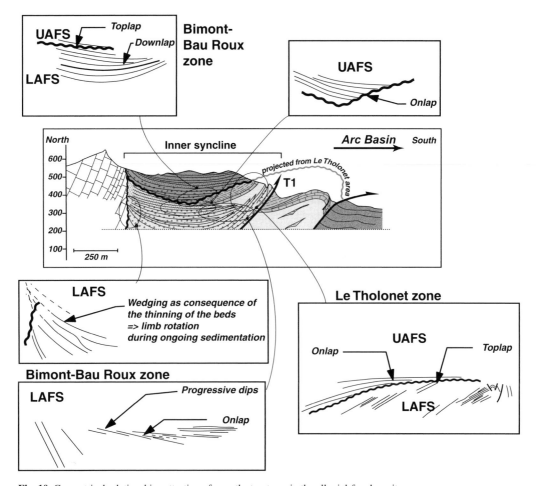

Fig. 10. Geometrical relationships attesting of growth structures in the alluvial-fan deposits.

active during deposition of the upper subsystem of the LAFS.

Growth structures in the (UAFS). Evidence for growth structures in the UAFS occurs in the Inner Syncline where conglomerate units thin and about northward against Jurassic strata. In the northern limb of the Inner Syncline conglomerates are associated with slump folds and with localized normal and reverse faults. These deformed zones show lateral and vertical transitions to well-stratified conglomerates. These structures are interpreted as soft-rock deformation features resulting from transient gravitational instabilities that formed during deposition of the UAFS during ongoing deformation and tilting in the northern margin of the basin (Fig. 11).

On the southern limb of the Inner Syncline, the conglomerates of the UAFS onlap a prominent erosional surface (Figs 5, 7b & 10). In the Roques Hautes area the depositional geometries do not show clear evidence for growth structures, which leads to the interpretation that the conglomerates were onlapping onto a palaeorelief. In the area of Le Tholonet further to the west, conglomerate beds of the UAFS seal the T1 thrust fault, implying that the T1 thrust was not active during deposition of the UAFS. In this area, the conglomerates of the UAFS also thin and onlap southwards onto red floodplain deposits overlying the LAFS (Figs 7c & 10). Because the uppermost conglomerate beds of the UAFS are in a subhorizontal position these strata are unlikely to have been tilted or deformed during deposition or later Eocene compression. Therefore, tilting of the southern limb of the Inner Syncline was related to synsedimentary thrusting along the T2 thrust fault. Activity along the T2 thrust fault probably stopped during deposition of the uppermost conglomerates in the UAFS. Unfortunately, the T2

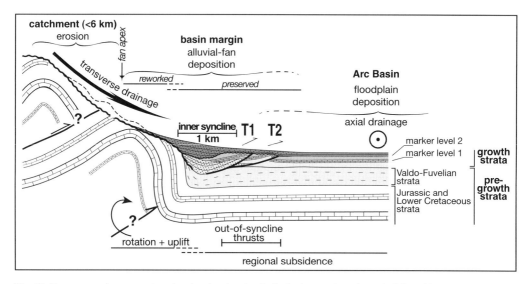

Fig. 11. Reconstructed cross-section showing the situation in the basin margin at the end of deposition of the UAFS before the onset of Eocene compression. The section shows the relationship between the depositional processes in the basin margin and the processes active in the catchment area (source) and the basin (sink).

thrust fault was reactivated during Eocene compression, as indicated by the subvertical dips of the conglomerate beds in the UAFS southward of the Inner Syncline. No growth structures related to the Eocene event were found.

The occurrence of growth structures on both the southern and northern limbs of the Inner Syncline clearly supports the idea that the LAFS and the UAFS formed contemporaneously with deformation and uplift in the northern margin of the Arc Basin. Because conglomerate bodies can be traced throughout the Inner Syncline, we can confidently link tilting related to the formation of the Inner Syncline with simultaneous shortening accommodated first along the T1 thrust fault before it localized along the T2 thrust fault.

Alluvial-fan development and implications for the basin-margin evolution

The sensitivity of alluvial fans to climate changes and tectonic processes makes them preserve an excellent record of the tectono-sedimentary evolution of basin margins. At the northern margin of the Arc Basin, tectonic processes interacted with climate changes controlling runoff, weathering, and other processes relevant for the erosion, transport and deposition of sediments from the source (the deformed catchment) towards their sink (mainly the Inner Syncline; Fig. 11). The bulk of the alluvial-fan conglomerates observed in the northern margin of the Arc Basin

represents superimposed and laterally amalgamated lobes, which were dominated by debris-flow deposition (Blair 1999; Gomez-Villar & Garcia-Ruiz 2000). Facies associations in both the LAFS and the upper UAFS reflect deposition on the mid–lower part of alluvial fans, i.e. downstream of the intersection point (Blair & McPherson 1994), whereas the proximal part of the lower UAFS has been preserved. Individual, 5–15 m-thick bodies represent periods of alluvial-fan aggradation (e.g. Harvey *et al.* 2003). Intervening calcimorph palaeosols represent the stabilization of alluvial fans by pedogenic calcretes (Alonso Zarza *et al.* 1998; Mack *et al.* 2000).

In the following sections the tectono-sedimentary evolution of the alluvial fans in the northern margin of the Arc Basin is examined, and the forcing factors controlling the deposition and preservation of alluvial fan systems are discussed.

Tectono-sedimentary evolution as recorded in the syntectonic alluvial-fan systems

Processes acting on a geological timescale are well recorded by the depositional architecture of the LAFS and UAFS. The observation of growth structure and aggradation in the northern margin of the Arc Basin indicate that deformation, uplift and erosion in the catchment area occurred simultaneously with subsidence and deposition in the Arc Basin to the south. Based on temporal and geometrical relationships between deformation structures and depositional sequences, the formation of growth structures in the alluvial fans is interpreted to be associated with thrusting along the T1 thrust during deposition of the upper subsystem of the LAFS, and later thrusting along the T2 thrust during deposition of the UAFS. Based on the position of the T1 and T2 thrust faults in the core of a km-scale synclinal structure and the observations that these faults are not rooted in the underlying Mesozoic limestones, the T1 and T2 faults are interpreted as 'out-of-syncline' thrusts (Rafini & Mercier 2002) forming during folding of the massive Mesozoic limestones. Out-of-syncline thrusting results from an excess of sedimentary deposits within the frontal hinge of a growth fold. This interpretation implies a direct relationship between the formation of the growth structures within the LAFS and UAFS, and large-scale deformation in the basin margin controlling structure formation and uplift in the catchment area and subsidence in the adjacent basin (Fig. 11). Based on the structural and sedimentological information recorded in the alluvial-fan systems the following tectono-sedimentary evolution of the northern margin of the Arc Basin is proposed (Fig. 12).

Pre-alluvial-fan development (Fig. 12a). The lack of Lower Cretaceous sediments over large parts of the Arc Basin, as indicated by Lower Campanian (Valdonian and Fuvelian) strata directly overlying Jurassic limestones, suggests that the Arc Basin formed within a previously elevated area (Aubouin & Mennessier 1963; Masse & Philip 1976) (Fig. 12a). Subsidence occurred during Campanian time, as indicated by the onlapping of Valdonian and Fuvelian sediments towards the north onto Mesozoic limestones. Deformation structures predating the alluvial fans are observed only within localized areas, bounding the Arc Basin to the north and to the south. In the Montagne Sainte Victoire area, folds containing Valdonian and Fuvelian sediments were formed and subsequently eroded (Tempier & Durand 1981), with incisions cut in Mesozoic limestones. This observation documents the first evidence for an emerging topography, which formed as a result of compressional deformation in the northern margin of the Arc Basin. Former rivers initially maintained a downstream gradient across the emerging E–W-trending ridge dissected by valleys. River courses were then abandoned when erosion rates became insufficient to keep pace with the rate of structural uplift of the ridge.

Tectono-sedimentary evolution during the deposition of the LAFS (Fig. 12b & c). In the lower subsystem of the LAFS, the laterally extensive conglomerate sheet (Fig. 6b) indicates coalescent alluvial fans analogous to a bajada system (Calvache et al. 1997) (Fig. 13a). Following abandonment of the previously

entrenched, pre-alluvial fan, antecedent drainage network, these deposits represent the response to the building of relatively steep topography and reflect the creation of a juvenile drainage network along the northern margin of the Arc Basin. In the upper part of the lower LAFS, the establishment of discrete conglomerate depocentres surrounded by floodplain deposits suggests a limited number of individual alluvial-fan apices interpreted as the result of a drainage network that was expanding by capture events in the catchment area. This change in the drainage network is associated with the onset of thrusting along the T1 thrust as recorded by the first growth structures in the southern limb of the Inner Syncline. Consequently, the T1 thrust noticeably post-dates the onset of the bajada-related topography in the catchment area (Fig. 12b, c). As deformations affecting the catchment area to the north and out-of-syncline thrusting in alluvial-fan deposits in the south (basin margin) are kinematically linked but structurally decoupled, a delay between the stratigraphic and tectonic record is documented in the alluvial system.

The evolution toward a better-integrated drainage occurred in tandem with a trend towards more 'muddy' debris flows. A second event of alluvial-fan deposition, associated with long-term aggradation, is recorded in the conglomerates of the upper subsystem of the LAFS (Fig. 12c). This event characterizes a period of relatively high sediment supply to discharge ratio leading to the infill of the palaeoincisions in the Mesozoic limestone within the catchment area. This is coeval with high aggradation in the basin (Westphal & Durand 1990). Alluvial fans and floodplain strata interdigitated above the active T1 thrust (Fig. 12c & 13b). Such conditions, which imply no topographic expression related to the activity of the out-of-syncline thrust, indicate a blind fault.

The lack of upper-fan deposits characterized by levée deposits suggests an indented mountain front with upper-fan facies restricted to subsequently reworked areas. Here, tilting and uplift recorded by growth structures at the basin edge resulted in a gradual, but continuous, drop in base level of erosion leading to an evolution dominated by the entrenchment of the successive fan surfaces and subsequent reworking of the upper fan deposits. In the uppermost levels of the LAFS, alluvial fans progressively became inactive and were gradually onlapped by floodplain and lacustrine sediments (e.g. the Calcaire de Rognac Formation, Marker Level 1; Fig. 2) (Figs 5 & 12d).

Evolution before the deposition of the UAFS (Fig. 12d, e). During the Upper Rognacian, following deposition of the LAFS, alluvial fans were partly inactive. Floodplain and lacustrine sediments aggraded onto the inactive fan surface (Fig. 12d). Despite the lack of growth structures, the formation of

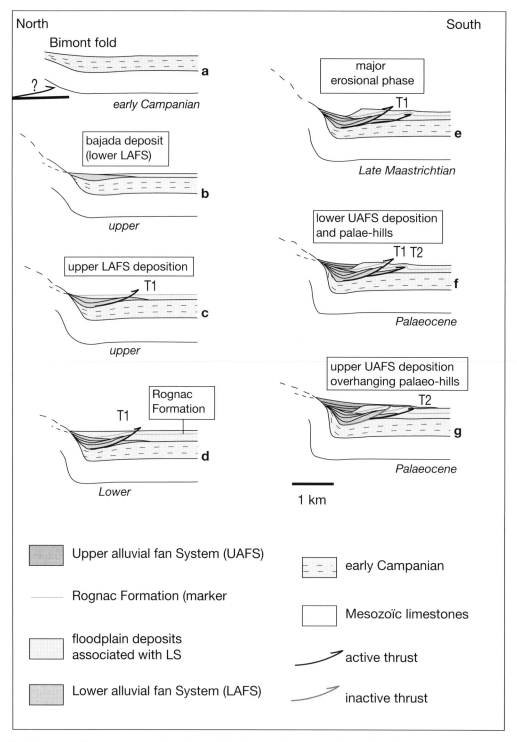

Fig. 12. Cartoon showing the temporal evolution of the basin margin sequence from initial deformation (**a**) to the deposition of the LAFS (**b** and **c**), the deposition of Marker Level 1 (**d**), major erosion phase (**e**) and final deposition of the UAFS (**f** and **g**). For discussion see the text.

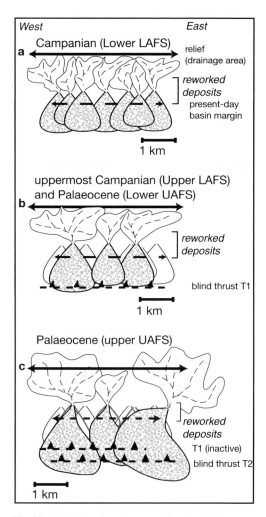

Fig. 13. Sketch showing the temporal and spatial evolution of the alluvial fan systems in a map view during the deposition of (**a**) the Lower LAFS, (**b**) the Upper LAFS and (**c**) the UAFS.

a palaeorelief associated with the T1 thrust is probably linked to tectonic activity, supported by the occurrence of two alluvial fan conglomerate beds below the Conglomérat de la Galante (Maastrichtian).

At the end of the Upper Rognacian, a major erosional event affected large parts of the basin margin (Fig. 12e). In the Inner Syncline, it is particularly well documented by scree deposition and incisions within the LAFS conglomerates and Maastrichtian floodplain deposits in both the northern and southern limbs. In the basin margin, erosion in places is at least partly synchronous with the Conglomérat de la Galante deposition. Sediments originating of the erosion of the Inner Syncline probably contributed to the sediment load of the fluvial system associated

with the Conglomérat de la Galante. In this case, the erosional surface in the basin margin is only partly time equivalent with the base of Marker Level 2 (Conglomérat de la Galante, Fig. 2).

Tectono-sedimentary evolution during the deposition of the UAFS (Fig. 12f, g). The architecture of the UAFS reveals that deposition was coeval with tectonic activity along the T1 thrust (growth structure) in the western zone (Le Tholonet). It is suggested that uplift of the southern limb of the Inner Syncline, related to the out-of-syncline process, promoted the deposition and aggradation of alluvial-fan sediments (e.g. Burbank *et al.* 1996, for an example related to a growing fold). This temporal relationship suggests that the sediments were trapped in the Inner Syncline to form the conglomerates of the lower UAFS. At the end of the lower UAFS deposition, transverse streams flowing across the Inner Syncline became entrenched as antecedent streams that incised through the uplifting hanging-wall palaeorelief formed by either T1 and/or T2 thrusts. They prevented the formation of a valley parallel to the basin margin in the axis of the Inner Syncline, and alluvial-fan sediments have been deposited in the basin (future monoclinal structure).

During the lower UAFS deposition, alluvial-fan apices approximately overlap those of the LAFS (Fig. 13b). Sealing of the T1 thrust fault by conglomerates of the uppermost lower UAFS and localization of the deformation along the T2 thrust fault indicates the migration of the deformation towards the basin associated with the out-of-syncline thrusts. At that time, alluvial fans prograded towards the basin depositing proximal sediments in the Inner Syncline, and alluvial-fan depositional surfaces overhang the palaeorelief forms inherited from the erosional stage. The progradation of the alluvial-fan deposits indicate a rejuvenation of the catchment area before the lower UAFS. As for the T1 thrust fault during the LAFS deposition, any topographic expression of the T2 thrust fault is subsequently recorded as alluvial deposits interfingered with floodplain deposit.

During deposition of the upper UAFS, fans were wider, depositing sediment lobes from the Inner Syncline area towards the south (future monoclinal structure). Increasing fan sizes (Fig. 13c), retrogradation of the proximal part of the alluvial fans towards the north, together with the occurrence of waterlain deposits and the large volume of silt and shale in alluvial-fan successions, suggest a coeval increase in catchment size reflecting the continuous integration of the drainage network in the catchment.

The UAFS shows a consistent thinning- and fining-upwards trend, coeval with the development of intervening muddy debris-flow deposits and floodplain facies. This late evolution of the UAFS is interpreted, using deformation located further to the north in the catchment area, as a major thrusting event (Fig. 11) causing the Liassic marls to be eroded. The occurrence of this deformation is coeval with the development of the calcimorph palaeosol accumulation before the upper UAFS which marks a delay between the deformation and the sedimentological response. The nature and the distribution of the upper UAFS suggest a diminishing topography in the catchment area, a migration of the basin margin towards north and a decline in tectonic activity during the uppermost UAFS.

During later Eocene compression, some of the pre-existing Late Cretaceous–Palaeocene structures were reactivated (e.g. T2 thrust and the Montagne Sainte Victoire thrusted anticline). No alluvial-fan deposits have been recorded related to the Eocene compression.

Forcing factors

The preservation of a thick alluvial-fan succession requires: (1) catchment surface uplift coeval with steady environmental conditions favourable for alluvial-fan development (Mack & Leeder 1999); and (2) a long-term rising stratigraphic base level, preventing erosional processes and related to regional subsidence and/or aggradation in the basin (Viseras *et al.* 2003). In this study, the catchment uplift is associated with large-scale folding leading to uplift, tilting and erosion. The occurrence of out-of-syncline thrusts implies uplift at the basin margin (Fig. 11). The latter factor controls the accommodation space for alluvial-fan deposition and preservation at the basin margin, which acts as a small-scale piggy-back basin soled by an out-of-syncline thrust.

The sequence of events as recorded by the two alluvial-fan systems comprises: (1) an erosional stage associated with the onset of the tectonic activity; (2) the development of aggrading alluvial fans, coeval with out-of-syncline thrusting and associated with the infilling of palaeotopography; and (3) the backstepping of the alluvial fans.

Differences between the two systems are the nature and zone of deformation in the catchment area, and the timing and the efficiency of the out-of-syncline thrusts. Out-of-syncline thrusting is delayed relative to the onset of the deformation in the catchment associated with the LAFS (T1 thrust fault), but efficient as soon as the rejuvenation and the erosion stage precedes the UAFS deposition (T2 thrust).

The occurrence of the two systems of alluvial fans has been claimed to reflect the sedimentary response subsequent to two distinct tectonic pulses in the Montagne Sainte Victoire (Durand & Tempier 1962; Corroy *et al.* 1964*a*, *b*; Chorowicz & Ruiz 1979). As shown above, the two alluvial-fan systems are associated in time and space with the development of

growth structures. In the following we discuss the major factors that controlled the evolution of the two syntectonic alluvial-fan systems.

Conglomerate accumulation appears coeval with warming trends in the basin (Fig. 2), possibly resulting in a stronger seasonality favouring floods or more active denudation rates. However, the occurrence of few debris-flow deposits in the more humid period (just below the Marker Sequence 2) does not fit with this hypothesis. Instead, the relationships between the tectonically induced catchment uplift and the folding of Mesozoic strata are strongly supported by the first-order architecture of alluvial-fan deposits. Growth structures preserved in the Inner Syncline are related to the tilting and uplift of the Mesozoic strata in the northern limb, and to the out-of-syncline thrusts in the southern limb. The growth structures associated with stratal wedging against the pre-growth strata, which occur in comparable tectono-sedimentary settings (e.g. Ford *et al.* 1997), indicate that folding was ongoing during sedimentation. They are recorded at all stratigraphic levels as early as the onset of alluvial-fan aggradation following the erosional stage or, alternatively, later, when the out-of-syncline T1 thrust was activated during the deposition of the upper subsystem of the LAFS. The activation of the latter depends on the balance between the wavelength of the fold affecting the Mesozoic strata, the rate of the fold-related tilting and the aggradation rate at the basin margin (Rafini & Mercier, 2002). In the northern Arc Basin, a rather tight fold and the localization of the conglomerate depocentre into the fold hinge probably enhanced the out-of-syncline process, which can be also accommodated by subordinate folds close to the limb of the major fold (e.g. Arenas *et al.* 2001). Furthermore, the abandonment of the T1 thrust and activation of the T2 thrust, corresponding to a normal-sequence thrust propagation, is dependent on the intrinsic wedge equilibrium in the thrusted sedimentary pile in the core of the major fold affecting the Mesozoic strata.

At the basin scale, the spatial distribution of the conglomerate was clearly controlled by tectonic processes, especially in those places where deformation and uplift rates lead to a steep topography. At the scale of the active basin margin, the gradual integration of the drainage network in the catchment was a key control on the location of conglomerate depocentres. The evolution from an emerging topography characterized by bajadas towards a more mature topography was reflected by distinct alluvial-fans apices associated with fans of greater size.

On the scale of an alluvial-fan system, displaying complex lithofacies, both long-term processes (backstepping profile) with high-frequency changes are superimposed (Fig. 9). More abundant muddy and cohesive debris flow occurred in the uppermost parts of the LAFS and UAFS, whereas gullying (lithofacies 7, Fig. 8), absent in the LAFS conglomerates, characterzes the upper UAFS in association with subordinate waterlain deposits. More mature clay-rich soil horizons may have sourced the muddy debris flows, whereas the onset of a drier climate associated with more violent floods during deposition of the UAFS may explain the rising water/sediment supply ratio.

High-frequency cycles within alluvial-fan successions (Fig. 9) represent periods of alluvial-fan aggradation interrupted by periods of stability. Based on the sedimentation rate of 20 m Ma^{-1}, each of these bodies reflects short-term (of the order millions of years) cyclic changes in the sediment supply to discharge ratio that should be classically related to a local tectonic control. The very high cycles recorded as alternation between fine conglomerates sheets and siltstones either represent the climatic record or autocyclic processes in these successions.

In conclusion, in the northern margin of the Arc Basin, the stratigraphic pattern in the Late Cretaceous–Palaeocene alluvial-fan systems mainly records deformation related to folding in the catchment, and subordinate out-of-syncline thrusts in response to shortening in the basin margin. On an intermediate timescale, the maturity of the drainage network controls the alluvial-fan spatial distribution and some changes in depositional facies. Climate changes have only been recorded in alternation of conglomerates–siltstones at the scale of interbedding, resulting in vertically superimposed alluvial-fan bodies comprising laterally amalgamated lobes corresponding to autocyclic events.

The authors wish to thank Trevor Elliott and Peter Friend for providing reviews of the original manuscript and they are grateful to Annie Bouzeghaia for the design of the final geological maps. This is a contribution of EOST, number 2005.501-UMR 7517.

References

ALONZO-ZARZA, A., SILVA, P.G., GOY, J.L. & ZAZO, C. 1998. Fan-surface dynamics and biogenic calcrete development: Interactions during ultimate phases of fan evolution in the semiarid SE Spain Murcia. *Geomorphology*, **24**, 147–167.

ARENAS, C., MILLÀN, H., PARDO, G. & POCOVÍ, A. 2001. Ebro Basin continental sedimentation associated with late compressional Pyrenean tectonics (north-eastern Iberia): controls on basin margin fans and fluvial systems. *Basin Research*, **13**, 65–89.

ASHRAF, A.R. & ERBEN, H.K. 1986. Palynologishe Untersuchung an der Kreide/Tertiär-Grenzze west-Mediterraner Regionen. *Paleontographica, Stuttgart*, **200**, 11–163.

AUBOUIN, J. & MENNESSIER, G. 1963. Essai sur la structure de la Provence. *In: Livre à la Mémoire du Professeur Paul Fallot, II. Mémoire Hors Série, Société Géologique de France*, **III**, 45–98.

BABINOT, J.F. & DURAND, J.-P. 1980a. Valdonien, Fuvélien, Bégudien, Rognacien, Vitrollien. Les étages français et leurs stratotypes. *Mémoire du Bureau de Recherche Géologique et Minière, Orléans*, **109**, 92–171.

BABINOT, J.F. & DURAND, J.-P. 1980b. Rognacien, Vitrollien. Les étages français et leurs stratotypes. *Mémoire du Bureau de Recherche Géologique et Minière, Orléans*, **109**, 184–192.

BABINOT, J.F. & DURAND, J.-P. 1984. Crétacé supérieur fluvio-lacustre. *In*: PHILIP J. (ed.) *Synthèse géologique du Sud-Est de la France, Crétacé Supérieur. Mémoires de Recherche Géologique et Minière, Orléans*, **125**, 362–367.

BIBERON, B. 1988. *Mécanismes et évolution de chevauchements à vergences opposées. Exemple de la Sainte-Victoire*. PhD Thesis, Université Joseph Fourier, Grenoble.

BILLEREY, A., DUGHI, R. & SIRUGUE, F. 1959. Les œufs de dinosaures et la datation des brèches de Sainte-Victoire. *Comptes Rendus de l'Académie des Sciences, Paris*, **248**, 272–274.

BLAIR, T.C. 1999. Sedimentology of the debris-flow-dominated Warm Spring Canyon alluvial fan, Death Valley, California. *Sedimentology*, **46**, 941–965.

BLAIR, T.C. & MCPHERSON, J.G. 1994. Alluvial fans and their natural distinction from rivers based on morphology, hydraulic processes, sedimentary processes, and facies assemblages. *Journal of Sedimentary Research*, **A64**, 450–489.

BURBANK, D., MEIGS, A. & BROZOVIC, N. 1996. Interactions of growing folds and coeval depositional systems. *Basin Research*, **8**, 199–223.

CHAMPION, C., CHOUKROUNE, P. & CLAUZON, G. 2000. La déformation post-Miocène en Provence occidentale. *Geodynamica Acta*, **13**, 67–85.

CALVACHE, M.L., VISERAS, C. & FERNANDEZ, J. 1997. Controls on fan development – evidence from fan morphometry and sedimentology; Sierra Nevada, SE Spain. *Geomorphology*, **21**, 69–84.

CHOROWICZ, J. & RUIZ, R. 1979. Observations nouvelles sur la structure des 'brèches' de Sainte-Victoire (Provence). *Comptes Rendus de l'Académie des Sciences, Paris*, **288**, D, 207–210.

CHOROWICZ, J., MEKARNIA, A. & RUDANT, J.-P. 1989. Inversion tectonique dans le massif de la montagne Sainte-Victoire (Provence, France). Apport de l'imagerie *Spot. Comptes Rendus de l'Académie des Sciences, Paris*, **C-II**, 1179–1185.

CHOUKROUNE, R. & MATTAUER, M. 1978. Tectonique des plaques et Pyrénées; sur le fonctionnement de la faille transformante Nord-Pyrénéenne; comparaison avec les modèles actuels. *Bulletin de la Société Géologique de France, Series 7*, **20**, 689–700.

CLAUZON, G. 1984. Evolution géodynamique d'une montagne provençale et de son piedmont; l'exemple du Luberon (Vaucluse, France). *In: Montagnes et piémonts; actes du colloque de geomorphologie, sur les relations entre les montagnes récentes et leurs piedmonts*. CNRS, France, 427–442.

COJAN, I. 1989. Discontinuités majeures en milieu continental. Proposition de corrélation avec des evenerments global (Bossin de Provence, S. France, Passage Crétacé/Tertiaire). *Compte Rendus de l'Académie des Sciences, Paris*, **309**, 1013–1018.

COJAN, I. 1993. Alternating fluvial and lacustrine sedimentation: tectonic and climate controls (Provence Basin, S. France, Upper Cretaceous/Palaeocene). *In:* MARZO, M. & PUIGDEFABREGAR, C. (eds) *Alluvial Sedimentation*. International Association of Sedimentologists, Special Publications, **17**, 425–438.

COJAN, I., MOREAU, M.-G. & STOTT, L.E. 2000. Stable carbon isotope stratigraphy of the Paleogene pedogenic series of southern France as a basis for continental-marine correlation. *Geology*, **28**, 259–262.

COJAN, I., RENARD, M. & EMMANUEL, L. 2003. Palaeoenvironmental reconstruction of dinosaur nesting sites based on a geochemical approach to eggshells and associated palaeosols (Maastrichtian, Provence Basin, France). *Palaeogeography, Palaeoclimatology, Palaeoecology*, **191**, 111–138.

COLSON, J. & COJAN, I. 1996. Groundwater dolocretes in a lake marginal environment: an alternative model for dolocrete formation in continental settings (Danian of the Provence Basin, France). *Sedimentology*, **43**, 175–188.

CORROY, G. 1957. *La Montagne Sainte-Victoire*. Bulletin du Service de la carte Géologique de de la France, **251**, 47.

CORROY, G., DURAND, J.-P. & TEMPIER, C. 1964*a*. Evolution tectonique de la montagne Sainte-Victoire en Provence. *Bulletin de la Société Géologique de France*, **7-VI**, 91–106.

CORROY, G., TEMPIER, C. & DURAND, J.-P. 1964*b*. Evolution tectonique de la montagne Sainte-Victoire en Provence (a). *Comptes Rendus de l'Académie des Sciences, Paris*, **258**, 1556–1557.

DEBROAS, E.J. 1990. Le flysch noir albo-cénomanien, temoin de la structuration albienne à sénonienne de la zone nord-pyrénéenne en Bigorre (Hautes Pyrénnées, France). *Bulletin de la Société Géologique de France, Series 8*, **6**, 273–285.

DECELLES, P.G., GRAY, M.B., RIDGWAY, K.D., COLE, R.B., PIVNIK, D.A., PEQUERA, N. & SRIVASTAVA, P. 1991. Controls on synorogenic alluvial fan architecture, Beartooth Conglomerate (Palaeocene), Wyoming and Montana. *Sedimentology*, **38**, 567–590.

DURAND, J.-P. & GUIEU, G. 1983. Cadre structural du bassin de l'Arc. *In:* ARCAMONE, J. *et al.* (eds) *Le gisement de charbon du Bassin de l'Arc (Provence occidentale)*. Mémoires du Bureau de Recherche Géologique et Minière, Orléans, **122**, 3–12.

DURAND, J.-P. & TEMPIER, C. 1962. Etude tectonique de la zone des brèches du massif de Sainte-Victoire dans la région du Tholonet (Bouches-du-Rhône). *Bulletin de la Société Géologique de France*, **7-IV**, 97–101.

FORD, M., WILLIAMS, E.A., ARTONI, A., VERGÉS, J. & HARDY, S. 1997. Progressive evolution of a fault related fold pair from growth strata geometries, Sant Llorenç de Morunys, SE Pyrenees. *Journal of Structural Geology*, **19**, 413–441.

GRADSTEIN, F.M., AGTERBERG, F.P., OGG, J.G., HARDENBOL, J., VAN VEEN, P., THIERRY, J. & HUANG, Z. 1994. A Triassic, Jurassic and Cretaceous time scale. *In:* BERGGREN. W.A., KENT, D.V., AUBRY, M.-P. & Hardenbol, J. (eds) *Geochronology, Time Scales and Global Stratigraphic Correlation*. Society for Sedimentary Geology, Special Publications, **54**, 95–126.

GOMEZ-VILLAR, A. & GARCIA-RUIZ, J.M. 2000. Surface sediment characteristics and present dynamics in alluvial fans of the central Spanish Pyrenees. *Geomorphology*, **34**, 127–144.

HARDY, S. & POBLET, J. 1995. The velocity description of deformation: paper 2. Sediment geometries associated with fault-bend and fault-propagation folds. *Marine and Petroleum Geology*, **12**, 165–176.

HARVEY, A.M. 1990. Factors influencing Quaternary alluvial fan development in southeast Spain. *In:* RACHOCKI, A.H. & CHURCH, M. (eds) *Alluvial Fans: A Field Approach*. Wiley, Chichester, 247–269.

HARVEY, A.M. 2002. The role of base-level change in the dissection of alluvial fans: case studies from southeastern Spain and Nevada. *Geomorphology*, **45**, 67–87.

HARVEY, A.M., FOSTER, G., HANNAM, J. & MATHER, A.E. 2003. The Tabernas alluvial fan and lake system, southeast Spain: applications of mineral magnetic and pedogenic iron oxide analyses towards clarifying the Quaternary sequences. *Geomorphology*, **50**, 151–171.

LUTAUD, L. 1935. Sur la genèse des chevauchements et écailles de la Provence Calcaire. *Comptes Rendus sommaires de la Société Géologique de France*, **16**, 261–263.

MACK, G.H. & LEEDER, M.R. 1999. Climatic and tectonic controls on alluvial-fan and axial-fluvial sedimentation in the Plio-Pleistocene Palomas half graben, southern Rio Grande Rift. *Journal of Sedimentary Research*, **69**, 635–652.

MACK, G.H., COLE, D.R. & TREVIÑO L. 2000. The distribution and discrimination of shallow, authigenic carbonate in the Pliocene-Pleistocene Palomas Basin, southern Rio Grande rift. *Bulletin of the Geological Society of America*, **112**, 643–656.

MASSE, J.P. & PHILIP, J. 1976. Paléogéographie et tectonique du Crétacé moyen en Provence : révision du concept d'Isthme Durancien. *Revue de Geographie Physique et de Geologie Dynamique*, (2), **XVIII**, 49–66.

MATTAUER, M. & SÉGURET, M. 1971. Les relations entre la chaîne pyrénéenne et le golfe de Gascogne. *In:* DEBYSER, J., LE PICHON, X. & MONTADERT, L. (eds) *Histoire structurale du golfe de Gascogne*. Technip, Paris, IV.4–1–IV.4–24.

MEDUS, J. 1972. Palynological zonation of the Upper Cretaceous in southern France and northeastern Spain. *Review of Palaeobotany and Palynology*, **14**, 287–295.

OLIVET, J.-L. 1996. La cinématique de la plaque ibérique. *Bulletin des Centre de Recherche d'Exploration–Production, Elf Aquitaine*, **20**, 131–195.

RAFINI, S. & MERCIER, E. 2002. Forward modelling of foreland basins progressive unconformities. *In:* MARZO, M., MUNOZ, J.A. & VERGÈS, J. (eds) *Growth Stata. Sedimentary Geology*, **146**, 75–89.

ROURE, F. & COLETTA, B. 1996. Cenozoic inversion structures in the foreland of the Pyrenees ans Alps. *In:* ZIEGLER, P.A. & HORVATH, F. (eds) *Peri-Tethys Memoir 2: Structures and Prospects of Alpine Basin and Forelands. Memoire du Museum National d'Histoire Naturelle*, Paris, **170**, 173–209.

SITTLER, C. & MILLOT, G. 1964. Les climats du Paléogène français reconstitués par les argiles néoformées et les microflores. *Geologische Rundschau*, **54**, 333–343.

TEMPIER, C. 1987. Modèle nouveau de mise en place des structures provençales. *Bulletin de la Société Géologique de France*, **8-III**, 533–540.

TEMPIER, C. & DURAND, J.-P. 1981. Importance de l'épisode d'âge crétacé supérieur dans la structure du versant méridional de la montagne Sainte-Victoire (Provence). *Comptes Rendus de l'Académie des Sciences, Paris*, **293**, 629–632.

VISERAS, C., CALVACHE, M.,L., SORIA, J.M. & FERNANDEZ, J. 2003. Differential features of alluvial fans controlled by tectonic or eustatic accomodation space. Examples from the Betics Cordillera, Spain. *Geomorphology*, **50**, 181–202.

WELLS, S.G., McFADDEN, L.D. & DOHRENWEND, J.C. 1987. Influence of Late Quaternary climatic changes on geomorphic and pedogenic processes on a desert piedmont, Eastern Mojave Desert, California. *Quaternary Research*, **27**, 130–146.

WESTPHAL, M. & DURAND, J.-P. 1990. Magnétostratigraphie des séries continentales fluvio-lacustres du Crétacé supérieur dans le synclinal de l'Arc (région d'Aix-en-Provence, France). *Bulletin de la Société Géologique de France*, **8-VI**, 609–620.

Index

Page numbers in italic, e.g. *42*, refer to figures. Page numbers in bold, e.g. **53**, signify entries in tables.